AF322735

DNA POLYMERASES

Discovery, Characterization and Functions in Cellular DNA Transactions

DNA POLYMERASES

Discovery, Characterization and Functions in Cellular DNA Transactions

Ulrich Hübscher

*Institute of Veterinary Biochemistry and Molecular Biology,
University of Zurich, Switzerland*

Silvio Spadari

*Institute of Molecular Genetics IGM-CNR,
National Research Council, Pavia, Italy*

Giuseppe Villani

*Institute de Pharmacologie et de Biologie Structurale,
CNRS-Université Paul Sabatier, Toulouse, France*

Giovanni Maga

*Institute of Molecular Genetics IGM-CNR,
National Research Council, Pavia, Italy*

World Scientific

NEW JERSEY · LONDON · SINGAPORE · BEIJING · SHANGHAI · HONG KONG · TAIPEI · CHENNAI

Published by

World Scientific Publishing Co. Pte. Ltd.

5 Toh Tuck Link, Singapore 596224

USA office: 27 Warren Street, Suite 401-402, Hackensack, NJ 07601

UK office: 57 Shelton Street, Covent Garden, London WC2H 9HE

British Library Cataloguing-in-Publication Data
A catalogue record for this book is available from the British Library.

DNA POLYMERASES
Discovery, Characterization and Functions in Cellular DNA Transactions

ISBN-13 978-981-4299-16-9
ISBN-10 981-4299-16-2

Typeset by Stallion Press
Email: enquiries@stallionpress.com

Printed by FuIsland Offset Printing (S) Pte Ltd. Singapore

Preface

Maintenance of the information embedded in the genomic DNA sequence is crucial for the survival of any living species. DNA polymerases play pivotal roles in this complex process since they are involved in all DNA synthesis events occurring in nature. Besides their essential tasks *in vivo*, DNA polymerases are now the workhorses in numerous important molecular biological and medical core technologies such as the widely applied polymerase chain reaction (PCR), cDNA cloning, genome sequencing, nucleic acids–based diagnostics and in techniques to analyze ancient and otherwise damaged DNA.

The history of DNA polymerases goes back over 53 years to the mid-1950s. With the discovery of a DNA polymerase (now known as DNA polymerase I) in the bacterium *Escherichia coli* by Arthur Kornberg and his colleagues, it was for the first time possible to synthesize the genetic material, the DNA, in the test tube. A manuscript reporting the discovery submitted to the *Journal of Biological Chemistry* was rejected with comments such as "the researchers were incompetent" and DNA polymerase was "a poor name" for this enzyme. What a terrible judgment by an incompetent editor! The observation by Kornberg and collaborators that a DNA polymerase synthesizes DNA according to the Watson–Crick base pair rule (A–T and G–C), needs activated bases (dNTPs), a template, a primer and $MgCl_2$ is still true today for most DNA polymerases in all organisms tested.

Bruce Alberts in a feature article in *Nature* stated in 2003[1]: "Knowledge of the structure of DNA enabled scientists to undertake the difficult task of deciphering the detailed molecular mechanisms of two dynamic processes that are central to life: the copying of the genetic information by DNA replication, and its reassortment and repair by DNA recombination. Despite dramatic advances towards this goal over the past five decades, many challenges remain for the next generation of molecular biologists."

A comprehensive book focusing on DNA polymerases appeared in 1986 and covered exclusively animal DNA polymerases.[2] At that time the animal DNA polymerase family was still small. Only three DNA polymerases, α, β and γ were known in detail and the scientific community just started to believe in a fourth DNA polymerase, called DNA polymerase δ. In the last 23 years, due to the analysis of various genomes, including the one from humans, we have witnessed the discovery of an abundance of novel DNA polymerases in the three kingdoms of life (bacteria, archaea and eukaryotes) with specialized properties whose physiological functions are only beginning to be understood. For several decades the dogma was widely accepted that very accurate DNA polymerases guarantee the faithful duplication of DNA, while their limited capacity of making mistakes might be one of the drivers for evolution and a cause of disease. When in 1999 the human genome was sequenced, nine newly discovered DNA polymerases appeared within three years in the literature (1999–2002). Many of these DNA polymerases appear to have distinct functions in translesion synthesis, in different DNA repair events or in immunoglobulin V(D)J recombination.

This book starts by presenting the history of the discovery of DNA and DNA polymerases, including the polymerase chain reaction (PCR) (Chapter 1), followed by the presentation of DNA polymerases from the three kingdoms of life: bacteria, archaea and eukaryotes (Chapter 2). Next, the structural and functional aspects of the different DNA polymerase families are described in prokaryotes (Chapter 3) and in eukaryotes (Chapter 4). Preventing genetic instability is of great importance in life. Temporal and spatial regulation of DNA polymerases is of paramount interest for the organism. This might occur via regulation of their expression, their stability and their localization. Many posttranslational modifications contribute to these properties. DNA polymerases are not autistic enzymes but rather work in a broader context within a cell and they can replace one another under certain physiological and even pathological situations (Chapter 5: Global functions of DNA polymerases). Many fundamental mechanistic properties have been elucidated, thanks to the study of DNA polymerases from bacterial and animal cells viruses. The most relevant ones will be described in Chapter 6. A recent field of Chemical Biology has developed techniques that allow the evolution of DNA polymerases in the test tube and thus a variety of novel applications may be ahead of us in the near future (Chapter 7). Many diseases have been correlated with malregulations and malfunctions of DNA polymerases (Chapter 8) and DNA polymerase inhibitors are used as chemotherapeutic agents (Chapter 9). The chapters are written so that they can be read and understood in their own and this will necessarily bring a certain redundancy. We have also made cross-references where appropriate. In summary, this book provides the arguments and evidence that characterize DNA polymerases as enzymes essential to life and also place them among the most important tools in modern chemistry, biology and medicine.

We would like to thank Ursula Hübscher, Ralph Imhof, Nicolas Tanguy Le Gac and Fabio Cobianchi for artwork and the following colleagues who revised chapters of this book: Paul Boehmer, Bill Copeland, Myron Goodman, Ghyslaine Henneke, Jean-Sébastien Hoffmann, Neil Johnson, Enni Markkanen, Andreas Marx, Kristijan Ramadan, Arthur Weissbach and George Wright.

Zurich (Switzerland), November 2009

References

1. Alberts B. 2003. *Nature* 421: 431–5
2. Fry M, Loeb LA. 1986. *Animal Cell DNA polymerases*. CRC Press, Inc. Baton Rouge, Florida.

Ulrich Hübscher
Institute of Veterinary Biochemistry and Molecular Biology,
University of Zurich, Switzerland

Silvio Spadari
Institute of Molecular Genetics IGM-CNR,
National Research Council, Pavia, Italy

Giuseppe Villani
Institute de Pharmacologie et de Biologie Structurale,
CNRS-Université Paul Sabatier, Toulouse, France

Giovanni Maga
Institute of Molecular Genetics IGM-CNR,
National Research Council, Pavia, Italy

Contents

4.   Structural and Functional Aspects of the Eukaryotic DNA Polymerase Families   **111**

5.   Global Functions of DNA Polymerases   **161**

CHAPTER 1

History and Discovery
of DNA Polymerases

1.1 Discovering DNA: A First Step Towards Understanding the Basis of Life

In retrospect, serendipity is often invoked as a major component of scientific discoveries and inventions although scientists and inventors might conceivably be reluctant to admit it. The accidental nature of some or many scientific discoveries does not in our opinion apply to DNA polymerases (pols), discovered in 1955–1957 by Arthur Kornberg (*1918–2007*) and co-workers. As a highly well-prepared and open-minded scientist, he was certainly more able than others to detect the importance of previous information that led to the discovery of DNA, to the demonstration that it carries genetic information, and finally to the elucidation of its structure and function in 1953.

But in his discovery of DNA pols Arthur Kornberg was undoubtedly guided by *"his unremitting fascination with enzymes, molecules which give the cell its life and personality and make things happen"*. He learned his love for enzymes from some of the best enzymologists of his time such as Severo Ochoa (*1905–1993*), Carl F. (*1896–1984*), and Gerty Cori (*1896–1957*). As we shall see later, prior to the discovery of *E. coli* DNA pol, he did in fact fundamental work on the elucidation of coenzymes (NAD, NADP, FAD) and nucleotide biosynthesis in a way that *"one thing had always led to the next one"*. This previous work, his deep belief that enzymes were the key to understanding biochemical processes, and the proposed structure of DNA by Watson and Crick, led him to discover DNA pols and to devote most of his scientific career to the enzymes that assemble DNA.

Although most exciting and fundamental discoveries in DNA research occurred over a period of 20 years in the middle of the twentieth century, research into DNA had begun some 80 years earlier. But the work of many DNA pioneers was completely obscured by the subsequent discovery by Watson and Crick of the double

1

helix structure of DNA and that work is now underappreciated by or even unknown to many young students and researchers. The structure of DNA proposed by Watson and Crick immediately appeared to everybody so naturally logical and unquestionable that there was no time for astonishment and for looking back at the times when the genetic secret of our life was unknown.

Thus, before discussing the discovery of DNA pols, we would like to mention and acknowledge the work of many scientists that led to the discovery of DNA, of its structure and function, and, as a consequence, to the search and discovery of DNA pols by Arthur Kornberg. For the first part of this chapter, references will be limited to few historical books.[1–3]

Following the discovery of the first practical Microscope in late 17th century by A. van Leeuwenhoek (*1632–1723*), research was dominated for several decades by cytologists doing observations of all kinds of material under the microscope. The English Robert Hooke (*1635–1703*) in his *Micrographia*, published in London in 1665, named *cells* the smallest living biological structures surrounded by walls observed in cork. A name derived by the latin *cella*, meaning a "small room" as those where the monks were living in.

Regarding the initial content of this chapter, the discovery of DNA, the way to the knowledge of *variation of individuals in certain characters, of the heredity of parental characters and finally of natural selection that allowed some individuals to survive and reproduce better than others*, was opened many years later when, in 1858, Charles R. Darwin (*1809–1882*) and Alfred Russel Wallace (*1823–1913*) published their joined communication "*On the tendency of species to form varieties and on the perpetuation of varieties and species by natural means of selection*",[4] thus suggesting the conception of the *struggle for existence and of the survival of the fittest*. This communication was followed in November 1859 by Darwin's book *On the Origin of Species by Means of Natural Selection* published in London. Then a few years later (1865–1869), Gregor Mendel (*1822–1884*) published in a local journal the results of his experiments with pea plants, made in the monastery garden of the convent at Brünn in Moravia, on the *nature of genetic inheritance of particular traits carried by units*. Almost unnoticed for some decades by the scientific world because his communications were to a local journal, Mendel's work defined the basic rules of heredity. In 1905 the Danish scientist W. L. Johannsen (*1857–1927*) called these *units*, residing in the chromosomes, *genes* and also introduced the terms *genotype, phenotype* and *biotype* that are now part of the common language of genetics. Historically, however, the terms "*genos*" (family or clan) and "*gonos*" (sperm or seed), forefathers of "*gene*", appeared for the first time in a text of *Corpus Hippocraticum* where Hippocrates stated that the so-called Sacred Disease (Epilepsy) was not a divine but a hereditary disease.

According to C.D. Darlington (*1903–1981*), an English biologist and geneticist, who significantly contributed to understand chromosomal crossing-over and its role in inheritance and evolution, it was the German botanist and biologist Wilhelm Friedrich Benedikt Hofmeister (*1824–1877*) who first observed, in 1848, small *rodlike bodies* in plant cells nuclei during mitosis, at least 30 years in advance of another German biologist, Walther Flemming (*1843–1905*). Thanks to developments in microscopy, Flemming, in 1879, observed and described within the nucleus *tiny thread-like structures,* which strongly absorbed basophilic dyes, and he called them *chromatin* from the Greek word for "color". Later, in 1888, they were called *chromosomes* — colored bodies — by the German anatomist Wilhelm von Waldeyer-Hartz (*1836–1910*) for their ability to absorb certain dyes. Flemming further investigated the process of cell division and the distribution of such chromosomes to the daughter nuclei, a process he called *mitosis* (from the Greek word *mitos* meaning "thread"). Flemming hypothesized that all cell nuclei came from another predecessor nucleus. He coined the phrase *"omnis nucleus e nucleo"* ("every nucleus from nucleus") after Virchow's (*1821–1902*) *"omnis cellula e cellula"* ("every cell from a cell"), but he did not realize the splitting of chromosomes into identical halves, the *daughter* chromatids. Unaware, as other men of science of his days, of Gregor Mendel's work on heredity, published, as we mentioned, in a local journal and left unnoticed for a long time, Flemming did not make the connection between his observations and genetic inheritance.

1.1.1 *Nuclein*

It was in 1869, almost a century before the Nobel Prize awarded to Watson, Crick and Wilkins for elucidating the structure of DNA, that a young Swiss physician, Friedrich Miescher (*1844–1895*) isolated a substance that he called *nuclein*, now known as *DNA*, as reviewed in Refs. 5 and 6. Just after receiving his M.D., encouraged by one of his Professors Ernst Felix Immanuel Hoppe-Seyler (*1825–1895*), Miescher chose scientific research rather than Medicine as a career.

Only few years earlier Gregor Mendel had finished a series of experiments with peas and made observations that turned out to be closely connected to the finding of *nuclein*. Mendel was in fact able to show that certain traits in the peas, such as their shape or color, were inherited in different units (later called *genes*), but his work went unnoticed for many years.

Thus at a time when scientists were still debating the concept of the *cell* and of its origin from another cell, Hoppe-Seyler, one of the pioneers in physiological chemistry, and his lab at the Faculty of Natural Sciences in Tübingen, were trying to isolate and study the molecules that made up cells and cellular organelles such as the *nucleus* (a latin name meaning *core* or central part of something, given in 1831 by the Scottish Robert Brown (*1773–1858*) best known for discovering and describing the random "Brownian" motion of molecules).

Miescher was thus given the task of researching the biochemical composition of lymphoid cells, the white blood cells, and of their organelles. He originally obtained the cells from lymph glands with the aim to isolate undamaged nuclei free of cytoplasm and to study their components. This had never been accomplished before, but he soon realized that it would be quite difficult to get enough of these cells for chemical analysis of their fundamental components. So he switched to leucocytes, the main cellular constituent of the pus that could be obtained fresh every day from used bandages in the nearby hospital. To avoid damaging the cells, before trying to separate the nuclei from the protoplasm, he washed them with various salt solutions but found that the cells swelled *giving rise to a highly viscous porridge that was very difficult to handle*. Miescher somehow realized that the substance could belong to the nuclei suspected, in those years, to contain the factors responsible for transmission of hereditary tracts. We now know that Miescher had indeed extracted high molecular weight DNA from damaged cells, and he did not loose the opportunity to study the chemistry and possibly the function of the nucleus.

So he soon developed protocols to better purify these cells and to isolate their nuclei and, after several attempts, he finally used a diluted solution of sodium sulfate to rinse the cells from the bandages without damaging them, removed tissue fibers by filtration, and left the cells to sediment to the bottom of the beaker (no laboratory centrifuges were available in those days). Following this procedure the leucocytes appeared morphologically intact at the microscope and showed no sign of damage. In the attempt to isolate the nuclei, he then treated the cells with warm alcohol to remove lipids, digested away the proteins of the cytoplasm with pepsin extracted from pig's stomach with dilute hydrochloric acid and thus lysed the cells, stripping off most of the cytoplasm. He obtained a greyish precipitate of cell nuclei which was subsequently redissolved in alkaline solutions, and on reacidification he again obtained a white, flocculent precipitate of gelatinous material that he therefore named *nuclein*. The quantity of isolated nuclear material was however too small for chemical analyses, and Miescher refined his purification procedure and applied it to other cells, in particular those of salmon sperm. Then by elementary analysis, one of the few methods then available to characterize an unknown compound, Miescher found that this new substance contained elements typical of organic molecules such as carbon, hydrogen, oxygen, nitrogen and sulfur (that indicates to us today that his preparation did contain contaminating proteins). Interestingly there was a unique ratio of phosphorus to nitrogen — 14 percent nitrogen, 3 percent phosphorus — distinct from that of known organic molecules (carbohydrates and proteins) and this made him confident that he had discovered a new type of organic molecule *perhaps serving to store phosphorus or acting as reservoir for other molecules.* Although Miescher did most of his work in 1869, his paper on nuclein was not published until 1871. Nuclein was such a unique molecule that his mentor Prof. Hoppe-Seyler was skeptical and wanted to confirm Miescher's results before publication.

In the next years Miescher found in many other tissues and cell types, including yeast cells, such phosphorus-containing substance and suggested that nuclei be defined by the presence of *nuclein* rather than by their morphological properties, getting quite close with his speculations to the concept that the nucleus is responsible for heredity although he remained uncertain about its function in the cell. It would be many years before the role of nucleic acids were recognized and Miescher himself, unable to conceive that a single substance might be responsible for the transmission of hereditary traits, like his contemporary scientists, believed that proteins were the molecules of heredity. He somehow established that all the phosphorus in nuclein was present as phosphoric acids, determined the P_2O_5 content to be 22.5% of the total mass (very close to the actual proportion of 22.9%), confirmed its acidic properties, concluded that it must be a molecule of high molecular weight and certainly laid the groundwork for the molecular discoveries that followed, encouraging chemists to further investigate nuclein.

1.1.2 *Nucleic Acid*

Ten years later in fact the German Albrecht Kossel (*1853–1927*), another physician much more interested in physiological chemistry than in medicine, who had as well served as assistant in Hoppe-Seylers's Institute of Physical Chemistry in Strasbourg from 1877 to 1881, where Miescher had isolated *nuclein* 10 years earlier, explored the chemistry of nuclein further. *Nuclein* was still thought to be a phosphorus-rich protein. With Richard Altmann (*1852–1900*), Kossel isolated a protein-free nuclein showing that Miescher's nuclein was actually composed of a protein contaminating portion and a non-protein portion. In 1889 Altmann, considering the protein a subcomponent of nuclein of yeast cells, proposed for the deproteinized material the more appropriate term *nucleic acid* that, by the 1930s, became *deoxyribose nucleic acid* and then *deoxyribonucleic acid* (DNA), thus serving to obscure Miescher's discovery of DNA.

Over the next 20 years Kossel and his research team made other groundbreaking discoveries about the structure of both the nucleic acids and cellular proteins. Using hydrolysis and other techniques to chemically analyze the nucleic acids, he discovered the components adenine, cytosine, guanine, thymine, and uracil, and the fundamental building blocks of nuclein, purine and pyrimidine bases, one sugar and phosphoric acid. He also determined their structures, showing that a pyrimidine has a single six-member ring, while a purine has a six-member ring that shares one side with a five-member ring. He isolated the two purines adenine and guanine, and showed that they existed in both ribonucleic acid (RNA) and deoxyribonucleic acid (DNA). He also isolated the three pyrimidines: thymine and cytosine, found in DNA, and uracil (present in RNA instead of thymine). Adenine was composed of carbon, hydrogen, and nitrogen; the others also contain oxygen.

In addition to these findings, Kossel was the first to isolate the protein histone and found that nuclein, together with histones, was a key component of chromatin, the structural material of the chromosomes that supports the DNA. He also discovered the amino acid histidine (1896), thymic acid, and other important biological compounds and in 1910 was awarded the Nobel Prize in Physiology or Medicine for his contributions to the knowledge of cell chemistry.

Kossel was always motivated to find the biological functions of the chemicals he studied and isolated. Several times he expressed his conviction about the importance of nucleic acids not as a source of energy or storage materials but rather in the synthesis of cytoplasm and nucleoplasm during cell multiplication. The terms *cytoplasm* and *nucleoplasm* had just been introduced in 1882 by E. Strasburger (*1844–1912*) to describe the protoplasm of the cell-body and of the nucleus, respectively. But the importance of nucleic acids remained obscure for several other decades although back in 1874 Leopold Auerbach (*1828–1897*) observed that fertilized worm oocytes contained two nuclei which fused prior to the first cell division. Oscar Hertwig (*1849–1922*) concluded that one of the nuclei came from a spermatozoon and the other one from the oocyte and that the fused nucleus gave rise to all subsequent nuclei during the animal development.

1.1.3 *Nucleic Acids Are Composed of Nucleotides*

Kossel's work was continued in particular by the physician and biochemist Phoebus Levene (*1869–1940*), who had once studied with him. Born in Russia, he left his country because of growing anti-Semitism and immigrated with his family to the United States. Between 1896 and 1905 he worked with a number of well-known chemists, including Albrecht Kossel and Emil Fischer (*1852–1919*), the nucleic acid and protein experts of the time, respectively, and was just introduced by Kossel to the study of nucleic acids at a time when still little was known about their structure and function. Levene did most of his nucleic acid work at the Rockefeller University, where in 1907 he was hired by the newly established Rockefeller Institute of Medical Research to head the biochemical laboratory and stayed there until his death.

In 1910, Levene found that the carbohydrate present in yeast nucleic acid, which differed from thymus nucleic acid for the presence of uracil instead of thymine, was the pentose sugar *ribose.* Following a chemical test for *deoxyribose*, developed by Robert Feulgen (*1884–1955*) during the 1920s, the first test capable of distinguishing DNA from RNA, he found in 1929 that the carbohydrate in thymus nucleic acid was also a pentose sugar but lacking one oxygen atom of ribose and was therefore called *deoxyribose.*

Although he understood that the nucleic acids components were linked together in the order *phosphate-sugar-base*, around 1910 Levene advanced the wrong hypothesis that a nucleic acid was composed of identical units containing 1A, 1T, 1G, and 1C (*tetranucleotide hypothesis*), units that were repeated in the DNA molecule to form a polynucleotide with the bases occurring in the same order throughout and their amounts the same in all DNA molecules, whatever their origin. He also concluded, incorrectly, that the linkage between the bases was a phosphomonoester linkage between the ribose residues (rather than the phosphodiester linkage subsequently found). Such a simple repetitive structure implied to Levene that DNA was too uniform to store a genetic code, and, therefore to contribute to complex genetic variation. For several additional decades, until Avery's and Chargaff's work in the 1940s, this opinion was widely accepted.

Because, like Miescher and many others, Levene was obviously studying the structure of DNA fragments, the extraordinary length of the DNA molecule as it appeared later in the 20th century escaped his attention. They believed it was a relatively small molecule, probably about 10 to 12 nucleotides long. So Levene, although successful in demonstrating the difference between RNA and DNA, and in showing that the molecule was a string of nucleotide units, imposed on them an incorrect, simple and repetitive tetranucleotide structure for which now he is mostly remembered. Despite this Levene published over 700 original papers and articles on the chemical structures of nucleic acids and many other biochemicals and he almost had the possibility to complete, a few decades before James Watson and Francis Crick, the revolution begun by Miescher in 1869 and by the many other biochemists, some of them we have mentioned here. Levene received a Nobel Prize in 1902 for his outstanding contribution to the chemistry of nucleic acids (composition of and the linkages in both DNA and RNA) but died in 1940 just before the true significance of DNA became clear. He remained still rather skeptical when he was informed, shortly before his death in 1940, of the preliminary results of Oswald Avery's work.

1.1.4 *DNA Is the Genetic Material*

Until this time and consequent to Levene's tetranucleotide hypothesis, proteins were recognized as the most important components of the cell (the name deriving from "*proteios*" meaning "*most important*"), and the general consensus was that the more complex proteins were the most logical candidate for the genetic material. Complex proteins contained 20 different amino acids shown by the German chemist Emil Fischer to be linked together in diverse sequences, versus the simplest structure of DNA containing only equimolar quantities of four bases. Therefore, proteins appeared at first to have a greater diversity and a more complex structure, certainly more suitable to assure the great diversity of the genes such as those mutations affecting color in flowers or change in the color of the eyes in Drosophila. The much simpler and uniform DNA was not considered to be a good candidate as carrier of genetic information and was rather thought to play a secondary role such as in chromosomal stability and maintenance. Even when the English physician A. Garrod (*1857–1936*), very interested on inborn errors of metabolism, in 1909 conceived *alkaptonuria*, *porphyria* and *albinism* as genetic diseases and the Swedish Torbjörn Caspersson (*1910–1997*) showed, in the 1930s, that DNA was a high molecular weight polymer, most people continued to believe in Levene's *tetranucleotide hypothesis*. The millions of nucleotides were still thought to be arranged in a monotonous and predictable way with no meaningful genetic information content.

A few decades later, however, novel findings pointed toward the conclusion that DNA was the genetic material. First it was the English army doctor Frederick Griffith (*1879–1941*) who, in 1928, inspired by Pasteur, in the attempt to make a vaccine against the bacterium *Streptococcus pneumoniae*, responsible of a type of pneumonia and much-feared in the days before antibiotics, discovered that a nonpathogenic **R** mutant (forming *rough* colonies because it lacked an enzyme required for the synthesis of the capsular polysaccharide) could be transformed into the pathogenic **S** form (forming *smooth* colonies). He found that a mixture of harmless heat-killed **S** bacteria and harmless live **R** bacteria was lethal when injected into mice, whereas live **R** or heat-killed **S** pneumococci were not lethal when injected alone. Furthermore **S**-type pneumococci were recovered from the blood of dead mice demonstrating that heat-killed **S** bacteria had transformed live **R** pneumococci into live **S** bacteria *in vitro*, suggesting that genetic material rescued from the mice had been transferred from dead to live bacteria.[7] Thus, Griffith's demonstration that genetic information could be passed from one bacterium to another must somehow be regarded as the first experiment of genetic engineering although he unfortunately failed to identify DNA as the specific molecule.

In 1941 George Beadle (*1903–1989*) and Edward Tatum (*1909–1975*) demonstrated in *Neurospora crassa* a precise relationship between genes and enzymatic proteins that subsequently led to the proposal of the *one gene–one enzyme hypothesis*.[8]

The definition of DNA as *transforming principle*, therefore genetic principle, was achieved in 1944 by Oswald Avery (*1877–1955*), Colin MacLeod (*1909–1972*), and Maclyn McCarty (*1911–2005*) at the Rockefeller Institute.[9] Inspired by the previous findings of Frederick Griffith, they made a more definitive experiment and demonstrated that DNA, and not proteins or other materials, was the transforming factor. They first extracted and partially purified nucleic acids, proteins and other materials from **S** bacteria. When they mixed each of them separately with **R** bacteria, they found that only **R** bacteria mixed with nucleic acids were transformed into **S** bacteria and killed mice. The transforming activity was resistant to proteases and ribonuclease but sensitive to deoxyribonuclease, thus providing definitive evidence that the genetic component of these cells was DNA. They called the uptake and incorporation of DNA by bacteria *genetic transformation*. In 1946–1947, Joshua Lederberg (*1925–2008*) and Edward Tatum (*1909–1975*) described exchange of genes in bacteria (*bacterial conjugation*)[10,11] and in 1952 N. Zinder and J. Lederberg[12] discovered gene transfer from phages to bacteria (*transduction*). For these studies J. Lederberg, at the age of only 33 years, shared the 1958 Nobel Prize in Physiology and Chemistry with G. Beadle and E. Tatum.

Further and definite support for the genetic role of DNA came in 1952 when Alfred Hershey (*1908–1977*) and Martha Chase (*1927–2003*) investigated the infection of *Escherichia coli* cells with a virus, *bacteriophage T2*, whose core of DNA is surrounded by a protein coat.[13] The year before Roger Herriot (*1908–1992*) suggested that "*the tail of the virus, that as such never enters the cell, may act like a little hypodermic needle contacting the host bacteria, making enzymatically a small hole through the outer membrane and allowing its nucleic acids to flow into the cell*".[14] Herriot's suggestions were tested and fully proved by Hershey and Chase by specifically labeling phage DNA with ^{32}P (phosphorus is present in DNA but not in proteins) and coat proteins with ^{35}S (sulfur is present in proteins but not in DNA) prior to infection. Following centrifugation, most of the ^{32}P phage DNA was found in the bacteria and the ^{35}S proteins in the supernatant. They could conclude that "*a physical separation of the phage T2 into genetic and non-genetic parts was achieved . . .*" and that "*the genetic material of bacteriophage T2 was DNA*".

Additional support then came of course from the studies on cell division and on the behavior of chromosomes during mitosis and meiosis, with daughter cells resembling parental cells. In a given species the DNA content is the same for the cells with a diploid set of chromosomes and *half* in haploid cells. The chromosomes are therefore the carriers of genes, and the genes are the basic units of heredity.

1.1.5 *Structure of DNA: The Watson–Crick DNA Double Helix and Mechanism of DNA Replication*

Once the genetic role of DNA was definitely established (the only exception being the case of RNA viruses, in which RNA is the only nucleic acid present in the virus and therefore the genetic material) biochemists started asking about the true structure of DNA and how DNA molecules were exactly copied during chromosomal replication to transfer the genetic information.

Most impressed by Avery's work was the Austrian chemist, Erwin Chargaff (*1905–2002*) who wrote "*I saw before me in dark contours the beginning of a grammar of biology. Avery gave us the first text of a new language, or rather he showed us where to look for it. I resolved to search for this text*". A pioneer in paper chromatography of nucleic acids, he thus studied the base compositions of DNAs from different species and demolished Levene's tetranucleotide hypothesis. He in fact found that each species differed in the amount of A, C, G and T whereas, within the species, the proportion of each is identical, no matter which tissue the DNA is extracted from. It was just what might be expected for a molecule that is the genetic material and thus a biological signature for the species. Even more significant was Chargaff's further discovery in 1950 that "*the ratio of Adenine to Thymine and of Guanine to Cytosine were nearly 1.0 in all species studied*",[15] a rule that became known as *Chargaff's ratios*. The rule turned out to be as meaningful and crucial as Franklin's X-ray diffraction pictures of DNA when the three-dimensional structure of DNA, based on the specificity of base pairing, was proposed three years later by Watson and Crick. Their announcement of the three-dimensional structure of DNA, from which its mechanism of replication could be immediately inferred, came in 1953, and the final phase of this puzzle relied on X-ray diffraction photographs of DNA fibers and on Chargaff's observations.

The X-ray technique was invented in 1913–1914 by William Bragg (*1862–1942*) and his son Lawrence (*1890–1971*), with whom he shared the Nobel Prize in Physics in 1915, and William Astbury (*1898–1961*), a student of William Bragg. It was used from the 1930s by Dorothy Hodgkin (*1910–1994*) to study and solve the structures of large biological molecules such as penicillin, lysozyme and vitamin B12, and by Max Perutz (*1914–2002*) with hemoglobin. In 1938, Astbury obtained X-ray pictures of DNA given to him by Caspersson, but they were hard to interpret.

In the late 1940s to early 1950s, separate groups were intensively working on DNA structure, in particular Maurice Wilkins (*1916–2004*) at King's College, London, joined in 1951 by Rosalind Franklin (*1920–1958*). A British physical chemist with a great reputation in X-ray crystallography, Franklin soon produced the best images of B-form DNA (the predominant form of DNA in solution), suggesting that maybe the DNA molecule was coiled into a helical shape. On the other hand, Linus Pauling (*1901–1994*), an American chemist who had discovered helical motifs in proteins structures, also begun to think about a possible helical structure for DNA but erroneously put the phosphate groups on the inside and the bases on the outside proposing a model of three intertwined chains.

In 1949, Francis Crick (*1916–2004*), an English physicist, had joined Max Perutz and John Kendrew (*1917–1997*) at the Cavendish Laboratory of Cambridge University, still under the general direction of Sir Lawrence Bragg, working on X-ray crystallography of proteins. He quickly learned the mathematical theory of X-ray diffraction by helical molecules, and when, always fascinated by grand designs, he joined the forces with James Watson, a younger chemist with expertise in the molecular biology of phages, they found themselves in the right place at the right time to reveal the molecular structure of DNA.[16,17]

As mentioned before, among the keys to developing and proposing the correct structure of DNA were: (1) the appearance of one of Franklin's photos of the so-called B form of DNA, much simpler and beautiful than those of the A form of DNA previously obtained by William Astbury; this photo definitely suggested to Crick the helical structure for the DNA molecule; (2) a visit by Erwin Chargaff to England in 1952 following his discovery that the *"the ratio of Adenine to Thymine and of Guanine to Cytosine were nearly 1.0 in all species studied"*, reinforcing the importance of this observation to Watson who persisted in building structural models (using metal plates for the nucleotides and rods for the bonds between them); and finally (3) a visit to the Cavendish Laboratory by the American chemist Jerry Donohue (*1920–1985*), who pointed out *"how hydrogen bonding allows A to bond to T and C to G"*.

All the above observations allowed Watson and Crick to propose a double helical structure for DNA, where the two strands had the hydrophobic bases packed and paired up by hydrogen bonds into the core (A with T and C with G so that all base pairs had the same shape and size), with their plane perpendicular to the common helix axis. This stabilized the double helix against the electrostatic repulsion between the negatively charged phosphate groups and, at the same time, provided a way to easily "unzip" the two complementary strands for replication. The hydrophilic phosphate-deoxyribose containing backbones of the nucleotide chains of DNA were positioned on the outside of the molecule so as to interact with water molecules and form right-handed helices. They could be imagined as the sides of a ladder whose rungs were the restricted base-pairs in the middle. Furthermore, the two chains had antiparallel orientation in the sense that they run in opposite directions: one in the direction P-3′-sugar-5′-P and the complementary one in the opposite direction. According to their model building the antiparallel orientation of the two strands favored the orientation of the base pairs in the center of the double helix.

As we said in the introduction, the discovery of the DNA structure proposed by Watson and Crick in 1953 was without precedent and it obscured all the previous work. Understanding the structure of this molecule meant understanding how it functioned (next section). Since then, the idea that every cell contains all the genetic information that allows its functioning and that our genetic destiny is all written in our chromosomes, became an integral part of our perception of life.

1.2 Imaging an Enzyme that Assembles the Nucleotides into DNA

The beauty of the DNA double helix model proposed by Watson and Crick was that the structure immediately suggested its function. As they hinted in their Nature paper: "It has not escaped our notice that the specific pairing we have postulated suggests a possible copying mechanism for the genetic material". The chemical complexity of the molecule was thought to be sufficient to store the genetic information and definitely settled a decades-long controversy over whether DNA or protein was the "life molecule". The authors had from the beginning the idea that the structure of DNA should allow the molecule to copy itself during cell division, so that an exact copy could pass into each new cell. Indeed the DNA molecule was self-replicating following the unwinding of the two complementary and antiparallel strands. "Each chain would act as a template for the formation on to itself of a new companion chain so that eventually we shall have two pairs of chains, where we only had one before". Moreover, the sequence of the pairs of bases will have been duplicated exactly because each base would attract its complementary one, by hydrogen bonding, so that two new double helices are assembled. Watson and Crick also proposed that one of the strand of each daughter molecule was newly synthesized and the other one was derived from the parental molecule. This *semiconservative mechanism of DNA replication* was soon (1958) demonstrated by Mathew Meselson and Franklin Stahl.[18]

The discovery of DNA structure started a new era in biology and the following years led to the complete elucidation of the genetic code, as reviewed in many articles and books such as those by Ycas[19] and Woese[20] and to the realization that DNA also directs the synthesis of proteins. This was enunciated by F. Crick in the so called "central dogma of molecular biology" (DNA $\rightarrow$ RNA $\rightarrow$ Protein) that implies a one-way flow of genetic information between macromolecules through DNA $\rightarrow$ DNA (*DNA replication*), DNA $\rightarrow$ mRNA (*transcription*) and mRNA $\rightarrow$ Protein (*translation*). Three kinds of RNA, synthesized by RNA polymerase, were found, i.e. messenger RNA (mRNA), transfer RNA (tRNA), and ribosomal RNA (rRNA). The mRNA transcript of DNA was indeed found to be the template for protein synthesis; tRNAs carried amino acids to the ribosomes and sequentially read on mRNA codons of three bases that specify an amino acid, and rRNA is the major component of ribosomes.

Robert W. Holley (*1922–1993*), Gobind Khorana and Marshall Nirenberg (*1927–2010*) shared the Nobel Prize in Physiology or Medicine in 1968 for their interpretation of the genetic code and its function in protein synthesis. However, other Nobel laureates such as Crick and S. Brenner, who found that amino acids are coded by groups of three bases, S. Ochoa, who isolated the polynucleotide phosphorylase, the enzyme that allowed the synthesis of many different RNAs, Francois Jacob and Jacques Monod (*1910–1976*), with their proposal of the mRNA hypothesis, contributed significantly to this outstanding accomplishment.

Going back to the DNA structure, the simplicity of the model for DNA replication, thanks to the complementarity of A to T and of G to C, even induced Watson and Crick to think that nucleotides, paired to the DNA template, might be zipped together without any enzyme action. Perhaps it was sufficient for the nucleotides to align spontaneously, because of their reciprocal affinity, along each of the two separated complementary strands, and the DNA was duplicated! But to a biochemist such as Kornberg this must have sounded like a heresy. Only a scientist who knew how nucleotides were synthesized and activated in the cell could imagine an enzyme that would assemble nucleotides into DNA.

As we have mentioned earlier, Kornberg, born in Brooklyn, N.Y., in 1918 and graduated in Medicine in 1941 from Rochester University School of Medicine, had learned *his unremitting fascination with enzymes* from some of the best enzyme biochemists of his time. He worked with Severo Ochoa in 1946 at New York University Medical School "learning the philosophy and practice of enzyme purification". He spent six months in 1947 with Carl F (*1896–1984*) and Gerty (*1896–1957*) Cori at Washington University Medical School in St. Louis, and was fascinated and inspired by their discovery of glycogen phosphorylase. They all contributed toward his scientific formation, made him fall in love with enzymes and convinced him that *enzymes were the key, the most effective way to understanding intracellular biochemical processes.*[21,22]

Thus, in the late 1940s to early 1950s, when it was becoming clear that pathways of degradation/energy production and biosynthesis were distinct, Kornberg returned to the NIH and started an "Enzymes and Metabolism Section" that included Leon Heppel and Bernard Horecker. He started to purify and to study enzymes that assembled various coenzymes such as NAD^+, NADP, FAD, thus greatly contributing to focusing the interest of biochemists in biosynthetic pathways. He first purified nucleotide pyrophosphatase from potatoes and found that the enzyme cleaved the pyrophosphate bond of NAD (nicotinamide adenine dinucleotide) releasing NMN (nicotinamide mononucleotide) and AMP. This led Kornberg to discover the enzyme NAD synthetase that utilized ATP to transform NMN into NAD with the release of pyrophosphate.

Then Kornberg discovered that FAD (flavin adenine dinucleotide) was synthesized by a similar mechanism as well. Thus, the mechanism involving nucleotidyl transfer from ATP (or any other nucleoside triphosphate such as UTP and CTP) and the release of pyrophosphate turned out to be a general mechanism for the synthesis of the precursors of each class of macromolecules in cells and tissues, proteins, lipids, carbohydrates, and, somehow, the nucleic acids. The building blocks, amino acids, acetic acid, sugar, nucleotides first react with ATP (or UTP or CTP) to become activated precursors before being added to the corresponding molecule through this mechanism that involves the nucleotidyl transfer from a nucleoside triphosphate with the release of pyrophosphate. The pyrophosphate is then hydrolyzed to inorganic orthophosphate by an efficient inorganic pyrophosphatase present in all cells, thus driving the reaction in the direction of the synthesis.

The synthesis of DNA and RNA, in fact, also resembled that of the coenzymes NAD and FAD, each nucleotide being incorporated as the monophosphate with release of pyrophosphate. This research earned Kornberg the Annual Paul-Lewis award in enzymology (now the Pfizer Award) and started attracting applications from young and promising biochemists. But the mechanism of the synthesis of nucleic acids was still far away.

Then in the late 1940s to early 1950s, with John Buchanan at the University of Pennsylvania and MIT, and Robert Greenberg at the University of Michigan, who were investigating the biosynthesis of purine nucleotides A and G, Kornberg focused on pyrimidine biosynthesis. He studied orotic acid and showed first how a liver enzyme could produce an activated form of ribose, phosphoribosyl pyrophosphate (PRPP), from ribose phosphate and ATP, and then how two yeast enzymes could transfer the activated ribose to orotic acid and then decarboxylate orotic ribose-phosphate to uracil ribose phosphate (UMP). UMP and other nucleoside monophosphates were then converted to di- and tri-phosphates by nucleoside monophosphate kinases that utilize ATP as the phosphoryl donor. Thus, UMP could be converted to UDP and UTP, and CTP could be formed by amination of UTP. Rather than uracil, DNA contains thymine, the 5'-methyl derivative of uracil; thus uracil deoxyribose phosphate (deoxyuridylate, dUMP) is methylated to thymine deoxyribose phosphate (deoxythymidylate, dTMP). Contrary to the biosynthesis of pyrimidine nucleotide, where ribose phosphate from PRPP is added to the pyrimidine ring, for example orotic acid, in the byosinthesis of purine nucleotides such as AMP,

Buchanan *et al.* found that PRPP reacts with smaller molecules, such as amino acids, in building the purine structure.

But soon Kornberg *et al.* found enzymes that could condense preformed adenine and guanine, such as those obtained from the breakdown of RNA and DNA by enzymes that degrade these molecules, with PRPP to make nucleotides directly. It thus became clear that there were two pathways for the biosynthesis of nucleotides: the *de novo pathway* in which the activated nucleotides are built from simpler molecules (amino acids, carbon dioxide, sugar phosphate etc.) and a *salvage pathway* in which A, G, C, U and T obtained from the hydrolysis of RNA and DNA are phosphorylated and recycled to produce activated nucleotides.

1.2.1 *DNA Polymerase Activity in Extracts of Escherichia coli*

In 1953 Kornberg resigned from NIH to become Professor and Chairman of the Department of Microbiology, Washington University, St. Louis. Having elucidated the biosynthesis of several coenzymes and learned how nucleotides are synthesized and activated in the cells, he was ready to try the synthesis of DNA in a broken-cell extract and to search for enzyme(s) that assemble the nucleotides into DNA. Of course Watson and Crick had just proposed in 1953 a model for DNA replication but also did not exclude a *spontaneous* assembly of nucleotides in the synthesis of a DNA chain directed by base pairing with each strand of the parental duplex. On the contrary, Kornberg, educated by Carl and Gerty Cori and by Ochoa, among the most famous enzymologists of that time, was well aware that one needed enzymes to understand biological events, and he was convinced that *a biochemist devoted to enzymes could reconstitute any metabolic event in the test tube as well as, or even better than, the cell does it.*

Kornberg thus hypothesized the need of an enzyme assembling the nucleotides into DNA in cell extracts, just like the synthesis of glycogen and fats were achieved outside living cells. And the search for such an enzyme, then named DNA polymerase (pol), was indeed soon successful in Kornberg's laboratory, thanks also to some young scientists such as I.R. Lehman, M. Bessman, and the technician E. Simms, later joined by J. Adler and S. Zimmermann. They made several appropriate observations and choices at the beginning and during earlier experiments with a broken-cell extract of *E. coli*, a rapidly growing bacterium and, therefore, a good source of the enzyme. A brief summary of these observations is given below:

(1) First, in December 1955, Kornberg's group obtained some [^{14}C]thymidine (a known constituent of DNA) from M. Friedkin, a colleague in the Pharmacology Department at the Washington University School of Medicine. They showed that after one hour of incubation with the bacterial cell extract, ATP, Mg^{2+} and buffer, a

very tiny amount of radioactivity was trapped in an acid-insoluble pellet, presumably DNA because of its sensitivity to deoxyribonuclease (DNase).

(2) Second, they identified dTMP in the reaction mixture as well as additional products of phosphorylation such as dTDP and dTTP. This suggested that they should look into *E. coli* extracts and purify an activity-converting thymidine into dTMP (thymidine kinase) useful for the synthesis of [32P]-TMP. The incorporation of [32P]-TMP, rather than [14C]thymidine, led to a dramatic increase in terms of counts/min in the acid-insoluble pellet, and also suggested that they synthesize [32P]-dTTP as a possible direct precursor for DNA synthesis.

Thymidine diphosphate (dTDP) was also a likely possibility because of the recent finding by Grunberg-Manago and Ochoa that ribonucleoside diphosphates, rather than triphosphates, were the substrates *in vitro* for a ribonucleotide-polymerizing enzyme (polyribonucleotide phosphorylase) from the bacterium *Azotobacter vinelandii*.[23] But there was no evidence of a template-directed synthesis of RNA, and the reaction was indeed reversible only *in vitro*. Thus polynucleotide phosphorylase *in vivo* turned out to be the enzyme involved in messenger RNA degradation rather than in RNA synthesis. Experimental evidence supporting the conclusion, in analogy with the synthesis of DNA, that the four rNTPs, a DNA template and a DNA-dependent RNA polymerase were required for RNA synthesis, came in the following years by the laboratories of Audrey Stevens, Sam Weiss, and Jerard Hurwitz.[24]

However, as we said previously, the fact that this enzyme, in the *in vitro* reverse reaction, was template-independent and the synthesized RNAs had a composition depending on the ratios of the ribonucleoside diphosphates present, proved quite useful for the *in vitro* synthesis of different RNAs that helped to define the genetic code.

This was a time when more was known about degradation rather than biosynthetic pathways. The glycogen phosphorylase of Carl and Gerty Cori, the enzyme which inspired Arthur Kornberg, could only elongate glycogen with glucose-1-phosphate *in vitro*. Later, in 1957, the Argentinian doctor and biochemist Luis Leloir (*1906–1987*), a 1970 Nobel laureate in Chemistry, demonstrated that the synthesis of glycogen *in vivo* was not the reverse of the degradation and was catalyzed by glycogen synthase, an enzyme that elongates glycogen using UDP-glucose. Thus *in vivo* synthesis and degradation of glycogen, as well as of other important biological molecules, are coordinated and one is switched off when the other one is active and *vice versa*.[25]

(3) Kornberg and his group correctly sensed that the DNA present in the extract was degraded to deoxyribonucleoside monophosphates (dNMPs) by endogenous endonucleases and that dNMPs were then converted to dNTPs by the corresponding

kinases in the presence of ATP. This was in agreement with the pathways of purine and pyrimidine biosynthesis leading to nucleoside-5′-phosphates and with his previous finding that ATP was the activated intermediate in the synthesis of coenzymes such as NAD, FAD and CoA. Thus, the Kornberg group purified the four [^{32}P]-dNMPs generated from [^{32}P]-labeled DNA (isolated from [^{32}P]-labeled *E. coli*) and converted them to the corresponding labeled dNTPs with ATP and the nucleotide kinases and nucleoside-diphosphate kinase they had partially purified from *E. coli*.

(4) Kornberg originally thought that DNA was needed in the reaction as a primer that was elongated by DNA pol, by analogy with the work of Carl and Gerty Cori on the elongation of glycogen by glycogen phosphorylase. He did not immediately think about DNA functioning as a template, as earlier suggested by Watson and Crick on the basis of its structure and of the opening of the double helix. (The initial lack of acknowledgment by Kornberg of the template function of DNA, and of the work of Watson and Crick, was probably one of the reasons for the caustic comments of one referee who rejected the JBC papers in 1957! See below). When Kornberg found that all four dNTPs were required, he admitted that the added DNA was serving as a primer and template, and indeed he subsequently demonstrated that the DNA synthesized by DNA pol was a faithful copy of the added DNA template (see next section).

(5) Finally, with the recognition that [α-^{32}P]-dNTPs were the true substrates for a DNA pol, the enzyme that converted them into an acid-insoluble product, (i.e. DNA), was finally purified in quantities that allowed its further characterization.

By analogy with Miescher's difficulties to publish his discovery of *nuclein* essentially because of the skepticism of his own mentor Hoppe-Seyler, who wanted to confirm himself the results before publication, the first two reports by Lehman *et al.* on the enzymatic preparation of the four dNTPs and the partial purification of *E. coli* DNA pol, and by Bessman *et al.* on the general properties of the reaction, raised much criticism by the reviewers and were declined by the *Journal of Biological Chemistry* in the fall of 1957. *"It is very doubtful that authors are entitled to speak of the enzymatic synthesis of DNA"* and *"Polymerase is a poor name"* etc. were some of the critical comments. The two papers[26,27] were accepted several months later by John Edsall, the new Editor-in-Chief.

Thus, Kornberg correctly imagined the DNA pol as an enzyme extending a DNA chain by successive additions of a properly activated nucleotide, but he added DNA to the reaction mixture *never imagining at the beginning to detect an enzyme that, unlike any others, took directions and instructions from its DNA substrate thus assembling nucleotides by Watson–Crick base pairing to create a complementary copy of the parental DNA chain.*

1.2.2 *Escherichia coli DNA Polymerase Can Synthesize DNA with Genetic Activity*: *Creating Life in the Test Tube*

In 1959 Arthur Kornberg moved to Stanford University School of Medicine as Professor and Chairman of the Department of Biochemistry and he also shared the Nobel Prize in Medicine or Physiology with Severo Ochoa. Among the referees who declined the JBC papers in 1957, one, *"whose caustic literary style unmistakably identified him"* to Kornberg, stated that *"the authors had to show genetic activity in the synthetic substance to have it qualify as DNA"*. *"Why?"* Said Kornberg: *"This criterion is met by less than 2 percent of the papers on DNA appearing in biochemical journals"*.

As we have mentioned, he withdrew the papers and waited for the next Editor-in-Chief. But certainly when he found that all four dNTPs were required, it became clear that the added DNA was serving not only as a primer but also as a template. Thus it was important to show that the DNA synthesized by the newly discovered DNA pol was a faithful copy of the added DNA template. This was proved by showing (1) that the newly synthesized DNA contained equal amounts of A and T, and of G and C; (2) using DNAs of various species with A + T/G + C ratios ranging from 0.5 to 1.9 as primers, the ratio in the product DNA closely corresponded to that of the added DNA and was independent of the relative concentrations of the dNTPs.

In summary, the bacterial DNA pol could copy any DNA, just for its ability to recognize the bases rather than their sequence in the template chain, and the above results suggested that *"the enzymatic synthesis of DNA by the DNA polymerase of E. coli represented the replication of a DNA template"*. It should be noted, however, that for some years the question of the "primer" for DNA pol remained open until it was found that all DNA pols also require a primer to initiate a DNA chain (see below).

Still after each of the many seminars he gave in those years, where he described the adequate performance as template-primers of DNAs from *Pneumococcus, Hemophilus* and *B. subtilis* but with a net loss rather than an increase of their measurable transforming activity, he was invariably asked the same question: *"Why have you been unable to replicate the genetic activity of the DNA template?"*

The solution came from studying the replication of a small bacterial virus ΦX174 containing a single-stranded circular DNA of about 5000 nucleotides. Robert Sinsheimer had found that the ΦX174 DNA was single-stranded for a part of its life cycle and, after entering *E. coli*, the DNA is converted by host enzymes to a double-stranded circle (a *replicative form* that serves as template for the synthesis of DNA of the progeny virus), and within 20 minutes the bacterial cells released hundreds of viral particles. Finally he showed that, although with a much lower efficiency, naked viral DNA was still infectious.

Kornberg and Mickey Goulian obtained some ΦX174 single-stranded DNA from Sinsheimer, and copied it with the purified *E. coli* DNA pol. By using appropriate radioactive and "heavy" precursors, (i.e. replacing deoxythymidine with 5-bromodeoxyuridine), and a novel enzyme DNA ligase (independently discovered in 1967, with different assays, by Martin Gellert at NIH,[28] Charles Richardson at Harvard,[29] Robert Lehman at Stanford,[30] Jerard Hurwitz at Albert Einstein Medical School,[31] and Kornberg and Nicholas Cozzarelli (*1938–2006*) at Stanford,[32] an enzyme able to join or ligate adjacent ends of a DNA chain, with a free OH group at the 3′ end and a phosphate group at the 5′ end to create a circle), they isolated fully synthetic circles of viral DNA and showed that they were infectious.

After a decade they had finally synthesized a DNA with genetic activity. The results published in the *Proceedings of the National Academy of Sciences* in 1967 as the 24th paper of the series "Enzymatic synthesis of DNA"[33] raised the interest of many newspaper and television reporters, not only in the United States but also in many other countries, essentially asking *"Is the DNA you have made a living molecule?"* It was quite difficult for Kornberg to explain that *"viral DNA has no life of its own, nor does the virus that bears it, in the sense of being able to grow and reproduce outside the cells of an organism. Yet the DNA has a potent life force. Once inside a bacteria, plant or animal cell, viral DNA can divert the host machinery to doing little else but making thousands of viruses identical to itself."*

1.2.3 *Bacteria Contain Many DNA Polymerases*

The DNA pol first discovered by Kornberg and co-workers in *E. coli* is now called DNA pol I, because other DNA pols (II , III , IV and V) were found later. As a matter of fact the simple mechanism of DNA replication suggested by the DNA structure proposed by Watson and Crick not only requires several DNA pols but also many other different enzymes and specific proteins to assure the extremely high number of polymerization steps on a single natural DNA template molecule and a high fidelity of DNA replication, thus avoiding a too high level of mutations unacceptable for individuals and species. Although the problem of DNA replication was solved intuitively by Watson and Crick in 1953, determining the actual molecular basis of DNA replication has required considerable more time and efforts.

Indeed, some ten years after Kornberg's discovery of DNA pol I, the DNA replication process became more complicated with the discovery that *E. coli* contained other DNA pols and the first one discovered by Kornberg was not the most important for the replication of the DNA molecule (see Chapter 3). In 1969 Paula De Lucia and John Cairns, then Director of the Cold Spring Harbor Laboratory, by analyzing thousands of colonies of mutagenized *E. coli* for DNA pol activity, found one with only 0.5 to 1 percent of the normal level of DNA pol. Impaired only by an increased sensitivity to ultraviolet light, the mutant, named polA1 mutant for its assonance with Paula, grew and multiplied at a normal rate.[34]

A year later, in 1970, it was just Arthur Kornberg's son, Thomas, who had joined Malcolm Gefter's laboratory in the Biology Department at Columbia University, who found in Cairns' mutant a DNA pol activity distinct from the one his father had discovered. He named the purified enzyme DNA pol II.[35,36] Similar results were reported the same year by Rolf Knippers[37] and by Robb Moses and Charles Richardson.[38]

In the course of chromatographic separation Tom Kornberg also noticed another band of DNA pol activity in the position where it would have been obscured by DNA pol I in wild type *E. coli*. He named it DNA pol III.[35,39] Gefter *et al.* located the gene for DNA pol III and were able to demonstrate that conditionally lethal mutations in this gene blocked DNA replication. Subsequent studies revealed that DNA pol III is the core of the DNA pol III holoenzyme.[22]

More bacterial DNA pols, named DNA pols IV and V were revealed almost 30 years later, in 1999, by studying inaccurate DNA replication across UV lesions during the so-called process of translesion DNA synthesis (see Chapter 3).

DNA pols I, II and III differed from each other in several properties, such as optimal template-primer, sensitivity to sulfhydryl-blocking agents, ethanol and salt, K_m for dNTPs, and interaction with DNA binding proteins. They, however, shared among themselves (and with DNA pols subsequently discovered not only in bacteria, but also in animal and plant cells and in viruses) basic and fundamental features such as: (1) requirement for a DNA template, guiding the DNA pol to select any deoxynucleotide by base-pairing to the one on the DNA template (a mismatch nucleotide at the 3′ terminus is usually removed by an intrinsic 3′ → 5′ exonuclease proofreading activity present in several DNA pols); (2) need for 2′-deoxyribonucleoside 5′-triphosphates (dNTPs), which are incorporated as monophosphate to the 3′-hydroxyl group at the growing end of a DNA strand (the *primer*) with formation of a new phosphodiester bond following nucleophilic attack of the α-phosphate of the incoming dNTP by the 3′-OH group on the primer terminus, with release of pyrophosphate (*no DNA pol is able to initiate new chains*); (3) chain growth exclusively in the 5′ → 3′ direction and antiparallel with respect to the template strand; (4) a 3′ → 5′ exonuclease activity, also called proofreading activity, and (5) a 5′ → 3′ nuclease activity, also called editing activity, important, for example, in the excision of pyrimidine dimers, is also present in DNA pol I as well as in other DNA pols. These and other features, such as binding sites for dNTPs, the template and the growing chains, will be subsequently discussed for each DNA pol as well as the ways they probably function *in vivo*.

We shall just mention here that Kornberg also demonstrated, by equilibrium dialysis, that dNTPs bind to a single site on DNA pol I with the triphosphate moiety being of primary importance for the specificity of binding. Thus he also found that $2'$, $3'$-dideoxyribonucleoside $5'$-triphosphates (ddNTPs), the analogs of the natural $2'$-deoxyribonucleotide substrates of DNA pol, but lacking the $3'$-hydroxyl group functioning as a primer, bind as well to the dNTPs binding site. In fact, ddTTP binds even better than dTTP. In the presence of an excess of DNA pol, ddNTPs are also incorporated and then function as DNA chain terminators. These results inspired Frederick Sanger to develop his chain-termination technique of determining the sequence of a DNA chain,[40] and for this in 1980, he received with Walter Gilbert and Paul Berg a second Nobel Prize in Chemistry, 22 years after earning his first Nobel Prize in Chemistry for determining the complete amino acid sequence of insulin.

1.2.4 *How Is a New DNA Chain Started? Discontinuous DNA Synthesis and the Need for an RNA Primer*

The replication of infectious circular single-stranded ΦX174 DNA, demonstrated in 1967, implied that DNA pol I could start a new chain. If this were the case the initiating nucleotide should retain the triphosphate group, but this was never found. A stimulatory effect of added boiled *E. coli* extract, removed by digestion with DNase, rather suggested that a fragment of *E. coli* DNA, later removed by the $3' \rightarrow 5'$ proofreading and $5' \rightarrow 3'$ editing activities associated to DNA pol I, initiated the reaction. But the problem of the *primer*, i.e. how a new DNA chain started *in vivo*, remained, because microscopic and genetic analysis indicated a synchronous replication of both daughter helices. How could one reconcile the synchronous replication of the two chains of the DNA helix, both serving as template, but oriented in opposite directions, with the limitation of DNA pols that only proceeded in the $5' \rightarrow 3'$ direction, just opposite that of the template?

The dilemma was resolved by Reiji Okazaki (*1930–1975*), a former visiting scientist in Kornberg's laboratory. He found that much of the newly synthesized DNA momentarily existed as small fragments, about 1000 nucleotides long (Okazaki fragments) near the replicating fork. These became covalently joined into longer daughter chains by DNA ligase as the replication process proceeded, suggesting that DNA replication is a discontinuous process.[41] The direction of DNA synthesis is thus always in the $5' \rightarrow 3'$ direction and might be continuous on the $3' \rightarrow 5'$ leading strand but needed to be discontinuous on the $5' \rightarrow 3'$ lagging strand.

1.2.5 *RNA Priming as a Mechanism for Initiation: DNA Primase*

The problem of how DNA chains are started was finally solved when Kornberg had the idea to study the conversion of the circular single-stranded chromosomal DNA of phage M13 into the double-stranded DNA replicative form. Such conversion, mediated by bacterial enzymes (including RNA polymerase), occurs upon phage entering the bacterial cell, and Kornberg hypothesized that RNA polymerase, which is able to start chains *de novo*, could make a short fragment of RNA on single strand M13 DNA that could be utilized by DNA pol to start a DNA chain. The small RNA fragment would then be removed by the proofreading and editing properties of the DNA replication machinery. The process should be inhibited by rifampicin, an inhibitor of RNA polymerase. Based on these hypotheses the results were obtained in a few days,[42] and such *RNA priming* was subsequently found to be the most general mechanism for initiation of DNA chains.

There are, however, exceptions to RNA priming such as: (1) a $3'$-hydroxyl primer can be created by a specific endonuclease, as in rolling circle replication of some phages and plasmids or by folding back of the $3'$OH end of a linear, single-stranded DNA (such as that of parvovirus); (2) a virus-encoded protein may serve as a primer by pairing a single deoxyribonucleotide (dCMP) with the end of a template strand as in the replication of the linear chromosomes of Adenovirus and *B. subtilis* phage Φ29.

The specialized RNA polymerases that synthesize minitranscripts serving as primers are called DNA primases. Their variety is fascinating: some read sequences of a few bases, others synthesize a primer of an exact length but are indifferent to their sequence, some even accept deoxyribonucleotides as well as ribonucleotides for elongation. Some are tightly bound to DNA pols such as in eukaryotic DNA pol α, others to a helicase (phage T4) and some contain the primase and helicase activities within a single polypeptide (phage T7). *E. coli* primase associates with many other proteins to form a mobile multifunctional primosome. In no instance it is clear how a primase, by analogy with RNA polymerase, recognizes its "promoter" or terminates "transcription".

The primers are ultimately removed, in *E. coli* by DNA pol I, in eukaryotes by RNase H, Dna 2 nuclease and flap endonuclease 1 when the DNA growing chain reaches the previous primer, just before another enzyme, DNA ligase, joins together the DNA fragments thus restoring the phosphodiester backbone.

1.3 Late 1960s to Early 1970s: DNA Replication Shows Its Complexity

In 1967 *in vitro* replication of a short circle of ΦX174 DNA appeared to be accomplished by just two enzymes, a DNA pol to start and extend the chain and a DNA ligase to join the ends. However, only a few years later the replication process definitely appeared much more complex, requiring the need of a complex network of interacting proteins and enzymes for the replication of both a simple bacteriophage DNA and the much more complex host bacterial chromosome.

We have already mentioned that the problem to reconcile the synchronous growth of both daughter chains with the $5' \rightarrow 3'$ direction of synthesis by all DNA pols was solved by Okazaki, whereas the need for a primer was solved again by Kornberg.

1.3.1 *DNA Structure Is Much More Complex, Rich of Conformational Flexibility and thus Full of Functional Potentialities than the One Proposed by Watson and Crick*

The need for a complex network of many proteins and enzymes interacting among themselves and with the DNA helices during DNA replication is not at all surprising. The structure proposed by Watson and Crick in 1953 immediately suggested the function of DNA and appeared to everybody so naturally logical and unquestionable that there was no time for astonishment and for looking back at the times when the genetic secret of our life was unknown.

However, the structure proposed by Watson and Crick in 1953 was a static, three-dimensional structure, and as time went by it became clear that the structure of DNA is much more complex, rich in conformational flexibility and full of functional potentialities compared to what it appeared at first sight. DNA is indeed a dynamic molecule whose movement and conformational changes are essential features for its function. The DNA molecule not only contains a type of genetic information written in the base sequence (nucleotide triplet) that, being a genetic code, in the end specifies the amino acids sequence, and, therefore, the protein structure. It also contains quite an important conformational type of genetic information, a kind of phenotypic code, that allows a specific interaction with proteins.

For the double helix itself, different variants were soon hypothesized and demonstrated, the best characterized being the A, B and Z structures, depending on the environment such as ionic strength, ions, relative humidity, organic solvents and temperature. The capacity to assume alternative conformational forms is further increased by the presence of particular sequences in the DNA, such as the highly repetitive AT-rich sequences of satellite DNA, the inverted repetitive "hairpin" sequences, the GC-rich telomeric sequences synthesized by a special enzyme (Telomerase), discovered in 1985 by Carol Greider and Elizabeth Blackburn,[43] that uses a short RNA, rather than a DNA, as a template, just to mention a few. Blackburn and Greider shared the 2009 Nobel Prize in Physiology or Medicine with Jack Szostak for the discovery of telomerase and of how chromosomes are protected by telomeres.

Furthermore, DNA has a kind of tertiary structure defined by the interchangeable terms of supercoiling, supertwisting, superhelicity, that define the spatial coiling or the twisting upon itself of a duplex DNA helix. This property was first observed and it is better understood in small circular molecules such as bacterial plasmids, small viral chromosomes, mitochondrial DNA. However, it is an integral feature of linear chromosomes as well as eukaryotic ones, when ends are united to a solid matrix to form circles or are too remote to rotate rapidly and free rotation of the ends of DNA is restricted. The degree of supercoiling is crucial in replication, transcription, recombination and repair and it is regulated by DNA topoisomerases (see next section).

1.3.2 *DNA Binding Proteins, DNA Helicases, DNA Topoisomerases*

While many groups began studying and finding several DNA pols in eukaryotic cells as well, another problem remained: what opens up the double helix giving rise to the replication fork in front of the advancing DNA pol(s)? We have seen that the structure of the double helix resembles a zipper that cannot be opened unless a slider is moved in the right direction. And, importantly, energy is needed for this.

The most obstinate hunter of the slider was Bruce Alberts who ultimately did not discover the slider but a new family of important proteins, the DNA binding or extending proteins, exemplified by T4 gp32 protein. These proteins promptly cover the bared single strands of DNA, protecting them against nuclease action, preventing them from reannealing and configuring them (by keeping them in an extended form) to serve as templates for DNA pols.[44–46]

The problem of the slider needed to open up the zipper was solved once again by Kornberg. In 1975 he found in *E. coli* a protein (the *rep* protein) that by utilizing ATP could move along the DNA opening up the double helix in front of the advancing DNA pol. It was a DNA-dependent ATPase and was named DNA helicase. The trick it used is the one utilized by the muscle motility proteins: ATP-dependent periodical changes in its shape enable the helicase to enter the double helix opening it to the advancing DNA pol(s).

If DNA binding proteins and DNA helicase solved one problem, another one immediately appeared: both the advancing of the replication complex and the fact that the daughter chains synthesized at the replication fork have to wind around each other to reform a double helix created topological problems downstream and upstream from the replication fork. For example, a great torsion would develop along the unreplicated DNA chains that should rotate continuously at the expense of a lot of energy. If we assume that a cell nucleus corresponds to a tennis ball, then its DNA would be a thread of 20 km, and it is easy to imagine what would happen if one tries to rotate a thread of 20 km in a tennis ball.

These problems were resolved with the discovery by James C. Wang of the first DNA topoisomerase, an enzyme that control the topological state of the DNA.[47] There are different types of DNA topoisomerases and with simple and economic mechanisms they can rotate DNA chains. As in Wangs's words,[48] *"DNA topoisomerases are magicians among magicians; they open and close gates in DNA without leaving a trace, and they enable two DNA strands or duplexes to pass each other as if the physical laws of spatial exclusion do not exist. Because the biological functions of the DNA topoisomerases are deeply rooted in the double helix structure of DNA, it should not be surprising that these enzymes participate in nearly all biological processes involving DNA"*.

1.4 Concluding Remarks, Parts 1.1–1.3

After the discovery of the DNA structure by Watson and Crick in 1953, the subsequent work of Kornberg and co-workers first led to the discovery of the DNA pols, enzymes that synthesize long chains of DNA, fill up gaps in DNA molecules and repair raw ends of DNA. Other important enzymes and proteins were then discovered and found essential for DNA replication (**Table 1.1**). Some of them allowed the

Table 1.1   DNA replication proteins in *E. coli* in 1975

Protein	Function
DNA polymerase I	Repair
DNA polymerase II	Repair
DNA polymerase III	Replicase
DNA primase	Synthesis of initiator RNA
DNA ligase	Replication — Repair (DNA joining)
DNA helicase	Unwinding of double-stranded DNA
SSB	Binding and protecting single-stranded DNA
DNA topoisomerases	Resolving topological issues of DNA

manipulation of DNA molecules such as *DNA ligases*, which join contiguous ends of DNA chains, various *DNA exonucleases*, and special DNA endonucleases, called *restriction nucleases,* that can modify DNA ends or cut DNA molecules at specific sequences respectively. (The latter studies earned Werner Arber, Hamilton Smith and Daniel Nathans (*1928–1999*) the Nobel Prize in 1978 for their applications to problems of molecular genetics). Finally a unique DNA template-independent DNA pol was found in calf thymus, *terminal transferase*, that can add nucleotides to the end of a DNA chain.[49]

The combined use of such enzymes has allowed the artificial manipulation of DNA molecules that soon led to important technical advances such as sequencing of DNA, that made it possible to sequence the genome of many organisms, including the human genome; gene cloning and genetic engineering, that have led and will lead to a cascade of progress and practical benefits in medicine, agriculture and chemical industry.

What are regarded as the first experiments of genetic engineering are those of Stanley N. Cohen, a physician in the Department of Medicine at Stanford University, and Herbert Boyer of the University of California, San Francisco. They were studying bacterial resistance to antibiotics treatments and wanted to understand how the genes of plasmids could made bacteria resistant to antibiotics. Back in 1972–1973, by using the restriction enzyme EcoR1 that cuts DNA molecules creating protruding complementary or cohesive ("sticky") ends which can be joined by DNA ligase, they were able to cut, combine and transplant genes in plasmids and, following the introduction of the plasmid into *E. coli*, to perpetuate an animal gene in the bacterium, opening the way for the introduction and propagation of any gene.[50] Furthermore, in 1976 Boyer, with the venture capitalist Robert Swanson, founded *Genentech* (a composite of *Genetic Engineering Technology, Inc.*), the leading biotechnology corporation that was the first to express human genes in bacteria such as the hormone *somatostatin* in 1977 and *insulin* in 1978.

1.5 Multiple DNA Polymerases in Eukaryotic Cells: DNA Polymerases α, β and γ as the First Ones

From what we have so far presented in a rather simple way, it is clear that, although the problem of DNA replication was solved intuitively by Watson and Crick in 1953, determining the actual molecular basis of replication has required considerable more time and effort. No wonder that the complexity of DNA replication forced things to be worked out first in prokaryotes, thanks to simple *in vitro* replication systems relying on bacteriophages pioneered by Kornberg (ΦX174, M13), Alberts (T4), and Richardson (T7). Such *in vitro* systems provided the means to identify the products of several genes involved in DNA replication, because DNA synthesis was observed in extracts prepared from wild-type strains but not in extracts of mutants. Using *in vitro* complementation and reconstitution assays, biochemists were able to fractionate the complex extracts and to define the individual enzymes and proteins necessary for replication, some of which have been mentioned above such as the main actors, the DNA pols, but also several accessory proteins, RNA polymerase and DNA primases, single-stranded DNA binding proteins, DNA helicase, DNA topoisomerases, RNase H, and DNA ligase.

The most important insight was that rather than carrying out a set of sequential individual reactions with one class of proteins completing its function, dissociating from DNA and the next protein further processing an intermediate, the proteins of the replication apparatus form a stable complex that remains associated with the DNA for thousands of nucleotides during chain growth. Some proteins are important to recognize an origin of replication, to alter its structural conformation and to lead to its further opening by the action of a helicase. Others are needed to suppress potential origins elsewhere on the chromosome, and finally others might appear more important or more specific to replicate the DNA such as the DNA pols, DNA primase, DNA helicase, DNA topoisomerase, single-stranded DNA binding proteins and DNA ligase because they are needed to prime and elongate DNA chains and to propel the two forks in bidirectional replication.

So if in prokaryotes different DNA pols are present and the one most directly involved in DNA replication has a higher complexity (See *E. coli* pol III holoenzyme in Chapters 2 and 3), it was quite likely that many DNA pols were also present in eukaryotes, and a similar or higher degree of complexity could be expected in functionally operating replicative DNA pols in eukaryotes. The excellent book *Animal Cells DNA Polymerases* published in 1986 by M. Fry and L.A. Loeb still contains the most complete list of references, including reviews, regarding DNA pols α, β, γ and δ up to 1985. Here, we will briefly review the most important steps leading to the discovery of these first DNA polymerases.

1.5.1 *DNA Polymerase* α

Following the first demonstration, in 1956, by Kornberg *et al.* of an activity (named *E. coli* DNA pol I) capable of incorporating radioactive thymidine in extracts of *E. coli*, Bollum and Potter in 1958 were the first to present evidence for a DNA pol activity in the high speed supernatant of homogenates from regenerating rat liver.[51] A similar activity was also present in nuclei, in agreement with two earlier reports in 1956 by Friedkin, Tilson and Roberts[52] and by Friedkin and Wood[53] on the incorporation of [^{14}C]-thymidine into DNA of chick embryos, bone marrow cells and isolated thymus nuclei, respectively.

But, in truth, back in 1951, Reichard and Estborn[54] had already demonstrated incorporation of [^{15}N]-thymidine into DNA, and in 1954 Friedkin and Roberts[55] had observed uptake of [^{14}C]-thymidine in Yoshida ascite tumor cells incubated *in vitro* and Lu and Winnick in embryonic tissue cultures.[56] Friedkin was the scientist who, in 1955, gave some [^{14}C]-thymidine to Kornberg for his first experiments in *E. coli* extracts, before it became commercially available in 1957. But Friedkin was no longer inclined to search for incorporation of thymidine into DNA with extracts from broken eukaryotic cells. Thus Bollum, certainly inspired by the preliminary experiments of Friedkin and then guided by the convincing and definite results of Kornberg in *E. coli*, can be considered as the pioneer of eukaryotic DNA pols. Upon the commercial availability of [^{3}H]-thymidine in 1958, Bollum reported the incorporation of thymidine into DNA in extracts of calf thymus, and, in 1960, just after the full characterization of *E. coli* DNA pol I by Kornberg's group in 1957, he also reported a partial description and purification in calf thymus of a DNA pol activity and its requirements for the four dNTPs, a primer DNA and Mg^{2+}.[57] An extensive purification of a high molecular weight DNA pol from calf thymus appeared in 1965 by Yoneda and Bollum.[58]

In the following years, a high molecular weight DNA pol activity was demonstrated to be ubiquitous in growing eukaryotic cells, such as human, murine, hamster, sea urchins, chick embryo, avian tissues and yeast, by various groups, including Bollum's and those of A. Weissbach, L.A. Loeb, D. Korn, R. Gallo, D. Baltimore, I.R. Johnston, S. Wilson, H.M. Keir, F. Chapeville, E. Wintersberger. A complete list of references can be found in the 1986 book of Fry and Loeb *Animal Cell DNA Polymerases*.[59]

This high molecular weight DNA pol activity showed a characteristic sensitivity to sulfhydryl group blockers such as N-ethylmaleimide and to NaCl concentration above 25 mM and has been referred to by Bollum as "high molecular weight cytoplasmic DNA polymerase" because most of the activity (>80%) was present in the cytoplasm of cells broken by an aqueous procedure. This was clearly an artifact due to the use of aqueous extraction, because in the following years different techniques, including mechanical dissection of nuclei and use of immunoperoxidase staining in fixed cells labeled with monoclonal antibodies against DNA pol α, confined this polymerase to the nuclei of all the examined cell types. Until a uniform nomenclature was adopted in 1975 (see later), it was called with different names such as "high MW cytoplasmic or maxi DNA pol" by Bollum, DNA pol II by Weissbach, cytoplasmic N2 by Korn, DNA pol I by Gallo, and DNA pol C by Baltimore.

1.5.2 *DNA Polymerase β*

For many years, by analogy with *E. coli* DNA pol I, the high molecular weight cytoplasmic DNA pol (6–8 S) was assumed to be the only type of DNA pol in eukaryotic cells. However, a low molecular weight DNA pol, much less abundant than the high molecular weight enzyme in proliferating cells, was first identified in 1971 in HeLa cells by Weissbach *et al.*[60] This enzyme was then found in rat liver nuclei by Baril *et al.*,[61] in calf thymus by Chang and Bollum,[62] and in rat liver by Haines *et al.*[63] and by Berger *et al.*[64] Other studies indicated that perhaps mitochondria contained a DNA pol that differed from these two high- and low-molecular weight DNA pols.

The new low molecular weight (3–4 S) DNA pol (or DNA pol I for Weissbach, "mini" for Bollum, DNA pol II for Gallo, cytoplasmic N1 for Korn and DNA pol N for Baltimore) was considered to be located exclusively in the nucleus and to be present in all eukaryotes examined. It showed resistance to sulfhydryl group blockers such as N-ethylmaleimide, to NaCl or KCl concentration up to 200 mM in 50 mM K-phosphate buffer, and to a number of chemical agents such as ethanol, acetone (20–25%), urea (5 M) and phosphonoacetic acid. An interesting study on the phylogeny of DNA pol β is that of L. Chang (1976) showing its wide distribution in multicellular animals and its absence in bacteria, plants and protozoa.[65]

Distinguishing among different DNA pols in eukaryotic cells was rather difficult because one had to completely rely on biochemical approaches, whereas we have previously seen that almost in the same years (1969–1972) conditional mutants of DNA replication allowed the unambiguous identification in *E. coli* of DNA pols II and III by T. Kornberg, M. Gefter, R. Knippers, R. Moses and C. Richardson, although they were much less abundant than DNA pol I.

1.5.3 *Lack of Relationship Between High- and Low-Molecular Weight DNA Polymerases*

However, the low-MW DNA pol was not immediately accepted as a distinct DNA pol by the scientific community. Immediately following its discovery, and with the possible identification of another different DNA pol in mitochondria, Chang and Bollum in 1972 reported an antigenic relationship between the high- and low-molecular weight DNA pols from a variety of cells and suggested that they could share common peptide sequences or even subunits[66] and others, in particular Hecht[67] and Lazarus,[68] in sedimentation studies of 1973, reported a salt-mediated conversion of the two DNA pols.

On the contrary, Spadari and Weissbach in 1974[69] and Smith *et al.* in 1975[70] with antibodies prepared against human high molecular weight DNA pol and Brun *et al.*

in 1975[71] with antibodies against avian high- and low-molecular weight DNA pols, showed an absence of cross-reaction between the two DNA pols. No cross-reactivity was also found with any of the oncornavirus reverse transcriptases, just discovered by Temin and Baltimore, or with another DNA pol found in Weissbach's laboratory (1972, 1973) in human cells and named R-DNA pol for is ability to copy, much better than the low molecular weight enzyme, a deoxyprimed homoribotemplate such as oligo(dT)/poly(A) (see section on DNA pol γ).

1.5.4 *1975: First Nomenclature System for Eukaryotic DNA Polymerases*

Because of the proliferation of systems for naming the DNA pols in various laboratories, A. Weissbach, D. Baltimore, F. Bollum, R. Gallo and D. Korn devised the nomenclature reported in **Table 1.2** at a meeting held at the Massachusetts Institute of Technology (MIT) on May 29, 1974.[72] In May 1975, the 48 scientists attending the international conference on Eukaryotic DNA pols organized by David Korn of Stanford University, at the Asilomar Conference Center, Pacific Grove, California, decided to accept this uniform system of nomenclature.

Thus, the size, cellular localization or template preference for naming DNA pols were abandoned in favor of a system of Greek letters, proposed by Weissbach, assigned to the DNA pols according to the order of discovery, a system acceptable to everybody because it was different from any of the systems previously used.

The high molecular weight DNA pol first discovered by Bollum (1958, 1960 and 1965) became DNA pol α; the low molecular weight DNA pol first discovered in 1971 in human cells in Weissbach's laboratory became DNA pol β; the other DNA pol, again described and characterized in Weissbach's laboratory in 1972–1974 and called R-DNA pol for its preference to copy a synthetic deoxyprimed polyribotemplate, such as oligo(dT)/poly(A), but not natural RNA, was called DNA pol γ.

All these pols, in particular DNA pols α and β, were active in copying "activated DNA" that is a DNA prepared by limited digestion of duplex DNA with DNase I,

Table 1.2   Nomenclature of eukaryotic DNA polymerases in 1975

	Formerly used systems (identified by the proposer's name)				
New system	Baltimore	Bollum	Gallo	Korn	Weissbach
DNA pol α	C	6-8S (maxi)	I	cytoplasmic N2	II
DNA pol β	N	3-4S (mini)	II	cytoplasmic N1	I
DNA pol γ	A		III		R

providing, in the DNA substrate, 3′OH primers and gaps to be filled up by DNA pols. Double-stranded and single-stranded DNAs, as expected, were inactive as template because of the lack of 3′-OH primers.

The retroviruses reverse transcriptases, discovered in 1970 by Howard Temin (*1942–1994*)[73] at the University of Wisconsin-Madison and independently by David Baltimore[74] at MIT, for which they shared the 1975 Nobel Prize in Physiology or Medicine with Renato Dulbecco, remained separate from this nomenclature because they represented distinct entities. Other viral-induced DNA pols such as Herpes and Vaccinia that could be referred to by the name of the virus, also remained separate.

Finally, terminal deoxynucleotidyltransferase (TdT), an unusual template-independent DNA pol, found in 1960 by Bollum in calf thymus, was also not included (see Chapter 2).

1.5.5 *DNA Polymerase γ*

If, at the Asilomar meeting of 1975, the attending scientists were all convinced of the existence of two distinct DNA pols, α and β, in eukaryotic cells, a word of reservation was still present about DNA pol γ, the last of the DNA pols described in eukaryotic cells included in **Table 1.2**. This reservation was shared by many scientists, whether either attending or not attending the meeting, because it represented only about 1% of the total cellular DNA pol activity and, most intriguingly, the preferred templates were deoxyprimed ribohomopolymers, in particular oligo(dT)/polyrA that in 1972 Chang and Bollum had found to be a substrate for DNA pol β as well.[75]

Rather than a novel cellular DNA pol, it was regarded more like a cellular "reverse transcriptase-like" entity with a possible function in normal cellular processes like cytodifferentiation or gene amplification. It was also suggested that it was an activity due to DNA pol β, because Chang and Bollum reported in 1972 that, in the presence of Mn^{2+} ions and salt, DNA pol β was able to copy oligo(dT)/poly(rA). After all, despite a marked preference for deoxyribohomopolymers, the *E. coli* DNA pol I was also able to use ribohomopolymers as templates.

At the Asilomar meeting, one of us (S. Spadari), visiting scientist in Weissbach's laboratory in 1973–1975, reported his results on DNA pol γ just after the previous speaker had said that DNA pols α and β represented approximately 95 and 5%, respectively, of the DNA pol activity in dividing cells, the "rest" or 0% being the other DNA pols! By modifying the previous purification procedure, Spadari and Weissbach purified DNA pol γ from cytoplasm (over 2000-fold) and from nuclei (almost 3500-fold). The two forms of DNA pol had a similar MW (110–120 000 daltons) as measured by sedimentation studies, much higher than the approximately 70 000 daltons of reverse transcriptases; they could use, although with lower efficiency, activated salmon sperm DNA as substrate and this activity was not inhibited by an antibody against DNA pol α, thus demonstrating that it was not due to a contamination with this DNA pol; they copied, in contrast with the less purified preparation first reported by Weissbach in HeLa cells in 1972,[76] all the homoribopolymers — poly(C), poly(I), poly(U) but with a preference for poly (A); contrary to DNA pol β, they utilized oligo(dT)/poly(rA) as primer-template in the presence of phosphate that completely inhibited the similar activity of DNA pol β; they were both unable, in contrast to reverse transcriptases, to copy natural RNA; they were not inhibited by antisera prepared against several reverse transcriptases as those obtained from Woolly monkey virus, Rauscher murine leukemia, or the Mason-Pfizer monkey virus. Interestingly, both forms showed K_m values for deoxyribonucleoside triphosphates (approximately 5×10^{-7} M) that were one order of magnitude or more lower than the K_m values of DNA pols α and β. Finally, similar forms of DNA pols γ were purified from the normal human diploid fibroblast cell line WI-38, suggesting that DNA pol γ was not unique to human cells derived from tumor tissues.[77]

Although the data presented by Spadari in Asilomar claiming that DNA pol γ was a distinct third DNA pol present in eukaryotic cells were not immediately accepted, subsequent studies very soon revealed, as we will see later, that DNA pol γ is the mitochondrial enzyme and that replicates mitochondrial DNA.

1.6   Early Attempts to Ascribe an *in vivo* Function to DNA Polymerases α, β and γ

The lack of eukaryotic conditional mutants in DNA replication made it difficult to assign a function to DNA pols α, β and γ. Even in prokaryotes, where mutants were available and the properties of enzymes and proteins involved in DNA transactions were much better known, assigning a functional role to a molecule was not without doubts. While it seemed unreasonable to assume that the DNA pols are geared for anything other that some form of DNA synthesis (replicative or repair synthesis), in prokaryotes there is no absolute division of labor. Thus, DNA pol III is responsible for chromosomal replicative synthesis and DNA pols I and II are mainly involved in repair-type synthesis. However, we know that DNA pol I is needed for the removal of the RNA primers from the Okazaki pieces and for the resynthesis of removed sequences, and DNA pol III is involved in some forms of repair synthesis including recombination. However, separation of main functions still remains in bacteria. Thus, following the identification of the first three DNA pols in eukaryotes, considered for several years as the sole eukaryotic DNA pols, scientists began wondering which of the animal DNA pols was more involved in either processes (see also global function of DNA pols in Chapter 5).

1.6.1 *Positive Correlation of DNA Polymerase α with Cellular DNA Replication and Development*

The catalytic properties of the first three DNA pols present in eukaryotic cells were different enough to allow, in the early 1970s, their simultaneous assay without appreciable interference. The assay for DNA pol β was absolutely specific because of its insensitivity to N-ethylmaleimide (NEM) and its high pH optimum and preference for high ionic strength. If enzyme samples were preincubated with 5 mM NEM at 0°C for 30 min to inactivate DNA pols α and γ, the β-assay was strictly specific for the DNA pol β. The γ-assay, using oligo(dT)/poly(rA) in the presence of 50 mM K-phosphate (pH 8.5) that completely inhibits the activity of DNA pol β on this template-primer, was also specific. Under the assay conditions used for DNA pol α, DNA pol γ was inactive and DNA pol β responded with 30% efficiency, and, therefore the DNA pol α results could be corrected for its contribution.

Thus, studies with cultures of mouse L cells in different growth conditions by Chang and Bollum in 1973,[78] in synchronized HeLa cells by Spadari and Weissbach in 1974,[79] and by Chiu and Baril in 1975,[80] with regenerating rat liver by Baril *et al.* in 1973,[81] in phytoaemoagglutinin-stimulated lymphocytes by Mayer *et al.* in 1975[82] and Bertazzoni *et al.* in 1976,[83] and many others[59] all showed a positive correlation between activity of DNA pol α and rate of DNA synthesis, with much lower changes of DNA pols β and γ. Because Okazaki fragments were also observed as intermediates during DNA synthesis in eukaryotes, Spadari and Weissbach in 1975[84] showed that only DNA pol α could synthesize DNA covalently bound to RNA. The RNA-DNA molecule could then be extended by either DNA pols α, β or γ indicating that the DNA pol α was capable of both RNA- and DNA-primed DNA synthesis, whereas DNA pols β and γ were capable of only DNA-primed DNA synthesis.

In the late 1970s when DNA pol δ was not yet widely recognized as a novel DNA pol (a destiny common to all DNA pols so far discovered after the identification of the major high MW DNA pol by Bollum), changes in DNA pols α, β and γ activities were studied in several systems to provide insights about their possible function. Among them fertilized eggs and developing embryos (from Xenopus, sea urchin, Drosophila) which showed very rapid DNA synthesis, mouse and rat testis, spleen, heart, brain and liver with important tissue changes during perinatal development and in senescence or, like in liver, a programmed proliferation of hepatocytes after partial hepatectomy. A positive correlation between DNA pol α activity and cell proliferation and development was always found.[59]

As an example, we report the impressive patterns of the three DNA pols α, β and γ observed by Hübscher and Spadari in 1977 in rat brain neurons (**Figure 1.1 A**),

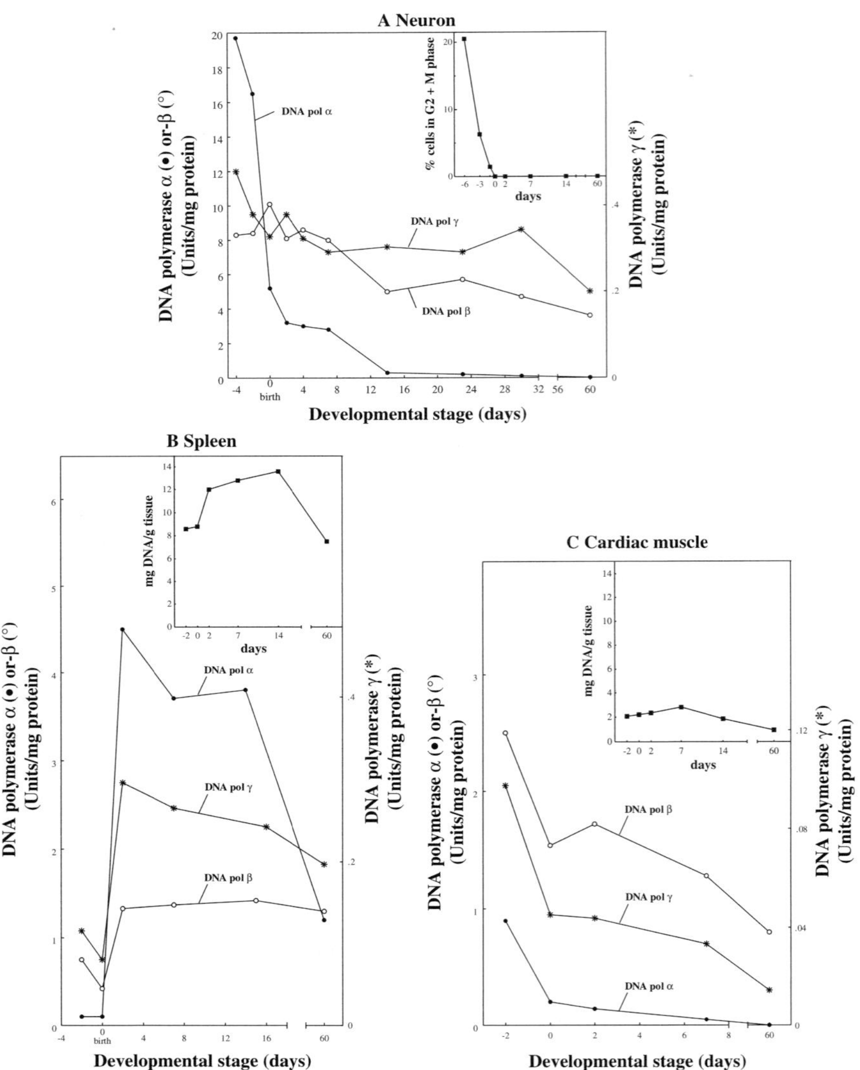

Figure 1.1 Developmental profiles of DNA pols α, β and γ in forebrain neurons (A), in spleen (B) and in ventricular cardiac muscle (C). The insert shows the percentage of cells in G_2 and M phase as determined by cell-cycle analysis (A) or the amount of DNA per g tissue chosen as a parameter of the rate of cell multiplication (B and C). Redrawn from Hübscher *et al.*[85]

spleen (**Figure 1.1 B**) and ventricular cardiac muscle (**Figure 1.1 C**) during perinatal development.[85] Comparison of the relative amounts of the three DNA pols present in late gestation in forebrain neurons showed that approximately 65–70% of the total activity is due to DNA pol α, 25–30% to β, and 2% to γ. After birth, the relative amounts of DNA pols β and γ increase to the adult proportions of 92–95% and 5–7%, respectively. The loss in DNA pol α activity correlates temporally very well with the decline in mitotic activity (**Figure 1.1 A, insert**), which drops to zero at birth. Later studies by Spadari *et al.* in 1988[86] showed that the DNA pols δ and ε levels during the course of neuronal development are coincident with those of DNA pol α, suggesting a coordinated function of these DNA pols in DNA replication. In spleen DNA pols α, β, and γ activities changed from 1%, 85%, and 14% prenatally (day −2) to 74%, 21%, and 5%, respectively, postnatally (day 2). In adults, DNA pol α is present in proportions (45%) much higher than in the other tissues studied (approximately 1% in neurons and cardiac muscle). The developmental variation of DNA pol α activity in spleen closely mirrors the rate of cell multiplication in this organ (**Figure 1.1 B, insert**). The high activity of DNA pol α in adults seems to be ascribable mainly to dividing and differentiating lymphocytes in the germinal centers.

Finally, the developmental profiles of DNA pols α, β and γ in ventricular cardiac muscle (**Figure 1.1 C**) were quantitatively similar to those in neurons and showed a good correlation with the virtually complete cessation of ventricular cellular proliferation before birth. All three DNA pols are already low by the second prenatal day where approximately 70% of the total activity is due to DNA pol β and only 26% to DNA pol α. DNA pol α declined more than β and γ during further development and can no longer be assayed by the seventh postnatal day, whereas the activities of DNA pols β and γ persist into adulthood. In summary, these experiments convincingly demonstrated the strong correlation between cell proliferation and DNA pol α.

1.6.2 *DNA Polymerase γ Is the Mitochondrial DNA Polymerase and Replicates Mitochondrial DNA*

The mitochondrial DNA pol, which had never been highly purified before 1977, was considered a fourth cellular DNA pol, distinct from DNA pol γ. Bolden *et al.*[87] in Weissbach's laboratory then purified a unique DNA pol from mycoplasma-free, HeLa cells mitochondria which was very similar to cytoplasmic and nuclear cellular DNA pol γ. This result suggested that the DNA pol γ and the mitochondrial DNA pol were identical.

In the same year, Hübscher and Spadari[88] aware in advance of this finding thanks to the personal communication of A. Weissbach, approached this problem by using an elegant system particularly suited to separating nuclear and mitochondrial compartments. Taking advantage of the fact that, in the brain, synaptic end knobs contain mitochondria anatomically away from and thus completely unassociated with nuclei, and that these structures can be isolated as distinct particles (synaptosomes) surrounded by a close membrane, they isolated mt-DNA pol from synaptosomes (**Figure 1.2 A and 1.2 B part A**) and the DNA pol γ from brain nuclei (**Figure 1.2 B part B**).

Almost 98% of the activity was recovered from synaptosomes, and comparison of the DNA pol activities purified from nuclei and mitochondria showed that a DNA pol with the properties of DNA pol γ occurred in both locations. Because DNA pol γ was the only DNA pol found in mitochondria, they concluded that DNA pol γ and mt-polymerase were identical. Bertazzoni *et al.* (1977) also made this observation on analogous enzymes from chick embryo nuclei and mitochondria.[89]

In the following two years (1978–1979) Hübscher and Spadari provided clear evidence that DNA pol γ replicated mt-DNA in purified synaptosomes, a *quasi in vivo* mitochondrial DNA synthesizing system.[90,91] They showed that

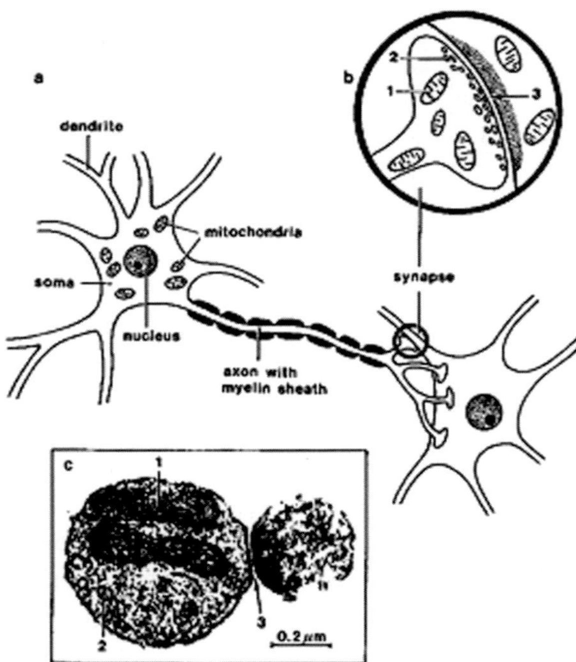

Figure 1.2A   Structure of neurons and their synapses simplified from electron micrographs. (a) Schematic representation of a neuron and its connections. (b) Enlarged synaptic end knob with mitochondria (1), synaptic vesicles (2), and synaptic cleft (3). (c) Electron micrographs of synaptosomes; identification of structures as in (b). Reproduced from Hübscher *et al.*[90]

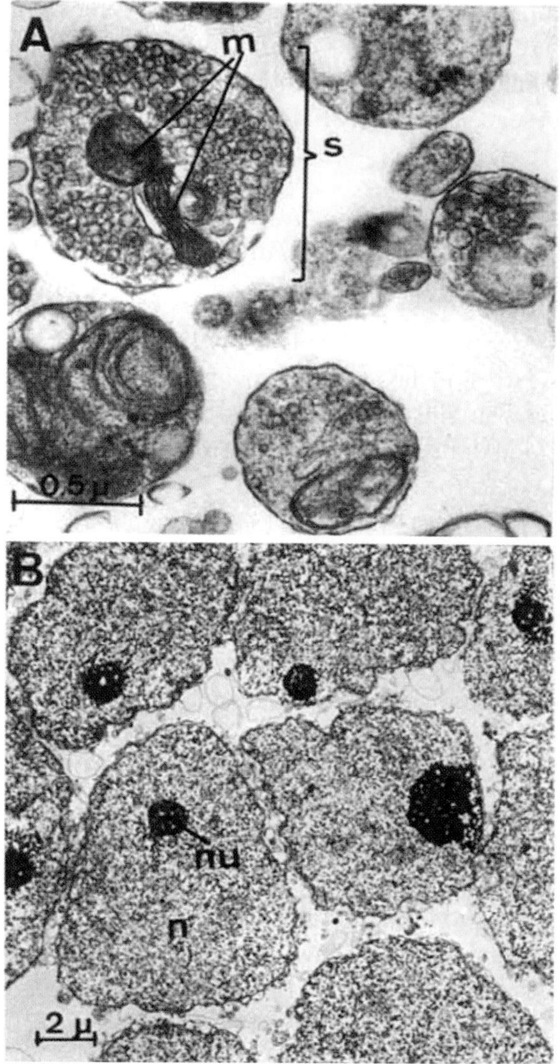

Figure 1.2B   Electron micrographs of synaptosomes (A) and nuclei (B) from rat brain cortex.
n = nucleus; nu = nucleolus; s = synaptosome; m = mitochondria. The bar represents 0.5 μm
(A) or 2 μm (B). Reproduced from Hübscher *et al.*[88]

synaptosomes permeabilized with 0.5% Brij 58 incorporated [^{3}H]dTTP into DNA
and that the labeled DNA was mitochondrial DNA by several criteria, such as:
(1) expected density following analytical ultracentrifugation in CsCl gradient, (2)
expected 6 bands upon digestion with Hind III, (3) presence of both fully duplex
circular DNA molecules and duplex circular DNA molecules containing replication

structures in electron micrographs, (4) a rate of DNA synthesis (around 10% of the input DNA in 1 hr) several thousand fold higher than the rate of repair synthesis in control nuclei and finally (5) a significant reduction of the number of replication structures in the presence of dideoxyTTP, an inhibitor of DNA pol γ.

1.6.3 *Further Evidence for a Major Involvement of DNA Polymerase α in DNA Replication and of DNA Polymerase β in DNA Repair*

In the years 1978–1980, other interesting evidence for a major involvement of DNA pol α in DNA replication and of DNA pol β in DNA repair came from the laboratories of Spadari and Hübscher. Following a preliminary report in 1973 that aphidicolin, a tetracyclic diterpenoid obtained from *Cephalosporium aphidicola*, inhibited incorporation of thymidine into DNA of cultured human embryonic lung cells and the growth of Herpes simplex and Vaccinia viruses, with no effect on RNA and protein synthesis,[92] it was found in 1978 that aphidicolin prevented mitotic division of sea urchin embryos,[93] which required replicative DNA synthesis. In particular Pedrali-Noy and Spadari in 1979 showed that aphidicolin selectively inhibited purified DNA pol α as well as DNA pols induced by Herpes simplex and Vaccinia viruses.[94] DNA pols β and γ were completely resistant to aphidicolin. The mechanism of inhibition of DNA pols by aphidicolin was also investigated by Pedrali-Noy and Spadari.[95]

With recognition that DNA pols δ and ε, later recognized as the 4th and 5th eukaryotic DNA pols with a critical role in DNA replication, were also found sensitive to aphidicolin, Spadari's group demonstrated that the selective inhibitory effect of aphidicolin on the replicative DNA pol α and the reversibility of its action *in vivo* allowed the synchronization of HeLa (and other eukaryotic cells) in culture.[96–98] They showed that the inhibition of DNA pol α by aphidicolin prevented G1 cells from entering the DNA synthetic period (S-phase), blocked cells in the S phase and allowed G2, M and G1 cells to continue the cell cycle and to accumulate at the G1/S border.

Aphidicolin thus appeared to be a more useful and physiological reagent than hydroxyurea and thymidine to synchronize cells because it did not affect cell viability or S phase duration. Furthermore, the 24 hours treatment with aphidicolin did not interfere with the synthesis of dNTPs or of DNA pols. Cells exposed to the drug continued to synthesize all three DNA pols α, β, and γ as well as all dNTPs which, when the block was removed, were present at levels optimal for DNA initiation and replication.[96] This simple technique could be applied to cells growing in suspension or monolayers (including plant cells — *see plant cell DNA polymerases section*) and allowed to harvest large quantities of synchronized cells.

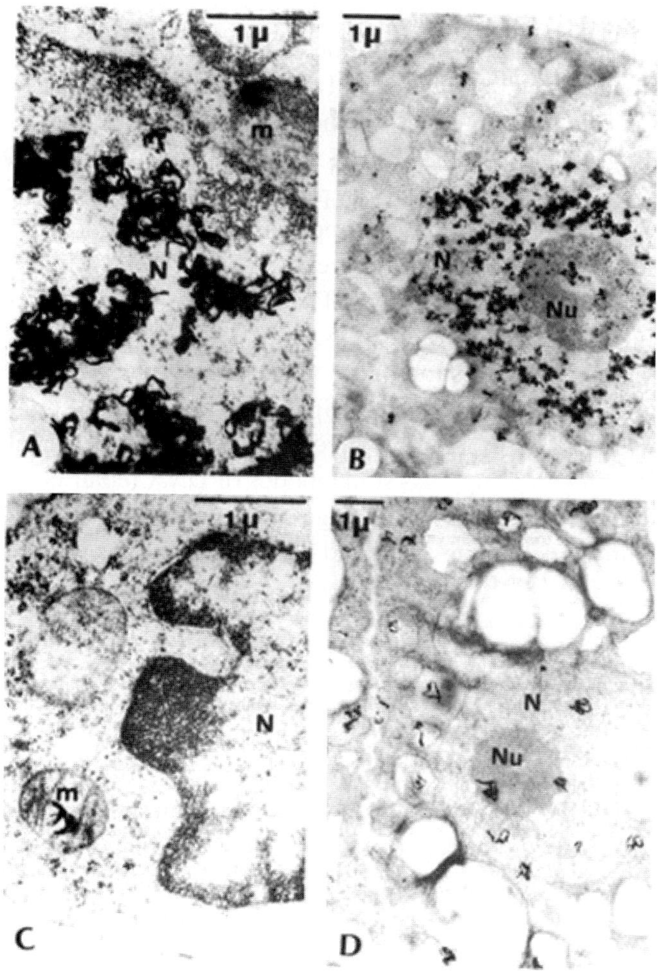

Figure 1.3   Electron microscopic autoradiographs of eukaryotic (A and C) and plant cells (B and D) incubated with labeled thymidine without (A and B) or with aphidicolin (C and D). Both types of cells, in the absence of aphidicolin, carry out nuclear and organellar (mitochondrial or chloroplast) DNA synthesis. The addition of aphidicolin completely inhibits the incorporation of thymidine into nuclear DNA but has no effect on organellar DNA synthesis. N = nucleus; Nu = nucleolus; m = mitochondria. Reproduced from Spadari *et al.*[98]

Plant cells α-like DNA pol was also found sensitive to aphidicolin by Spadari's group. **Figure 1.3** depicts electron microscope autoradiographs of mammalian and plant cells incubated with aphidicolin, showing that aphidicolin stopped nuclear DNA synthesis in both types of cells while replication of mitochondrial or chloroplast DNA, respectively, was not affected.[97-99]

In 1978–1979 a major role of DNA pol β in DNA repair was assigned more by exclusion, following the demonstration of the role of DNA pol α and γ in the replication of nuclear and mitochondrial DNA, respectively, rather than by a direct proof. Hübscher and Spadari clearly demonstrated, however, that rat nuclei from adult nondividing neurons that they had shown to contain virtually no DNA pol other than β (β, 99.2%; γ, 0.8%) were able to repair UV-damaged DNA. Incorporation of [³H]dTTP in such nuclei was stimulated approximately 7–10 fold by irradiation with UV light, and this stimulation was suppressed when DNA synthesis was allowed to proceed in the presence of dideoxyTTP, an inhibitor of DNA pol β. These data for the first time suggested a role of DNA pol β in DNA repair.[91,100] In 1981 a clear radioautographic demonstration of DNA repair synthesis in HeLa cells in the presence of aphidicolin was obtained by Spadari's group, again suggesting a major role of DNA pol β in DNA repair.[101] Of course the above experiments did not rule out that aphidicolin could suppress some modes of DNA repair due to DNA pol α and that some other types of DNA damage might require DNA pol α. The extent of damage, nature of damaging agent and many other factors might determine the relative contribution of DNA pols β and α to repair of DNA.

1.7   DNA Polymerases δ and ε

The father of DNA pol δ was Antero So. In 1976, a year after the proposal to name eukaryotic DNA pols with Greek letters in the order of discovery, Byrnes *et al.* purified from erythroid hyperplastic bone marrow a new species of DNA pol.[102] In contrast to DNA pols α, β and γ, the novel DNA pol was associated with a very active $3' \rightarrow 5'$ proofreading exonuclease activity, similar to the one associated with prokaryotic DNA pols and not separable from the DNA pol activity by several chromatographic steps and by sucrose gradient centrifugation. Both exonuclease and DNA pol activities had identical rate of heat inactivation, suggesting the coexistence of the two activities in a single polypeptide chain. The novel DNA pol was named δ but for 4–5 years it was neglected by the scientific community, its uniqueness being only the inseparability from the $3' \rightarrow 5'$ exonuclease. All other activities and properties (such as reaction requirements, sensitivity to NEN and aphidicolin, and resistance to dideoxy-NTPs) resembled those of DNA pol α which was the predominant DNA pol activity in the cells.

DNA pol δ was generally assayed with polydA/oligoT as template-primer because of a rather feeble activity observed with primed single-stranded DNA, and this property did not help to immediately recognize it as a distinct DNA pol. In the 1980s however all doubts about DNA pol δ as a polymerase structurally and immunological distinct from DNA pol α were dispelled, and its fundamental role as a complex with several auxiliary proteins in DNA replication in conjunction with DNA pol ε and DNA pol α/primase was then widely recognized (see Chapter 4).

A distinctive feature of DNA pol δ, revealed more than 10 years later in 1987 by Tan, again in So's laboratory,[103] was its dependence on proliferating cell nuclear antigen (PCNA), a protein that increased the DNA pol δ processivity by 40 fold (see also Chapter 2). Thus the high processivity of DNA pol δ and the much lower processivity of DNA pol α, which was found to possess a primase activity, suggested a role for DNA pol δ in continuous replication of the leading strand and for DNA pol α/primase for the discontinuous replication of the lagging strand.

Soon, however, Focher *et al.* in Hübscher's laboratory in 1988 established a novel procedure to purify calf thymus DNA pol δ from cytoplasmic extracts that allowed the simultaneous isolation and separation of DNA pol δ, DNA pol α, a DNA-dependent ATPase and PCNA.[104] The final preparation had typical properties of a DNA pol δ including a $3' \rightarrow 5'$ exonuclease activity, and efficiently replicated naturally occurring genomes such as primed single-stranded M13 DNA and single-stranded porcine circovirus DNA, this last one thanks to an associated or contaminating primase activity. Processivity of at least a thousand nucleotides was evident for this pol and even in the absence of PCNA, a property that was in disagreement with the dependence of the processivity of DNA pol δ on PCNA.

A similar PCNA-independent DNA pol was identified in a variety of cells, e.g. HeLa cells,[105] purified in several ways and recognized as a DNA pol distinct from δ and, therefore, named DNA pol ε. Recent experiments indicate that DNA pol δ is responsible for the replication of the lagging strand of DNA while DNA pol ε replicates the leading strand (see Chapter 4).

1.8 1985: Polymerase Chain Reaction (PCR), a Concept with Tremendous Practical Applications

It was at about this period when, thanks to the properties elaborated with DNA pols, the paramount technology of polymerase chain reaction (PCR) was developed in a Californian Company named Cetus Inc. We therefore introduce here a section about the PCR development so that especially young readers can feel the time when this important discovery was made.

> *In hindsight — basic science at a medically oriented faculty of Zürich in 1975–1980.* When Spadari and Hübscher were elaborating the function of mammalian DNA pols α, β, and γ between 1974–1978 (for details see above and **Figures 1.1 and 1.2**), the latter, who was a graduate student at the Department of Veterinary Pharmacology and Biochemistry of the University of Zürich, was told often by

(*Continued*)

(*Continued*)

> pathologists, microbiologists, parasitologists, physiologists and nutritionists not to work on "submolecular bullshit". The comments, not made by clinicians, were as follows: "It is impossible that a simple biochemical molecule such as an enzyme called DNA polymerase can ever be useful for a practical application." Here is the story in Hübscher's words: "At this time (1974) I tried to promote my modest career (almost a decade before PCR was discovered) as a simple veterinarian interested in basic science and in particular in an enzyme named DNA pol. I realized that I felt in love with this enzyme. With this idea I was immediately attacked by some considered as opinion leaders at that time at our Faculty saying: "Hübscher, are you crazy to work on an issue that *will never* be of any practical use?" Fortunately, my supervisors let me do my work. 25 years later we do not have to apologize as DNA pol scientists. The practical development of the use of a DNA pol in PCR documents in a convincing way that the scientific community has always to be open to issues that are not obvious to many people. In other words the DNA pol studies were leading not only to the polymerase chain reaction (PCR), but also to the DNA sequencing technology invented by Fred Sanger some years earlier (see above). Nowadays the DNA pols serve millions of people and animals as diagnostic and prophylactic tools and have certainly saved thousands of lives. Moreover, I am sure that the successors of the mentioned Departments are now daily using PCR and DNA sequencing."

Concept of PCR. When the concept of PCR was proposed by Kary Mullis in 1983, the basic properties of DNA pols were known. A DNA pol requires a template, a primer to start DNA synthesis, the four deoxyribonucleoside triphosphates, magnesium, and certain additives such as buffer, salt and others. The concept of PCR was not to invent novel properties of DNA pols but rather to use them in a reaction that does not generally occur in nature, namely amplification. When Mullis first brought the PCR idea in 1983 to Cetus, a biotech company founded in 1971 in the San Francisco Bay area, the enthusiasm was not overwhelming. Nevertheless, in collaboration with the hard working scientists and technicians Henry Ehrlich, Norman Arnheim, Randall K. Saiki, Glen Horn, Corey Levenson, Steven Scharf, Fred Faloona and Tom White, the company achieved within two years reproducible amplification of DNA. This led to the first patent application on March 28th, 1985. On September 20th 1985 a paper was sent to *Science* and appeared on 20th December 1985 entitled "Enzymatic amplification of beta-globin genomic sequences and restriction site analysis for diagnosis of sickle cell anemia" by R K. Saiki, S. Scharf, F. Faloona, K.B. Mullis, G. Horn, H.A. Ehrlich and N. Arnheim.[106] In this paper an application but not the technique of PCR *per se* was the main topic.

Mullis tried to publish at the same time the PCR method in *Nature* and *Science* and both journals rejected his work as insufficient for a top journal. Under the title "Specific synthesis of DNA *in vitro* via polymerase-catalyzed chain reaction" Mullis and Faloona published their results in 1987 in *Methods in Enzymology*,[107] meaning that one of the most revolutionary technology concepts was not published via a regular referee process! Nevertheless, the inventor of the PCR concept Kary Mullis was eventually awarded the Noble Prize in Chemistry in 1993.

The three steps of the PCR cycle are: first, denature the DNA to be amplified at 95°C; second, anneal primers between 50–70°C depending on their sequences; and, third, synthesis of DNA by the DNA pol. Here came the initial problem: because *E. coli* DNA pol I was used in the early days of PCR, fresh enzyme had to be added after each cycle, as it was inactivated upon denaturation and hybridization of DNA. This was not only laborious and time-consuming, but also enhanced the danger of carryover contamination, a problem that in the early days of PCR was common to many laboratories.

Around 1985 the idea came up at Cetus to purify a DNA pol from the heat-stable organism *Thermus aquaticus*. Because Kary Mullis apparently refused to isolate this DNA pol, David Gelfand and his technical assistant Susanne Stoffel purified in a very short time a DNA pol of about 90 000 daltons that was heat-stable and free of contaminating endo- and exo-nucleases. In addition, the enzyme was unexpectedly very efficient in adding dNTPs at 70–75°C when non-ionic detergents were added. With this enzyme, now called *Taq* DNA pol, they were quickly in a position to develop and to optimize single tube PCR for over 30 cycles. Moreover, Gelfand and Stoffel soon cloned and successfully over-expressed *Taq* DNA pol in *E. coli* so that large amounts of extremely pure enzyme became available to the scientific community, and the PCR technology spread over the globe in a very short time.[108,109] It was this great discovery of heat-stable DNA pol for PCR that exponentially increased the scientific publications using PCR after 1990. The last December 1989 issue of *Science* praised *"DNA polymerase as the molecule of the year"*[110] and this attribute was due to the spreading of the PCR technology as a result of its ease, thanks to heat-stable DNA pols.

PCR has deeply changed the daily practices in many laboratories. Thousand, of scientists very quickly began using PCR for various purposes. Many novel applications appeared, such as nested PCR, single-molecule amplification, inverse PCR, multiplex amplification, direct DNA sequencing, quantification of virus and bacterial load, single gamete typing, reverse transcriptase PCR (known as single enzyme RT-PCR), dUTP incorporation to prevent carryover, combinatorial libraries, aptamers, amplification of ancient DNA, *in situ* PCR, sequence tagged sites, long PCR, accurate PCR by using a combination of DNA pol and $3' \rightarrow 5'$

exonuclease. These applications led to a "golden age" of research on thermophiles, because many DNA pols appeared with different properties. Moreover, site directed mutagenesis allowed engineering of DNA pols with desired properties (see also Chapter 7: Synthetic evolution of DNA polymerases for novel properties). In conclusion, PCR has changed the daily work not only in scientific laboratories of topics from basic to applied, but also in practical terms, because many diagnostic and forensic laboratories now rely on PCR.

1.9 Yeast DNA Polymerases

Yeasts are eukaryotic microorganisms with about 1500 species. Commercial bakers' yeast *S. cerevisiae* is a very important model organism in cell biology research, and for a long time has been used by researchers to obtain information about the biology of eukaryotic cells. Thus, following the discovery of a DNA pol activity in bacteria by Kornberg and in eukaryotic cells by Bollum, the abundance of yeast and the power of yeast genetics and molecular analysis immediately led researchers to look for a similar DNA synthesizing enzyme in yeast.

Erhard Wintersberger, at the Institut fur Krebsforschung der Universitat of Wien, started to look for a DNA pol activity in yeast mitochondria in 1966,[111] organelles in eukaryotes that have their own replicating machinery. Later, in 1970, he wondered whether the enzyme he had found in mitochondria was different from the DNA pol active in the yeast nucleus.[112] Following the report by Eckstein *et al.* in 1967 from Hamburg that extracts from synchronized cells of *S. cerevisiae* exhibited a DNA pol activity oscillating during the cell cycle with a maximum just before the onset of DNA synthesis,[113] Ulrike and Erhard Wintersberger partially purified and characterized from a mitochondria-free cell extract of *S. cerevisiae* two high MW DNA pols: a major one, called **A**, and a minor one, called **B**, both apparently distinct from the high MW mt-DNA pol they had further purified at the same time both from wild type and from a *"petite"* mutant.[112] A low MW DNA pol activity similar to DNA pol β found in those years in higher eukaryotic cells was never detected in *S. cerevisiae*.[114]

A better purification and characterization of the two DNA pols present in extracts of commercial bakers' yeast and wild type *S. cerevisiae* grown aerobically to late log phase was done by Chang in 1977 in order to compare them with the properties of the well- characterized bacterial and mammalian DNA pols. Thus, she purified two immunologically distinct DNA pols and called them I and II, corresponding to the previously described **A** and **B** enzymes.[115] DNA pol I carried out pyrophosphate exchange and pyrophosphorolysis reaction and had no associated nuclease activity, in general resembling DNA pol α. However, DNA pol II had an associated 3′ → 5′ exonuclease activity and could excise mismatched 3′ nucleotides from suitable template systems, resembling more prokaryotic DNA pols II and III.

Like other DNA pols so far discovered, neither of the two yeast DNA pols was capable of initiating *de novo* DNA synthesis but Chang had shown that only DNA pol I could use oligoribonucleotides as initiators. Plevani and Badaracco, following the finding that DNA pol α from several eukaryotes was associated with a DNA primase, isolated a complex of yeast DNA pol I and DNA primase in 1984.[116] In addition, Johnson *et al.* in Campbell's laboratory clearly demonstrated in 1985 that the gene coding for this enzyme was essential, because disruption of the gene in germinated spores gave a terminal phenotype consistent with a defect in the S phase.[117]

The discovery of another yeast nuclear DNA pol, biochemically and immunologically distinct from DNA pols I and II, came in 1988, from the laboratory of Burgers at the Washington University School of Medicine, St. Louis.[118] By using yeast cells from a protease-deficient strain, Burgers and co-workers purified a complex containing three subunits, one of them with a DNA pol and a 3′ → 5′ exonuclease activity that could excise single nucleotide mismatches providing a base-paired primer-template that could be elongated by the DNA pol, consistent with a proofreading function during *in vivo* DNA replication. Thus this novel DNA pol, that was called DNA pol III, appeared functionally related to mammalian DNA pol δ. With appropriate mutants Campbell's group in 1989 showed that not only DNA pols II and III were genetically distinct, but also that DNA pol II, the one present in lower amounts, was clearly distinct from pol I. Highly purified DNA pol II contained a 3′ → 5′ exonuclease but not DNA primase activity.[119,120]

Thus, four DNA pols as well as their corresponding genes had been identified in yeast cells: DNA pol I was analogous to DNA pol α; DNA pols II and III resembled DNA pols ε and δ, and a mitochondrial DNA pol was akin to DNA pol γ. Still another DNA pol was inferred by Morrison *et al.* in 1989 from the sequence of gene REV3, a nonessential gene required for the mutagenic repair of UV damage[121] (see also Chapter 5).

1.9.1 *Revised Nomenclature for Eukaryotic DNA Polymerases*

Burgers *et al.* in 1990 proposed a revised nomenclature for eukaryotic DNA pols (**Table 1.3**) and the use of a separate letter of the Greek alphabet to designate each genetically distinct yeast DNA pol.[122] Later, the availability of yeast genome sequence has allowed the identification of other DNA pols for a total of at least 8 enzymes (see **Table 1.4** and Chapter 2).

Table 1.3   Nomenclature of eukaryotic DNA polymerases in 1990

DNA polymerase	Yeast enzyme	Yeast gene	Mammalian enzyme
α	I	*POL1*[a]	α
β	—	—	β
γ	m	*MIP1*	γ
δ	III	*POL3*[b]	δ
ε	II	—	ε

[a]Former designation, *CDC17*; [b]Former designation *CDC2*; m = mitochondrial

1.10 Plant Cell DNA Polymerases

Following the discovery of multiple DNA pols in prokaryotes (DNA pols I, II and III) and eukaryotes (DNA pols α, β and γ), several groups started to study DNA pol activities in plant cells in the mid-1970s. Litvak's group at the Université de Bordeaux II was one of the most active, studying the DNA pol activities in ungerminated or germinating wheat embryos. In 1975 they purified and partially characterized three cytoplasmic DNA pols (**A**, **B** and **C**) from ungerminated wheat embryos and another DNA pol from purified mitochondria that behaves like DNA pol **B**.[123] Considering the possibility that cytoplasmic DNA pol **B** was a release product of broken mitochondria, on the basis of chromatographic separation, template-specificity and enzyme properties, the conclusions emerging from their and others' studies[124–128] were that: (1) plants might contain two high MW DNA pols; (2) such enzymes possessed associated exonuclease activities like those of bacterial DNA pols, and (3) no low MW DNA pol analogous to animal DNA pol β was found in plant cells.

In an attempt to gather experimental data on the DNA pols of plant cells, which could allow a better and meaningful comparison with mammalian cells, Spadari's group in Pavia in 1978 used plant cells in culture where the relationship between DNA synthesis and DNA pols could be more easily examined under strictly controlled experimental conditions (proliferating or resting cells, normal or tumorous cells, mutant cell lines). The major DNA pol activity they purified from cultured *Oryza sativa L.* cells shared most of the properties of animal DNA pol α, and, at the stage of purification employed, possessed an associated 3′ → 5′ exonuclease activity.[129] Because in *Oryza sativa L.* cells as well as in other plant cell lines like *Parthenocissus tricuspidata*, *Acer pseudoplatanus L.* and *Medicago sativa L*, the purified DNA pol was the most abundant in proliferating cells and responded to changes in the rate of cell multiplication, they postulated that this activity corresponded functionally to the animal DNA pol α and propose to call it α-*like* DNA polymerase.

In 1980 the same group[130] showed that spinach leaves contained at least two DNA pols, one of them localized in the chloroplasts. The one present and predominant in crude leaf extracts and practically absent from purified chloroplasts was the previously described α-*like* DNA pol. They purified a distinct DNA pol from purified chloroplasts whose properties significantly resembled those of the animal mitochondrial DNA pol γ and thus proposed to name it γ-*like* DNA pol. The chloroplast γ-*like* DNA pol differed from the mitochondrial DNA pol described by Litvak's group in wheat embryos.

Because, by analogy with animal DNA pols, the plant cell α-*like* and the γ-*like* enzymes were sensitive and resistant, respectively, to aphidicolin,[131] they studied the effect of aphidicolin on the synthesis of DNA in the nucleus, chloroplast and mitochondrion of cultured rice cells, as assessed by autoradiography by light and electron microscopy (see **Figure 1.3**). Their results[99] clearly showed that aphidicolin specifically prevented the synthesis of nuclear DNA, while it had no effect on the synthesis of organellar DNA, thus proving that the α-like DNA pol played an essential role in the replication of nuclear DNA and that this DNA pol was not involved in the replication of plastid and mitochondrial DNA. The replication of plant organellar DNA required a different DNA pol, the γ-*like* DNA pol described in chloroplast or a mitochondrial DNA pol, if different from the γ-*like* one, both of them able of strand displacement synthesis of the organellar DNAs.

Once again, by analogy with the results obtained with mammalian cells, Spadari *et al.* demonstrated by autoradiography in 1983, that aphidicolin, by inhibiting the α-*like* DNA pol, and thus nuclear DNA replication, could be used to synchronize cultured *Daucus carota* cells by inducing accumulation of cells at the G1/S boundary of the cell cycle. A single blockage with aphidicolin synchronized 95% of the cell population (the remaining nonsynchronized 5% cells most probably representing the fraction of nonmultiplying cells), and this was reversible, resulting in immediate and synchronous resumption of nuclear DNA synthesis following removal of the drug.[132] Obviously a small fraction of the cells was blocked in the process of DNA synthesis; if necessary these cells, which were estimated to represent <10% of total population, could be forced to follow the main stream population by an appropriate second exposure to aphidicolin.

We previously described how Hübscher and Spadari demonstrated in 1979 a role of DNA pol β in DNA repair by exploiting the unique property of brain neurons which, in adult rats, are devoid of DNA pol α and do not carry out nuclear DNA replication, thus allowing a study of UV-induced DNA repair synthesis. Similarly Spadari's group in Pavia used plant protoplasts prepared from leaves of *Nicotiana sylvestris*, which, under the experimental conditions used, did not carry out nuclear DNA replication, to study UV-induced DNA repair synthesis which also occurs in cells of higher plants.[132] They clearly demonstrated by autoradiography that in this system UV-induced DNA repair synthesis was resistant to aphidicolin and thus was not performed, at least under these physiological conditions, by the α-*like* DNA pol. As mentioned above, there was no evidence as to the existence of a low MW β-*like* DNA pol in plant cells and such repair synthesis could be due to a yet unknown DNA pol. As a matter of fact the genome sequence has recently allowed the identification in plant cells of several other DNA pols and the plant model organism *Arabidopsis thaliana* could contain 10 DNA pols (see also Chapter 2).

1.11 Virus-Induced DNA Polymerases

Viruses are leading cause of many human diseases. Some oncogenic DNA viruses, such as polyoma and SV40, are known to induce cellular DNA synthesis upon infection and to cause a rise in various cellular DNA pols, in particular DNA pol α. But often a virus-induced DNA pol is required for viral DNA replication and this has led to isolation of viral DNA pols and to a search for antiviral compounds that might selectively inhibit the DNA pol and viral DNA replication. Here we shall limit the history of the discovery of viral DNA pols to some human viruses such as Herpes, Vaccinia, Hepatitis B and retroviruses for their medical interest (see also Chapters 5, 6 and 9).

1.11.1 *Herpes Virus DNA Polymerase*

Among human viruses, Herpes viruses (whose name is derived from the Greek word *herpein* which means to creep, reflecting the creeping and spreading nature of the skin lesions they usually cause) are second only to influenza and cold viruses, in causing overt disease. Once a patient has become infected by a herpes virus, the infection remains for life because the virus remains silent (*latent*) in peripheral sensory ganglia (Herpes simplex-1 (HSV-1) or 2 (HSV-2), Varicella Zoster virus (VZV)) or in monocytes or lymphocytes (Epstein-Barr virus (EBV), cytomegalovirus (CMV), Herpes lymphotropic virus, Human herpes virus-7 (HHV-7)) or in unknown target cells such as Human herpes virus-8 (HHV-8) and Kaposi's sarcoma associated herpes virus (KSHV). Latency is often followed by subsequent reactivation and recurrence of infection.

Herpes viruses encode their own DNA-dependent DNA pols. In addition, most herpes viruses encode thymidine kinase (TK) that activates several nucleoside analogs to triphosphates that are incorporated by viral DNA pol thus inhibiting replicating viruses (see Chapter 9). Herpes viruses replicate their DNA in the cell nucleus, and Keir *et al.* in 1966 observed that the infective process in BHK or

Hep-2 cells by HSV-1 was accompanied by an increase, among other enzymes, of a DNA pol activity with physical and immunological properties distinct from host cellular DNA pols. They also investigated the template specificity of the induced DNA pol in salt-fractionated crude extracts.[133] The first extensive purification and characterization, however, was reported by Weissbach *et al.* in 1973 from nuclei of HSV-1 HeLa infected cells. The enzyme had a high MW (180 000) and was optimally active at 0.15 M K_2SO_4 or 250 M KCl, salt concentrations that inhibited the HeLa DNA pols α and β by 90%. The best primer/template was oligodG/polydC, approximately 10 times better than activated DNA, perhaps a reflection of the 67% G+C base composition of HSV-1 genome.[134] These results opened the way to the identification and isolation of a novel virus-induced DNA pol in all other members of the herpes group of viruses and to the demonstration that the HSV-induced DNA pol is required for viral replication (see Chapter 6).

1.11.2 *Vaccinia Virus DNA Polymerase*

Vaccinia virus, belonging to the poxvirus family and responsible for smallpox, contains a linear, double-stranded DNA genome of approximately 190 kbp and it is unique among DNA viruses because it replicates only in the cytoplasm of the host cell. Berns *et al.* (1969) and Citarella *et al.* (1972), in Weissbach's laboratory, were the first ones to partially purify and characterize a vaccinia-directed DNA pol from the cytoplasmic fraction of virus-infected HeLa cells.[135,136] Subsequently purified to homogeneity by Chalberg and Englund in 1979,[137] the enzyme is a 110 kDa single polypeptide with an active $3' \rightarrow 5'$ proofreading exonuclease activity. Like DNA pol α and Herpes virus DNA pol, vaccinia DNA pol is strongly inhibited by aphidicolin.[94]

1.11.3 *DNA Polymerase Activity in Hepatitis B Particle*

Whereas the DNA pols induced during infection by Herpes or Vaccinia viruses are localized in the cell, a DNA pol activity within the virion of the Hepatitis B Dane particle was first reported by Robinson's group in 1973.[138] They also showed that the particle contained a small circular double-stranded DNA with a MW of 1.6×10^6 daltons, which served as a primer-template for the DNA pol. The DNA pol was a specific marker for Dane particles and Robinson used the DNA pol reaction also to radiolabel the Dane particle core for a specific and sensitive radioimmunoprecipitation assay for antibodies against hepatitis B core antigen (anti-Hbc).[138–141] Hepatitis B virus replicates through reverse transcription of a pregenomic RNA by viral pol that possesses protein-priming, reverse transcriptase, DNA pol and RNase

H activities (see Chapter 6). Thus, the reverse transcription activity of the viral polymerase can be the target of nucleoside analogs inhibiting viral replication and the polymerase can be used for *in vitro* screening of novel anti-HBV compounds directed against wild-type and mutants of the polymerase (see Chapter 9).

1.11.4 *Retroviruses Reverse Transcriptase*

Reverse transcription, the flow of genetic information from RNA to DNA, typifies retroviruses more than any other properties. Howard Temin in the 1960s was the first to suggest that some RNA viruses could replicate through a DNA intermediate, subsequently integrated into the host genome. His hypothesis was based upon the observation that replication of retroviruses was sensitive to inhibitors of DNA synthesis. As mentioned earlier, the discovery in purified virions of a novel polymerase, a RNA-dependent DNA pol (reverse transcriptase), was made in 1970 by Howard Temin at the University of Wisconsin- Madison[73] and independently by David Baltimore at MIT.[74] This discovery astonished the scientific world because it violated the central dogma of molecular biology, that is the flow of genetic information from DNA to RNA to proteins. The AIDS pandemic later stimulated renewed interest in reverse transcriptase (see Chapter 6) as a therapeutic target (see Chapter 9).

1.12 1999–2000: Nucleotide Sequence Analysis of Eukaryotic Organisms Allowed the Identification of Many Novel Specialized DNA Polymerases

As soon as the human genome sequence was published in 1999, a "golden age" of DNA pols followed. More than 30 years ago it was hypothesized by Miroslav Radman that error-prone DNA pols might exist in bacteria as well as in eukaryotes that could, like terminal deoxynucleotidyl transferases, insert random nucleotides in a template-independent fashion, and that DNA damage-induced mutagenesis might be a specific cellular response to genotoxic damage. This event is known as SOS response and in *E. coli* more than 30 inducible genes were identified (reviewed in Refs. 142 and 143). The isolation of mutants defective exclusively in SOS mutagenesis led to the identification of the *umu*C, *umu*D and *din*B genes in *E. coli*. These genes are now known as the two translesion DNA pols IV (*din*B) and V (*umu*C and *umu*D) (see Chapter 3).

1.12.1 *DNA Polymerase ζ, the Lesion Extender*

Over 30 years ago Christopher Lawrence discovered that in yeast Rev 3 and Rev 7 genes are indispensable for UV mutagenesis and for mutagenesis derived from translesion DNA synthesis (TLS) through abasic sites.[142] The two subunits form the heterodimer of DNA pol ζ.[144] This enzyme belongs to the B family, but lacks a $3' \rightarrow 5'$ exonuclease proofreading activity and is essential, because mice without the DNA pol ζ gene die during embryogenesis.[145] Its function is not to incorporate opposite a lesion but to *extend* from a variety of lesions (for details see Chapters 4 and 5).

1.12.2 *DNA Polymerases* λ *and* μ, *Two Family X DNA Polymerases*

In 2000 Luis Blanco reported DNA pol λ as a novel eukaryotic DNA pol with a potential role in meiosis.[146] Soon thereafter he also identified a second family X polymerase that he called DNA pol μ, homologous to TdT, that could act as a DNA mutator in eukaryotic cells.[147] These two DNA pols were subsequently studied by various laboratories. Both of these enzymes appear to have important DNA repair functions such as base excision repair, translesion DNA synthesis, double-stranded break repair and, in a physiological sense, they have important roles in the hypermutation of immunoglobulin genes, a recombination event leading to the diversity of antibodies (for details see Chapters 4 and 6).

1.12.3 *The Complex Y Family of DNA Polymerases*

DNA pols of the Y family belong to the five subfamilies which include: (i) the *umu*C and *umu*D (pol V) branch that is only found in prokaryotes; (ii) the *din*B branch which is found in *E. coli* (pol IV), in archaea (Dbp/Dpo4) and in eukaryotes (DNA pol κ). The other three subfamilies found exclusively in eukaryotes are (iii) DNA pol η, (iv) Rev1 and (v) DNA pol ι.

The discoveries of novel DNA pols of the Y family were mostly reported in the year 1999–2000 (reviewed in Ref. 148). In *E. coli* the groups of Goodman, Woodgate and Livneh simultaneously published in 1999 that DNA pol V is a dimer of the post-translationally modified *UmuD* protein in complex with *UmuC*.[149,150] Soon thereafter Matsui and collaborators identified *DinB* as a DNA pol and called it DNA pol IV.[151] Also in 1999 Hanaoka *et al.* purified a DNA pol from human Xeroderma pigmentosum variant (XP-V) cells and called it DNA pol η.[152] Ohmori's[153] and Friedberg's[154,155] groups described the human *DINB1* gene product as DNA pol κ. Lin *et al.*[156] identified in human cells Rev1, a homolog of the *S. cerevisiae* REV1, which also codes for a dCMP transferase that is dependent upon a DNA template. Finally, the most astonishing discovery in these years was the fact that Woodgate *et al.* identified DNA pol ι,[157] that has such a low fidelity that it even violates the Watson-Crick base pair rule!

Table 1.4   Nomenclature of eukaryotic DNA polymerases in 2001

Greek name	HUGO name	Class	Other names	Proposed main function
alpha	POLA	B	POL1	DNA replication
beta	POLB	X		Base excision repair
gamma	POLG	A	MIP1	Mt replication
delta	POLD1	B	POL3	DNA replication
epsilon	POLE	B	POL2	DNA replication
zeta	POLZ	B	REV3	Bypass synthesis
eta	POLH	Y	RAD30, XPV	Bypass synthesis
theta	POLQ	A	mus308	DNA repair
iota	POLI	Y	RAD30B	Bypass synthesis
kappa	POLK	Y	DinB1	Bypass synthesis
lambda	POLL	X	POL4	Base excision repair
mu	POLM	X		Non-homologous end joining
sigma	POLS	X	TRF4	Sister chromatid cohesion
	REV1L	Y	REV1	Bypass synthesis
	TDT	X		Antibody diversity

Adapted from Burgers *et al.*[158]

1.12.4 *Present Nomenclature for Eukaryotic DNA Polymerases*

In 2001 Burgers *et al.* proposed a novel nomenclature for eukaryotic DNA pols[158] reported in **Table 1.4**.

1.13 Concluding Remarks, Parts 1.5–1.12

With few exceptions (such as the protists *yeast* and *amoebae* which are individual, unicellular organisms), eukaryotes are complex multicellular organisms made up of many specialized cells containing DNA/chromosomes enclosed by a membrane defining the nucleus, whose presence gives the name to these organisms. Eukaryotic cells are also much larger and have a greater degree of structural organization and functional complexity than prokaryotic cells. Furthermore, nearly all of them possess self-replicating organelles such as *mitochondria* (plant cells and various algae have *plastids*, among which the most prominent and well studied are the *chloro-plasts* present in green parts of the plant). Both types, mitochondrion and plastid, have probably evolved from aerobic bacteria. They retain their own bacteria-like chromosomal DNA and are self-replicating organelles that apparently divide independently of the cell, although over time most of the original genes were transferred to the host cell and therefore most of their proteins are now coded for by the nuclear DNA. Because of this they are unable to live on their own if isolated from the eukaryotic cell.

The basic mechanisms of DNA synthesis are similar in prokaryotes and eukaryotes, but DNA replication is much more complicated in eukaryotes. It starts at a specific point in the cell cycle, the beginning of the S phase and at several points in the chromosome leading to chromosome duplication first and then to nuclear and cell division. The surface/volume area ratio in eukaryotic cells is smaller than in prokaryotes and explains the lower metabolic rate and longer generation time in eukaryotes.

All eukaryotes, including plants, are vulnerable to the effects of DNA damage either due to metabolic processes inside the cell or to environmental factors. They thus possess different types of DNA repair mechanisms such as base excision repair (BER), nucleotide excision repair (NER), mismatch repair (MMR), nonhomologous end joining (NHEJ), recombinational or homologous recombination repair (see Chapters 2 and 5). When necessary, they are also able to tolerate DNA damage allowing the DNA replication machinery to replicate past a DNA lesion (translesion DNA synthesis, TLS).

No wonder that, by analogy with multiple DNA pols (at least 5) found in *E. coli*, we have just seen that eukaryotes as well contain many DNA pols, e.g. at least 8 in yeast and at least 15 in mammalian cells. The model plant organism *Arabidopsis thaliana* could contain 10 DNA pols. To replicate and maintain genome integrity (including the organellar genome) many DNA pols are thus required, and some cooperate with multiple proteins induced/expressed in response to DNA damage and involved in the various types of error-free DNA repair mechanisms as well as in DNA damage checkpoint. Sometimes specialized DNA pols (e.g. the Y family translesion DNA pols) allow greater tolerance to damage leading to enhanced mutagenesis as well as to an increased rate of survival (see Chapters 2, 3, 4 and 5).

Viruses are intracellular parasites, and for their replication strategies they must either operate within limits imposed by the host cells or circumvent these limits. Several viruses, such as parvoviruses, papillomaviruses, polyomaviruses, adenoviruses and herpes viruses, enter the cell nucleus to replicate their DNA and either induce and use a cellular DNA pol (parvoviruses, papillomaviruses, polyomaviruses), or encode their own DNA pol (adenoviruses and herpes viruses). Among cytoplasmic viruses, vaccinia viruses are of historic interest and might be used as possible agents of bioterrorism. As we have seen in the case of vaccinia, they possess their own DNA pol. Hepatitis B virus differs from the other DNA viruses in that it is more similar to the oncornaviruses (RNA tumor viruses) in its mode of replication. As described above it encodes its own DNA pol that possesses protein-priming, reverse transcriptase, DNA pol and RNase H activities, all required for its replication (see Chapter 6).

Finally, the RNA tumor viruses (retroviruses) are an important class with an RNA genome that must be copied to DNA by an RNA-dependent DNA pol (reverse transcriptase) prior to integration in the host cell chromosomes. A member of these viruses, the *human immunodeficiency virus* (HIV) has in the last decades caused the AIDS pandemic, and its reverse transcriptase is the therapeutic target of many inhibitors described in Chapter 9.

As in the case of bacterial infections, some human/animal pathological states such as autoimmune diseases, cancer and viral infections are attributable to uncontrolled DNA replication, and cellular/viral DNA pols are therapeutic targets for a variety of therapeutic agents that will be discussed in Chapters 8 and 9.

The discovery of DNA pols has been a long and hard work, spanning more than half a century, and took the best efforts of some of the most talented scientists of several disciplines. As always in science, it did not follow a straight line, but rather it was a winding and climbing trail, with dead ends, crevices and pitfalls. **Figure 1.4** ideally represents this extraordinary effort.

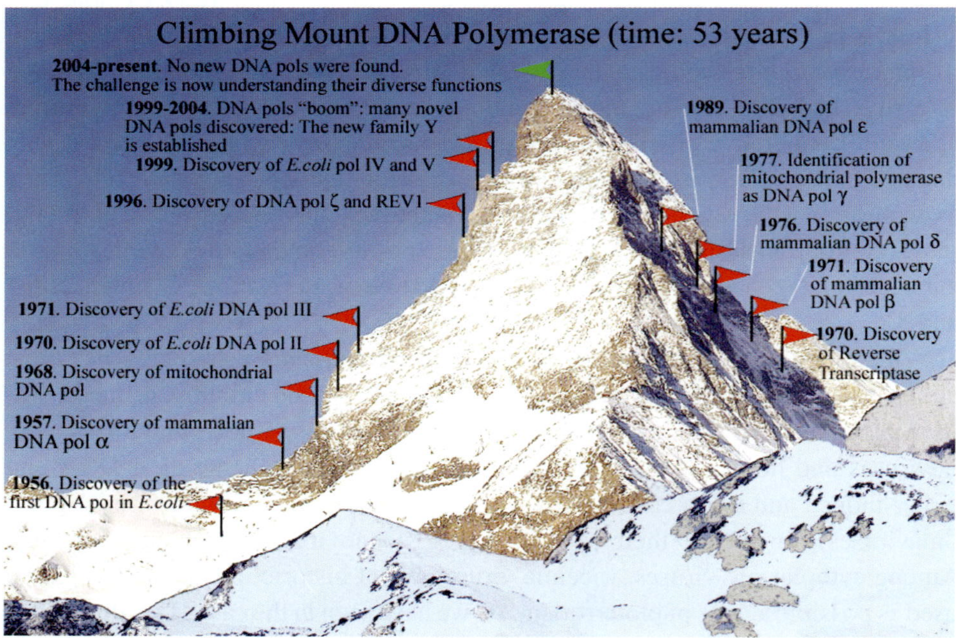

Figure 1.4   Climbing Mount DNA polymerase. The most significant achievements are indicated in chronological (ascending) order by the red flags. The background picture shows Mt. Cervino (Italy)/Mt. Matterhorn (Switzerland) (4478 m/14730 ft), in the Swiss-Italian Alps.

References

1. Singer C. 1950. *A History of Biology*. New York: Henry Schuman, Inc.
2. Portugal FH, Cohen IS. 1977. *A Century of DNA*. Cambridge: MIT Press.
3. Wilson EB. 1959. *The Cell in Development and Heredity*. New York: The Macmillan Company.
4. Darwin CR, Wallace AR. 1858. *Journal of the Proceedings of the Linnean Society of London. Zoology*: 46–50.
5. Dahm R. 2005. *Dev Biol* 278: 274–88.
6. Dahm R. 2008. *Hum Genet* 122: 565–81.
7. Griffith F. 1928. *Journal of Hygiene* 27: 113–59.
8. Beadle GW, Tatum EL. 1941. *Proc Natl Acad Sci U S A* 27: 499–506.
9. Avery OT, MacLeod CM, McCarty M. 1944. *Journal of Experimental Medicine* 79: 137–57.
10. Lederberg J, Tatum EL. 1946. *Nature* 158: 558.
11. Tatum EL, Lederberg J. 1947. *J Bacteriol* 53: 673–84.
12. Zinder ND, Lederberg J. 1952. *J Bacteriol* 64: 679–99.
13. Hershey AD, Chase M. 1952. *J Gen Physiol* 36: 39–56.
14. Herriot RM. 1951. *J Bacteriol* 61: 752–4.
15. Chargaff E, Magasanik B, Vischer E, Green C, Doniger R, Elson D. 1950. *J Biol Chem* 186: 51–67.
16. Watson JD, Crick FH. 1953. *Nature* 171: 737–8.
17. Watson JD, Crick FH. 1953. *Nature* 171: 964–7.
18. Meselson M, Stahl FW. 1958. *Proc Natl Acad Sci U S A* 44: 671–82.
19. Ycas M. 1969. *The Biological Code*. Amsterdam & London: John Wiley & Sons.
20. Woese CR. 1967. *The Genetic Code*. New York: Harper and Row.
21. Kornberg A. 1989. *For the Love of Enzymes*. Cambridge, Massachusetts: Harvard University Press.
22. Kornberg A, Baker TA. 1992. *DNA Replication*. New York: WH. Freeman and Company.
23. Grunberg-Manago M, Oritz PJ, Ochoa S. 1955. *Science* 122: 907–10.
24. Hurwitz J. 2005. *J Biol Chem* 280: 42477–85.
25. Leloir LF. 1971. *Science* 172: 1299–302.
26. Lehman IR, Bessman MJ, Simms ES, Kornberg A. 1958. *J Biol Chem* 233: 163–70.
27. Bessman MJ, Lehman IR, Simms ES, Kornberg A. 1958. *J Biol Chem* 233: 171–7.
28. Gellert M. 1967. *Proc Natl Acad Sci U S A* 57: 148–55.
29. Weiss B, Richardson CC. 1967. *Proc Natl Acad Sci U S A* 57: 1021–8.
30. Olivera BM, Lehman IR 1967. *Proc Natl Acad Sci U S A* 57: 1426–33.
31. Gefter ML, Becker A, Hurwitz J. 1967. *Proc Natl Acad Sci U S A* 58: 240–7.
32. Cozzarelli NR, Melechen NE, Jovin TM, Kornberg A. 1967. *Biochem Biophys Res Commun* 28: 578–86.
33. Goulian M, Kornberg A. 1967. *Proc Natl Acad Sci U S A* 58: 1723–30.
34. De Lucia P, Cairns J. 1969. *Nature* 224: 1164–6.
35. Kornberg T, Gefter ML. 1971. *Proc Natl Acad Sci U S A* 68: 761–4.
36. Kornberg T, Gefter ML. 1970. *Biochem Biophys Res Commun* 40: 1348–55.
37. Knippers R. 1970. *Nature* 228: 1050–3.
38. Moses RE, Richardson CC. 1970. *Biochem Biophys Res Commun* 41: 1565–71.
39. Kornberg T, Gefter ML. 1972. *J Biol Chem* 247: 5369–75.
40. Sanger F, Nicklen S, Coulson AR. 1977. *Proc Natl Acad Sci U S A* 74: 5463–7.
41. Okazaki R, Okazaki T, Sakabe K, Sugimoto K, Sugino A. 1968. *Proc Natl Acad Sci U S A* 59: 598–605.

42. Brutlag D, Schekman R, Kornberg A. 1971. *Proc Natl Acad Sci U S A* 68: 2826–9.
43. Greider CW, Blackburn EH. 1985. *Cell* 43: 405–13.
44. Alberts BM, Amodio FJ, Jenkins M, Gutmann ED, Ferris L. 1968. *Cold Spring Harb Symp Quant Biol* 33: 289–305.
45. Alberts BM. 1970. *Fed Proc* 29: 1154–63.
46. Alberts BM, Frey L. 1970. *Nature* 227: 1313–8.
47. Wang JC. 1971. *J Mol Biol* 55: 523–33.
48. Wang JC. 1991. *J Biol Chem* 266: 6659–62.
49. Bollum FJ. 1962. *J Biol Chem* 237: 1945–9.
50. Cohen SN, Chang AC, Boyer HW, Helling RB. 1973. *Proc Natl Acad Sci U S A* 70: 3240–4.
51. Bollum FJ, Potter VR. 1958. *J Biol Chem* 233: 478–82.
52. Friedkin M, Tilson D, Roberts D. 1956. *J Biol Chem* 220: 627–37.
53. Friedkin M, Wood HI. 1956. *J Biol Chem* 220: 639–51.
54. Reichard P, Estborn B. 1951. *J Biol Chem* 188: 839–46.
55. Friedkin M, Roberts D. 1954. *J Biol Chem* 207: 257–66.
56. Lu KH, Winnick T. 1954. *Exp Cell Res* 7: 238–42.
57. Bollum FJ. 1960. *J Biol Chem* 235: 2399–403.
58. Yoneda M, Bollum FJ. 1965. *J Biol Chem* 240: 3385–91.
59. Fry M, Loeb L.A. 1986. *Animal cell DNA polymerases*. Boca Raton, Florida, US: CRC Press, Inc.
60. Weissbach A, Schlabach A, Fridlender B, Bolden A. 1971. *Nat New Biol* 231: 167–70.
61. Baril EF, Brown OE, Jenkins MD, Laszlo J. 1971. *Biochemistry* 10: 1981–92.
62. Chang LM, Bollum FJ. 1971. *J Biol Chem* 246: 5835–7.
63. Haines ME, Holmes AM, Johnston IR. 1971. *FEBS Lett* 17: 63–7.
64. Berger H, Jr., Huang RC, Irvin JL. 1971. *J Biol Chem* 246: 7275–83.
65. Chang LM. 1976. *Science* 191: 1183–5.
66. Chang LM, Bollum FJ. 1972. *Science* 175: 1116–7.
67. Hecht NB. 1973. *Biochim Biophys Acta* 312: 471–83.
68. Lazarus LH, Kitron N. 1973. *J Mol Biol* 81: 529–34.
69. Spadari S, Muller R, Weissbach A. 1974. *J Biol Chem* 249: 2991–2.
70. Smith RG, Abrell JW, Lewis BJ, Gallo RC. 1975. *J Biol Chem* 250: 1702–9.
71. Brun GM, Assairi LM, Chapeville F. 1975. *J Biol Chem* 250: 7320–3.
72. Weissbach A, Baltimore D, Bollum F, Gallo R, Korn D. 1975. *Science* 190: 401–2.
73. Temin HM, Mizutani S. 1970. *Nature* 226: 1211–3.
74. Baltimore D. 1970. *Nature* 226: 1209–11.
75. Chang LM, Bollum FJ. 1972. *Biochemistry* 11: 1264–72.
76. Fridlender B, Fry M, Bolden A, Weissbach A. 1972. *Proc Natl Acad Sci U S A* 69: 452–5.
77. Spadari S, Weissbach A. 1974. *J Biol Chem* 249: 5809–15.
78. Chang LM, Brown M, Bollum FJ. 1973. *J Mol Biol* 74: 1–8.
79. Spadari S, Weissbach A. 1974. *J Mol Biol* 86: 11–20.
80. Chiu RW, Baril EF. 1975. *J Biol Chem* 250: 7951–7.
81. Baril EF, Jenkins MD, Brown OE, Laszlo J, Morris HP. 1973. *Cancer Res* 33: 1187–93.
82. Mayer RJ, Smith RG, Gallo RC. 1975. *Blood* 46: 509–18.
83. Bertazzoni U, Stefanini M, Pedrali-Noy G, Giulotto E, Nuzzo F, *et al.* 1976. *Proc Natl Acad Sci U S A* 73: 785–9.
84. Spadari S, Weissbach A. 1975. *Proc Natl Acad Sci U S A* 72: 503–7.
85. Hubscher U, Kuenzle CC, Spadari S. 1977. *Nucleic Acids Res* 4: 2917–29.
86. Spadari S, Focher F, Hubscher U. 1988. *In Vivo* 2: 317–20.
87. Bolden A, Pedrali-Noy G, Weissbach A. 1977. *J Biol Chem* 252: 3351–6.

88. Hubscher U, Kuenzle CC, Spadari S. 1977. *Eur J Biochem* 81: 249–58.
89. Bertazzoni U, Scovassi AI, Brun GM. 1977. *Eur J Biochem* 81: 237–48.
90. Hubscher U, Kuenzle CC, Limacher W, Scherrer P, Spadari S. 1979. *Cold Spring Harb Symp Quant Biol* 43 Pt 1: 625–9.
91. Hubscher U, Kuenzle CC, Spadari S. 1979. *Proc Natl Acad Sci U S A* 76: 2316–20.
92. Bucknall RA, Moores H, Simms R, Hesp B. 1973. *Antimicrob Agents Chemother* 4: 294–8.
93. Ohashi M, Taguchi T, Ikegami S. 1978. *Biochem Biophys Res Commun* 82: 1084–90.
94. Pedrali-Noy G, Spadari S. 1979. *Biochem Biophys Res Commun* 88: 1194–202.
95. Pedrali-Noy G, Spadari S. 1980. *J Virol* 36: 457–64.
96. Pedrali-Noy G, Spadari S, Miller-Faures A, Miller AO, Kruppa J, Koch G. 1980. *Nucleic Acids Res* 8: 377–87.
97. Geuskens M, Hardt N, Pedrali-Noy G, Spadari S. 1981. *Nucleic Acids Res* 9: 1599–613.
98. Spadari S, Sala F, Pedrali-Noy G. 1982. *Trends in Biochemical Sciences* 7: 29–32.
99. Sala F, Galli MG, Levi M, Burroni D, Parisi B, *et al.* 1981. *FEBS Lett* 124: 112–8.
100. Waser J, Hubscher U, Kuenzle CC, Spadari S. 1979. *Eur J Biochem* 97: 361–8.
101. Hardt N, Pedrali-Noy G, Focher F, Spadari S. 1981. *Biochem J* 199: 453–5.
102. Byrnes JJ, Downey KM, Black VL, So AG. 1976. *Biochemistry* 15: 2817–23.
103. Tan CK, Castillo C, So AG, Downey KM. 1986. *J Biol Chem* 261: 12310–6.
104. Focher F, Gassmann M, Hafkemeyer P, Ferrari E, Spadari S, Hubscher U. 1989. *Nucleic Acids Res* 17: 1805–21.
105. Syvaoja J, Linn S. 1989. *J Biol Chem* 264: 2489–97.
106. Saiki RK, Scharf S, Faloona F, Mullis KB, Horn GT, *et al.* 1985. *Science* 230: 1350–4.
107. Mullis KB, Faloona FA. 1987. *Methods Enzymol* 155: 335–50.
108. Lawyer FC, Stoffel S, Saiki RK, Myambo K, Drummond R, Gelfand DH. 1989. *J Biol Chem* 264: 6427–37.
109. Saiki RK, Gelfand DH, Stoffel S, Scharf SJ, Higuchi R, *et al.* 1988. *Science* 239: 487–91.
110. Guyer RL, Koshland DE, Jr. 1989. *Science* 246: 1543–6.
111. Wintersberger E. 1966. *Biochem Biophys Res Commun* 25: 1–7.
112. Wintersberger U, Wintersberger E. 1970. *Eur J Biochem* 13: 20–7.
113. Eckstein H, Paduch V, Hilz H. 1967. *Eur J Biochem* 3: 224–31.
114. Wintersberger U. 1974. *Eur J Biochem* 50: 197–202.
115. Chang LM. 1977. *J Biol Chem* 252: 1873–80.
116. Plevani P, Badaracco G, Augl C, Chang LM. 1984. *J Biol Chem* 259: 7532–9.
117. Johnson LM, Snyder M, Chang LM, Davis RW, Campbell JL. 1985. *Cell* 43: 369–77.
118. Bauer GA, Heller HM, Burgers PM. 1988. *J Biol Chem* 263: 917–24.
119. Budd ME, Sitney KC, Campbell JL. 1989. *J Biol Chem* 264: 6557–65.
120. Sitney KC, Budd ME, Campbell JL. 1989. *Cell* 56: 599–605.
121. Morrison A, Christensen RB, Alley J, Beck AK, Bernstine EG, *et al.* 1989. *J Bacteriol* 171: 5659–67.
122. Burgers PM, Bambara RA, Campbell JL, Chang LM, Downey KM, *et al.* 1990. *Eur J Biochem* 191: 617–8.
123. Castroviejo M, Tarrago-Litvak L, Litvak S. 1975. *Nucleic Acids Res* 2: 2077–90.
124. Sahai Srivastava BI, Grace JT, Jr. 1974. *Life Sci* 14: 1947–54.
125. Tarrago-Litvak L, Castroviejo M, Litvak S. 1975. *FEBS Lett* 59: 125–30.
126. Gardner JM, Kado CI. 1976. *Biochemistry* 15: 688–97.
127. Mory YY, Chen D, Sarid S. 1975. *Plant Physiol* 55: 437–42.
128. McLennan AG, Keir HM. 1977. *Biochem Soc Symp*: 55–73.
129. Amileni A, Sala F, Cella R, Spadari S. 1979. *Planta* 146: 521–7.
130. Sala F, Amileni AR, Parisi B, Spadari S. 1980. *Eur J Biochem* 112: 211–7.

131. Sala F, Parisi B, Burroni D, Amileni AR, Pedrali-Noy G, Spadari S. 1980. *FEBS Lett* 117: 93–8.
132. Sala F GM, Nielsen E, Magnien E, Devreux M, Pedrali-Noy G. Spadari S. 1983. *FEBS Letters* 153: 204–8.
133. Keir HM, Subak-Sharpe H, Shedden WI, Watson DH, Wildy P. 1966. *Virology* 30: 154–7.
134. Weissbach A, Hong SC, Aucker J, Muller R. 1973. *J Biol Chem* 248: 6270–7.
135. Berns KI, Silverman C, Weissbach A. 1969. *J Virol* 4: 15–23.
136. Citarella RV, Muller R, Schlabach A, Weissbach A. 1972. *J Virol* 10: 721–9.
137. Challberg MD, Englund PT. 1979. *J Biol Chem* 254: 7812–9.
138. Kaplan PM, Greenman RL, Gerin JL, Purcell RH, Robinson WS. 1973. *J Virol* 12: 995–1005.
139. Robinson WS, Greenman RL. 1974. *J Virol* 13: 1231–6.
140. Robinson WS, Clayton DA, Greenman RL. 1974. *J Virol* 14: 384–91.
141. Greeman RL, Robinson WS, Vyas GN. 1975. *Vox Sang* 29: 77–80.
142. Lawrence CW, Christensen R. 1976. *Genetics* 82: 207–32.
143. Friedberg EC, Wagner R, Radman M. 2002. *Science* 296: 1627–30.
144. Lawrence CW. 2004. *Adv Protein Chem* 69: 167–203.
145. Wittschieben J, Shivji MK, Lalani E, Jacobs MA, Marini F, *et al.* 2000. *Curr Biol* 10: 1217–20.
146. Garcia-Diaz M, Dominguez O, Lopez-Fernandez LA, Lain de Lera T, Saniger ML, *et al.* 2000. *J Mol Biol* 301: 851–67.
147. Dominguez O, Ruiz JF, Lain de Lera T, Garcia-Diaz M, Gonzalez MA, *et al.* 2000. *Embo J* 19: 1731–42.
148. Prakash S, Johnson RE, Prakash L. 2005. *Annu Rev Biochem* 74: 317–53.
149. Reuven NB, Arad G, Maor-Shoshani A, Livneh Z. 1999. *J Biol Chem* 274: 31763–6.
150. Tang M, Shen X, Frank EG, O'Donnell M, Woodgate R, Goodman MF. 1999. *Proc Natl Acad Sci U S A* 96: 8919–24.
151. Wagner J, Gruz P, Kim SR, Yamada M, Matsui K, *et al.* 1999. *Mol Cell* 4: 281–6.
152. Masutani C, Kusumoto R, Yamada A, Dohmae N, Yokoi M, *et al.* 1999. *Nature* 399: 700–4.
153. Ogi T, Kato T, Jr., Kato T, Ohmori H. 1999. *Genes Cells* 4: 607–18.
154. Gerlach VL, Aravind L, Gotway G, Schultz RA, Koonin EV, Friedberg EC. 1999. *Proc Natl Acad Sci U S A* 96: 11922–7.
155. Gerlach VL, Feaver WJ, Fischhaber PL, Richardson JA, Aravind L, *et al.* 2000. *Cold Spring Harb Symp Quant Biol* 65: 41–9.
156. Lin W, Xin H, Zhang Y, Wu X, Yuan F, Wang Z. 1999. *Nucleic Acids Res* 27: 4468–75.
157. Tissier A, McDonald JP, Frank EG, Woodgate R. 2000. *Genes Dev* 14: 1642–50.
158. Burgers PM, Koonin EV, Bruford E, Blanco L, Burtis KC, *et al.* 2001. *J Biol Chem* 276: 43487–90.

CHAPTER 2

DNA Polymerases in the Three Kingdoms of Life: Bacteria, Archaea and Eukaryotes

2.1 Synthesis and Maintenance of DNA in Nature Need DNA Polymerases

In all three kingdoms of life (bacteria, archaea and eukaryotes) DNA constitutes the genetic material. Even many viruses consist of DNA. Universally in any organism in nature, a DNA molecule is built up of adenine (A) pairing to thymine (T) and cytosine (C) pairing to guanine (G). As initially suggested by Watson and Crick in 1953, by this base pairing rule the mother DNA strand encodes the instruction for the synthesis of the new daughter DNA strand in DNA replication during the cell division. The mother strand also serves as a copy to replace the damaged strand in different DNA repair events.[1] Ever since the first discovery in 1956, the enzymes that catalyze the synthesis of DNA are called DNA polymerases (pols)[2,3] (for their history and discovery see Chapter 1). DNA replication and the various DNA repair events all rely on DNA synthesis carried out by those DNA pols (**Figure 2.1**).

Nature has evolved a system where DNA can be faithfully transmitted from one generation to the next by replication. However, a few mistakes performed in a DNA replication cycle might be critical to guarantee changes that possibly allow the evolution of species. A DNA molecule in a single mammalian cell has a length of 2 m, contains about 6×10^9 bases and faces up to $100\,000$ alterations per day. Such alterations include depurination, depyrimidation, oxidation, non-enzymatic methylation, bulky adduct formation, DNA deletions, DNA expansions, recombinations and breaks of the phosphodiester backbone, all of which can lead to single-stranded or even double-stranded breaks. The enormous and difficult joint task of the various DNA repair mechanisms is to keep these alterations at a "forgivable" level so that the continuity of the cellular functions is guaranteed at any time. Whenever mistakes accumulate, the chances increase that those alterations stably establish themselves in the DNA. If DNA repair is not keeping up in time, the consequences of such alterations in the DNA sequence can be cell death or changes of the cellular physiology. Programmed cell death, called apoptosis, is a physiological event to

59

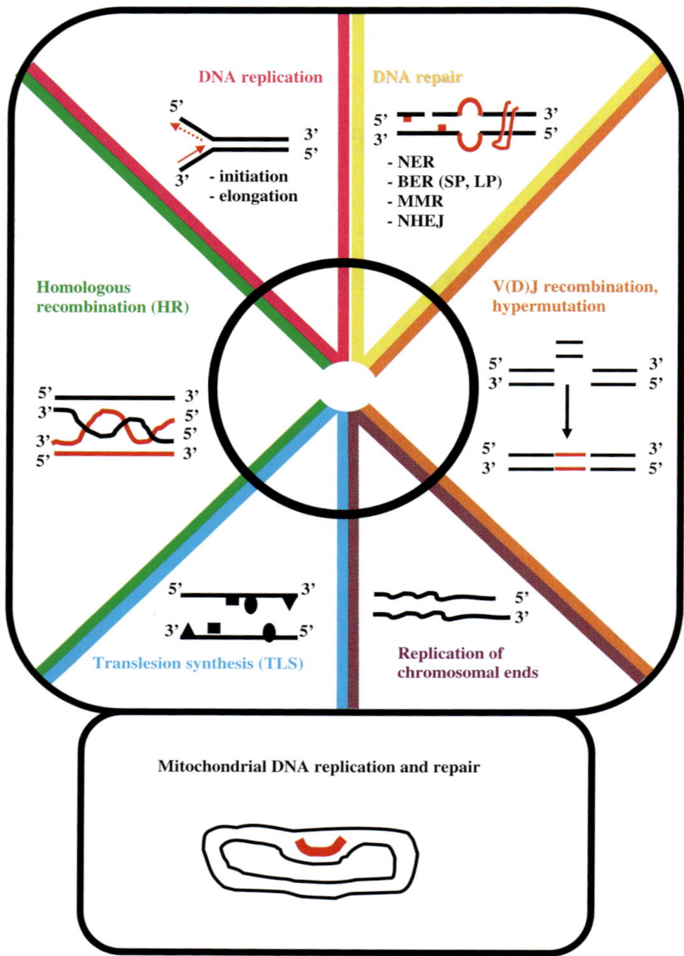

Figure 2.1   Cellular events that need DNA polymerases. NER: nucleotide excision repair; **BER**: base excision repair; **SP**: short patch; **LP**: long patch; **MMR**: mismatch repair; **NHEJ**: non-homologous end joining; **HR**: homologous recombination; **TLS**: translesion synthesis. For more details see text.

eliminate useless cells that contain altered DNA coding for proteins with reduced, enhanced or ablated functions. However, if those cells cannot undergo apoptosis, even more mutations can accumulate. These can lead to a gain of unlimited replication potential and uncontrolled cell division, a process than can eventually give rise to cancer. Still, as deleterious as mutations may seem, an intermediate level of accumulating DNA mutations might provide a driving force for the evolution of species.

Depending on the organism, DNA replication relies on two to three different DNA pols[4,5] (for details see also Chapters 3, 4 and 5). DNA repair events need DNA pols capable of short-gap filling as in short-patch base excision repair

(SP-BER) or DNA pols that can synthesize longer patches of DNA (in long patch (LP-) BER, nucleotide excision repair (NER), mismatch repair (MMR) and non-homologous end-joining[4,6]). Moreover, long patch DNA synthesis is also required for homologous recombination and physiological processes occurring in the context of immunoglobulin recombination.[7] As indicated above, the DNA pols also have to face the many alterations continuously being generated in the DNA. Specifically selected for those tasks a variety of DNA pols that act as translesion DNA pols has evolved.[8,9] Translesion synthesis (TLS) can either be faithful (error free) or mutagenic (error prone).

DNA pols have a unique polarity and synthesize DNA only in the $5' \rightarrow 3'$ direction (see below). This leads to a fundamental problem to fully replicate the chromosomal ends, the telomeres, giving rise to a phenomenon called telomere shortening. A specialized DNA pol, called telomerase, can fulfil this function and prevent the shortening of telomeres.[10] Finally, DNA pols are not only simple working horses to synthesize DNA but they can also be involved in control mechanisms such as checkpoint control during the cell cycle.[11]

2.2 The DNA Polymerase Reaction

The basic reaction mechanism of DNA pols is universal (**Figure 2.2**). A DNA pol needs (i) the four bases in their activated form as deoxyribonucleoside triphophates (dATP, dTTP, dCTP and dGTP) as substrate, (ii) a DNA template that instructs the DNA pol which bases to incorporate (A to T and C to G and vice versa), (iii) a primer containing a $3'$-OH group at its end which is hybridized to a template (there are only a few exceptions to this, e.g. the so called protein priming in bacteriophage φ29 and in adenovirus), (iv) the divalent cation magnesium as cofactor, and (v) in some cases other proteins called DNA pol auxiliary proteins (see below for further details).

DNA pols incorporate a base by the following five-step-mechanism[12,13]: first, a correct hydrogen bonding between the incoming dNTP and the templating base is established (A to T and G to C and vice versa); second, water is excluded from the active site; third, the geometric selection in the active site takes place; fourth, the dNTP binding affinity leads to an induced conformational change in the active site and fifth, the phosphodiester bond is formed between the last base of the primer and the new nucleotide by the release of pyrophosphate.

DNA pols, as everything else in nature, can make mistakes and occasionally incorporate wrong bases. To reduce the amount of those mistakes many, but not all DNA pols, contain a second activity, the $3' \rightarrow 5'$ exonuclease.[14] This so-called proofreading exonuclease activity removes non-properly paired bases. Such incorrectly base paired nucleotides can either derive from a wrong pairing of two normal nucleotides or be caused by altered or missing bases that prevented

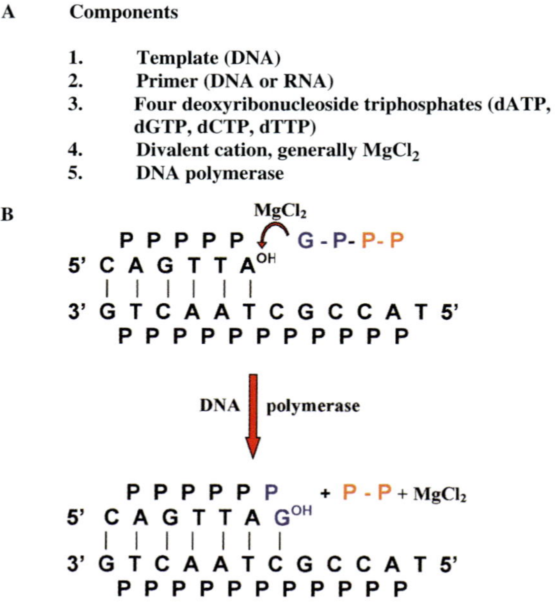

Figure 2.2   The DNA polymerase reaction. For details see text.

the DNA pol active site from choosing the correct nucleotide to be incorporated. If an incorrect base pair is overlooked by the DNA pol, the so called post-replication mismatch repair (MMR) system can remove the incorrectly incorporated base (for details see later).[15]

How accurate is then the overall DNA synthesis conducted by DNA pols? While spontaneous correct base pairing in the absence of a DNA pol can occur at a frequency of 10:1 correct vs. incorrect incorporation events, DNA pols display frequencies of correct incorporation that range from 10:1 to 100 000:1. The reasons why DNA pols have different fidelities and what are the advantages of having low fidelity DNA pols will be discussed in later Chapters. By the intrinsic $3' \rightarrow 5'$ exonuclease, DNA pols can improve their fidelities up to 1 000 000:1. However, since the higher eukaryotic genomes, including humans, contain 6×10^9 base pairs, a mistake every 1 000 000 base pairs would lead to the occurrence of 6×10^3 mistakes per cell division, a number which is unacceptable for the preservation of the integrity of genomic information. This problem is taken care of by DNA pol auxiliary proteins such as proliferating cell nuclear antigen (PCNA),[16] replication factor C (RF-C)[17] and replication protein A (RP-A),[18] which improve the processivity and the fidelity of DNA pols by a great extent (see DNA pol holoenzymes below). Finally, the above mentioned post-replication MMR system can discriminate between the old and newly synthesized DNA strand and thus remove the incorrectly incorporated nucleotide.[14] The auxiliary proteins and the post replicative mismatch repair system

Table 2.1   Fidelity of DNA synthesis

Step	Fidelity
Spontaneous base pairing without DNA pol	10^{-1}
A "sloppy" DNA pol	10^{-1} to 10^{-4}
A faithful DNA pol with $3' \rightarrow 5'$ exonuclease	10^{-5} to 10^{-6}
A faithful DNA pol with $3' \rightarrow 5'$ exonuclease and auxiliary proteins	10^{-7} to 10^{-8}
A faithful DNA pol with $3' \rightarrow 5'$ exonuclease, auxiliary proteins and the post-replication mismatch repair system	10^{-9} to 10^{-10}

together contribute to an increase in fidelity of 1000–10 000. In summary these four steps, DNA pols, their $3' \rightarrow 5'$ exonuclease activity, DNA pol auxiliary proteins and the MMR system, can account for the overall DNA replication fidelity required in nature (**Table 2.1**).

2.3 The Universal Structure of a DNA Polymerase Resembles a Human Right Hand

So far, the crystal structures that have been obtained for many DNA pols from all kingdoms of life have revealed a quite common structure.[19,20] The overall architecture of a DNA pol resembles a human right hand and consists of three domains: palm, fingers and thumb (**Figure 2.3**). Upon binding of the DNA pol to the template/primer and the dNTP, conformational changes in the DNA pol as well as in the DNA itself lead to the assembly of a DNA pol active site. These changes allow a catalytically competent DNA pol to exert its function. The finger domain interacts with the incoming dNTP and single-stranded DNA. The thumb on the other hand binds to double-stranded DNA. This double-stranded DNA is either a primer hybridized to a template or represents the newly synthesized DNA with its templating strand. The palm domain harbors the catalytic residues that bind magnesium ions which are required for the phosphoryl transfer reaction (**Figure 2.2**). The phosphoryl transfer reaction includes the attachment of the dNMP to the 3'-OH group of the primer or the growing DNA chain and the release of pyrophosphate. The palm domain, where, as indicated, the catalytic reaction occurs, is extremely conserved in most DNA pols analyzed so far. Exceptions are: (i) the pol β-like nucleotidyltransferase superfamily X enzymes such as DNA pol β,[21] (ii) the DNA pols III from *E. coli* and *Thermus aquaticus*[22] and the DNA pol X from *Deinococcus radiodurans*[23] (see Chapters 3 and 4 for details). Despite these differences, all DNA pols utilize the same two-metal (mostly magnesium)-ion mechanism for dNTP addition. In contrast to the widely conserved palm domain, the fingers and the thumbs differ in the various DNA pol families (see below and Chapters 3 and 4 for details).

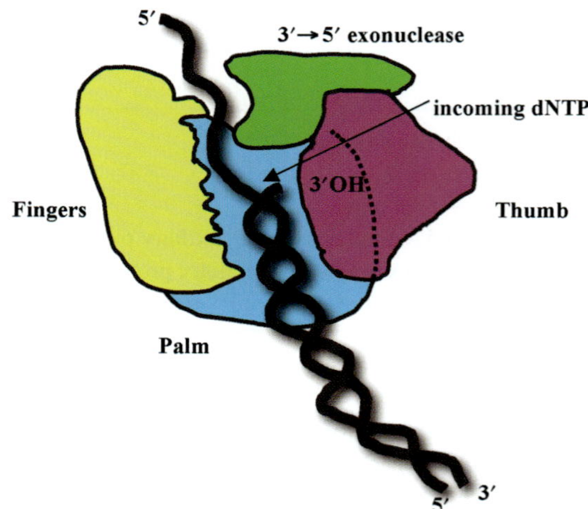

Figure 2.3   Structure of a DNA polymerase. Schematic representation of a typical DNA pol containing a $3' \rightarrow 5'$ exonuclease proofreading activity. Yellow: fingers; pink: thumb; blue: palm; green: $3' \rightarrow 5'$ exonuclease domain.

As already mentioned, some DNA pols contain an associated $3' \rightarrow 5'$ exonuclease activity (**Figure 2.3**). The exonuclease domains are structurally conserved in all DNA pols tested so far.[14] These DNA pols contain two active sites — one for polymerization and one for exonucleolytic proofreading. The proofreading DNA pols must have a mechanism to balance between DNA synthesis and proofreading exonuclease action. It is likely that the structure of the DNA at the primer terminus and a few bases upstream of the active site, is responsible for the switch between "going forward" (DNA synthesis) and "going back" (proofreading). It is thought that a non-properly base paired primer terminus becomes single-stranded and is frayed out to bind to the exonuclease domain, while the correctly base paired primer terminus forms perfectly double-stranded DNA that preferentially binds to the DNA pol active site. It seems that when a DNA pol is incorporating nucleotides, the exonuclease domain is in closed conformation that does not allow the binding of single-stranded DNA. However, upon incorporation of a wrong nucleotide, the single-stranded DNA primer end allows the formation of an open exonuclease domain conformation ready for proofreading.

2.4   The Seven DNA Polymerase Families and Their Functions: An Overview

So far, seven different DNA pol families have been grouped based on sequence homologies. They are called family A, B, C, D, X, Y and RT (**Table 2.2**).[4] None of

Table 2.2   The seven DNA polymerase families

Family	Viral	Bacteria	Archaea	Eukaryotes
A	T3 DNA pol T5 DNA pol T7 DNA pol	DNA pol I *E. coli* DNA pol I *T. aquaticus*		DNA pol γ DNA pol θ DNA pol υ
B	T4 DNA pol T6 DNA pol RB69 DNA pol Adeno[a] DNA pol HSV-1[b] DNA pol Vaccinia DNA pol Phi29 DNA pol	DNA pol II *E. coli*	DNA pol BI DNA pol BII DNA pol BIII	DNA pol α DNA pol δ DNA pol ε DNA pol ζ/Rev3
C		DNA pol III *E. coli* DNA pol III *B. subtilis* DNA E *B. subtilis* DNA pol III *T. aquaticus*		
D			DNA pol D	

X	ASFV[c] DNA pol	DNA pol X
		   D. radiodurans
		DNA pol X
		   B. subtilis
		
		DNA pol X
		   L. monocytogenes
		DNA pol X
		   S. saprolyticus
		DNA pol X
		   S. aureus
		DNA pol X
		   D. reducens
		DNA pol X
		   A. aeolicus
		DNA pol X
		   T. thermophilus
		DNA pol X
		   T. denitrificans
		DNA pol X
		   T. aquaticus
			DNA pol X
			   M. mazei
			DNA pol X
			   M. thermauto-
			   trophicus
			DNA pol X
			   T. volcanium
			DNA pol X
			   F. acidarmanus
				DNA pol β
				   (pol IV S.c.[d])
				DNA pol λ
				   (pol LSP S.p[e])
				
				DNA pol μ
				
				DNA pol σ
				
				TdT[f]

Y		DNA pol IV *E. coli* DNA pol V *E. coli*	Dpo4 DNA pol[g] Dbh DNA pol[g]	DNA pol η DNA pol κ DNA pol ι Rev1
RT	RT[h]			Telomerase

a: Adenovirus; b: Herpes simplex virus; c: African swine fever virus; d: *Saccharomyces cerevisiae*; e: *Schizosaccaromyces pombe*; f: Terminal deoxyribonucleotidyl transferase; g: From *Sulfolobus solfataricus*; h: Reverse transcriptases from retroviruses and lentiviruses.

the three kingdoms of life contains all the seven families. Prokaryotes contain five, namely A, B, C, X and Y, Archaea harbor four (B, D, X and Y) and eukaryotes are composed of the five families A, B, X, Y and RT. Moreover, the DNA pols from a variety of viruses of prokaryotes and eukaryotes belong to the three families A, B and X.

As noted above, the overall structures of DNA pols are very similar. It appears that the different DNA pol families have evolved in evolution as an adaptation of an organism to certain survival strategies. Viruses contain 1–2 different DNA pols (the African swine fever virus codes for an own X family repair DNA pol in addition to its replicative enzyme, see also below),[24] while bacteria and archaea, as simple single cell organisms, code for 4–5 different DNA pols. In contrast to prokaryotes, eukaryotic cells possess two- to four-fold more distinct DNA pols, a fact that might reflect the higher complexity of DNA maintenance and replication when a nucleus is present. However, it is remarkable that all those eukaryotic DNA pols, with the exception of telomerase (RT family), belong to the same DNA pol families that already exist in prokaryotes. This means, that, apart from the telomerase, no new DNA pol families have evolved with the development of eukaryotic cells. The telomerase obviously evolved due to the need of eukaryotes to solve the replication problem at the end of their chromosomes. Thus, the strategy of the eukaryotes was rather to multiply DNA pols *within* a family. In eukaryotes there are 3 family A (DNA pol γ, DNA pol θ and DNA pol υ); 4–5 family B (DNA pol α, DNA pol δ, DNA pol ε, DNA pol ζ/Rev3, possibly DNA pol φ); 5 family X (DNA pol β, DNA pol λ, DNA pol μ, DNA pol σ and TdT) and 4 family Y members (DNA pol η, pol κ, pol ι, and Rev1). This multiplication within the families asks for an explanation, which we will try to provide in the following.

In order to understand the reason for a multiplication of family members, we have to consider first their functional tasks. For many decades the scientific community believed that each single DNA pol had a unique and particular function. In higher eukaryotes, DNA pol α was considered as the replicative enzyme, DNA pol β as the repair enzyme and DNA pol γ as the mitochondrial replicase (for the rationale see Chapter 1). Eventhough this statement is still true today, the overall picture concerning the functional tasks of those and all the recently discovered DNA pols is much more complex and far from being solved in detail. What follows is an attempt to group proposed DNA pol functions in viruses, bacteria, archaea and eukaryotes:

(i) *Viruses* (**Table 2.3**): DNA viruses either rely on cellular DNA pols or have their own replicative DNA pols (e.g. T4 and T7 in bacteria and adenovirus, herpes simplex virus and vaccinia virus in eukaryotes). The African swine fever

Table 2.3   Functions of viral and bacterial DNA polymerases[a]

Family	DNA polymerase	Main function
A	T3 DNA pol	replication
	T5 DNA pol	replication
	DNA pol I *E. coli*	replication, repair
	DNA pol I *T. aquaticus*	replication
B	T4 DNA pol	replication
	T6 DNA pol	replication
	DNA RB69 pol	replication
	Adeno[b] DNA pol	replication
	HSV-1[c] DNA pol	replication
	Vaccinia DNA pol	replication
	Phi29 DNA pol	replication
	DNA pol II E. coli	repair, replication?
C	DNA pol III *E. coli*	replication
	DNA pol III *B. subtilis*	replication
	DNA pol E *B. subtilis*	replication
	DNA pol III *T. aquaticus*	replication
X	ASFV[d] DNA pol	repair
	DNA pol X *D. radiodurans*	repair (double-stranded breaks)
	DNA pol X *B. subtilis*	repair
	DNA pol X *L. moncytogenes*	repair
	DNA pol X *S. saprolyticus*	repair
	DNA pol X *S. aureus*	repair
	DNA pol X *D. reducens*	repair
	DNA pol X *A. aeolicus*	repair
	DNA pol X *T. thermophilus*	repair
	DNA pol X *T. denitrificans*	repair
	DNA pol X *T. aquaticus*	repair
Y	DNA pol IV *E.coli*	translesion synthesis
	DNA pol V *E. coli*	translesion synthesis
RT	reverse transcriptase	DNA synthesis from RNA

a: For more details and references see Chapter 3; b: Adenovirus; c: Herpes simplex virus; d: African swine fever virus

virus, the causative agent of African swine fever, is a large double-stranded DNA virus, which replicates in the cytoplasm of infected cells. It is the only member of the Asfarviridae family. It is an exception to other viruses in the context of DNA pols since it contains a pol X DNA pol, likely with properties in DNA repair. Such a repair DNA pol allows better survival of the virus under conditions in the infected cell that might otherwise lead to genomic instability.

(ii) *Bacteria* (**Table 2.3**): Bacteria have sophisticated replication and repair machineries with at least *one replicative* DNA pol, belonging to the C family (pol III) and five DNA pols for repair (A family DNA pol I, B family DNA pol II, C family DNA pol III, and Y family DNA pols IV and V). It is the repair task that obviously became important for the evolution of bacteria. For BER, NER, MMR and recombinational repair they rely on DNA pol I, DNA pol II and DNA pol III. Moreover, translesion synthesis over damaged DNA requires DNA pol IV and DNA pol V (**Table 2.3**). For replication, on the other hand, it appears that the family C pol III forms a *triple DNA pol replisome* containing three times the *same* DNA pol (see also holoenzyme below).

(iii) *Archaea* (**Table 2.4**): For DNA replication, archaea appear to have DNA pols of two different families: B and D. This is the first time in evolution that an organism relies on DNA pols belonging to *two different* families for replication. Archaea have to survive harsh conditions such as high temperatures, high salt, high sulfur and high metals just to mention a few. For repair they rely on family X and B and for TLS on family Y DNA pols.

(iv) *Eukaryotes* (**Table 2.5**): It appears that eukaryotes contain at least 15 different DNA pols. As already indicated, with the exception of telomerase, *no new families* have evolved in comparison with prokaryotes. In contrast, eukaryotic cells even miss the C and the D families present in prokaryotes and archaea. However, to compensate, the four families A, B, X and Y in eukaryotes *each* contain 3-5 different DNA pols. Studies in many laboratories can be summarized by saying that redundant functions of DNA pols seem to be an important survival strategy for eukaryotes (see also Chapter 5). Moreover, new DNA synthesis events evolved, such as: (i) mitochondrial DNA replication and repair, (ii) physiological repair in immunoglobulin recombination and in antigen receptor diversity generation and (iii) in sister chromatid cohesion.

Table 2.4   Functions of the archaeal DNA polymerases[a]

Family	DNA polymerase	Main function
B	DNA pol BI	Replication, repair?
	DNA pol BII	Replication, repair?
	DNA pol BIII	Repair (BER), replication?
D	DNA pol D	replication
X	DNA pol X *M. mazei*	repair
	DNA pol X *M. thermautotrophicus*	repair
	DNA pol X *T. volcanium*	repair
	DNA pol X *F. acidarmanus*	repair
Y	Dpo4 DNA pol[b]	translesion synthesis
	Dbh DNA pol[b]	translesion synthesis

a: For more details see Chapter 3; b: From *Sulfolobus solfataricus*

Table 2.5   Functions of the eukaryotic DNA polymerases[a]

Family	DNA polymerase	Main function
A	DNA pol γ	replication in mitochondria
	DNA pol θ	replication of cross links, base excision repair, hypermutation of immunoglobulin genes
	DNA pol υ	repair
B	DNA pol α	replication (priming)
	DNA pol δ	replication (lagging strand), repair
	DNA pol ε	replication (leading strand), repair
	DNA pol ζ/Rev3	translesion synthesis (extension)
X	DNA pol β (pol IV S.c.[b])	base excision repair
	DNA pol λ (pol LSP in S.p[c])	base excision repair, double-strand break repair, immunoglobulin recombinational repair, translesion synthesis
X	DNA pol μ	immunoglobulin recombinational repair
	DNA pol σ[d]	sister chromatid cohesion
	TdT[e]	antigen receptor diversity
Y	DNA pol η	translesion synthesis (incorporation)
	DNA pol κ	translesion synthesis (incorporation)
	DNA pol ι	translesion synthesis (incorporation)
	Rev1	translesion synthesis (incorporation)
RT	Telomerase	replication of chromosome ends

a: For more details and references see Chapters 4 and 5; b: *Saccharomyces cerevisiae*; c: *Schizosaccaromyces pombe*; d: DNA pol σ was later shown to be a poly (A) RNA polymerase[25]; e: Terminal deoxynucleotidyl transferase

In DNA replication *three different* DNA pols are required (DNA pols α, δ and ε). DNA repair events each require several DNA pols such as DNA pols β, λ, θ, δ and ε in BER, DNA pols δ, ε, κ and η in NER, DNA pols δ in MMR, DNA pols λ and μ in DSBR and DNA pols λ, μ and TdT in immunoglobulin recombination. Often in BER, NER and MMR, the cell borrows the replication enzymes DNA pols δ and ε for repair. To be able to cope with modified, frequently miscoding nucleotides that are produced by DNA damage, TLS has been established. It is a synthesis event aimed at decreasing the genetic alterations after DNA damage. For this purpose, eukaryotes have evolved an entire set of specialized DNA pols from the three families Y, X and B. While the Y family DNA pols are responsible for incorporating nucleotides opposite certain lesions (DNA pols η, κ, ι, and Rev1), the B family DNA pol ζ/Rev3 can extend from the often mis-incorporated nucleotides opposite such lesions. 8-oxo-G is a frequent lesion caused by reactive oxygen species (ROS) and poses a big problem to the genomic integrity, as replicative DNA pols preferentially incorporate a wrong A instead of the correct C opposite the lesion. This preferential misincorporation is due to the fact that an A pairing to 8-oxo-G forms a Hoogsteen base pair (which is, in contrast to the Watson-Crick base pairing, an incorrect way to base pair two nucleotides) that creates no single-stranded DNA that could activate the closed exonuclease conformation (see also Chapter 4). Therefore, the mispairing of A to 8-oxo-G goes unnoticed by the proofreading function of replicative DNA pols and can lead to the generation of GC to TA transversion mutations. However, the X family DNA pol λ has been shown to act as repair enzyme for this lesion by faithfully incorporating a C opposite an 8-oxo-G, instead of the wrong A that is preferentially incorporated by the replicative DNA pols. Thus, DNA pol λ seems to play a role in preventing GC to TA transversions that occur due to 8-oxo-G.

In summary, the DNA pol number increased concomitantly with the evolution towards more complex cellular structures and their functions are closely linked to the complexity of those organisms. In subsequent Chapters we will gain more insight into the fascinating areas of expertise of all the individual DNA pols (Chapters 3 and 4) and an attempt is made to give a global picture how they work in concert in a living eukaryotic cell (Chapter 5).

2.5 DNA Polymerase Holoenzymes

More than 30 years ago the laboratory of the late Arthur Kornberg (*1918–2007*), a pioneer in the field of DNA pols, discovered that the replicative DNA pol III was inefficient in DNA replication compared to DNA pol I *in vitro*.[26] The subsequent isolation of an efficient replication complex from crude *E. coli* extracts soon indicated that a more complex form of DNA pol III was present in the cell. From this finding it was realized that DNA pol III functions in the context of a multipolypeptide complex, which was thereupon called DNA pol III holoenzyme.[26] DNA pol III holoenzyme contains 10 different polypeptides (**Table 2.6**).[27,28] These proteins are grouped into the three major functional units: (i) the DNA pol III core consisting of three subunits, (ii) the sliding clamp consisting of a homodimeric ring and (iii) the clamp loader/polymerase connector complex consisting of six different subunits (one subunit, τ, being a dimer). A fourth functional unit is the single-stranded DNA binding protein (SSB), a necessary component for a DNA pol III holoenzyme to efficiently synthesize DNA. This unit is mentioned here since there is a lot of evidence in the literature that SSB, as well as its eukaryotic counterpart replication protein A (RP-A), both physically interact with the holoenzymes from *E. coli* and eukaryotes, respectively. The DNA pol III holoenzyme contains two copies of the

Table 2.6   Components of DNA polymerase holoenzymes[a]

Components	*E. coli* (M_r, kDa)	Human (M_r, kDa)	Function
DNA polymerase	DNA pol III core (166)	DNA pol ε (350.3)	replication (leading strand)
	α (129.9)	p260 (261.5)	
	ε (27.5)	p59 (59.5)	
	θ (8.6)	p17 (17.0)	
		p12 (12.3)	
DNA polymerase	DNA pol III core (166)	DNA pol δ (238.7)	replication (lagging strand)
	α (129.9)	p125 (123.6)	
	ε (27.5)	p66 (51.4)	
	θ (8.6)	p50 (51.3)	
		p12 (12.4)	
Sliding clamp	Dimer (81.2)	Trimer (86.1)	processivity factor for DNA polymerases
	β-subunit (40.6)	PCNA (28.7)	
Clamp loader	γ/τ complex (291.1)	RF-C (314.9)	pentameric clamp loader
	γ/τ (47.5/71.1)	p140 (128.2)	
	δ (38.7)	p40 (39.7)	
	δ′ (36.9)	p38 (38.5)	
	χ (16.6)	p37 (39.2)	
	ψ (15.2)	p36 (40.6)	

a: Only the two examples of *E. coli* and human are exemplified. For other DNA pol holoenzymes see Chapter 4 (*S. cerevisiae and S. pombe*) and Chapter 6.1 (bacteriophage T7 and T4).

DNA pol III core which are connected together by the clamp loader. Both DNA pol III core enzymes themselves are attached to a homodimeric ring like protein called the β-clamp. In such a conformation, the dimeric holoenzyme can act in a coordinated way at the replisome (see Chapters 3 and 5 for details about the dimeric replication model proposed by Alberts *et al.* in 1980 and the dynamics at the replication fork[29]). In summary the DNA pol III holoenzyme is a complex formed by 17 polypeptides: two DNA pol III core enzymes consisting of 6, two β-clamps consisting of 4 and the clamp loader/connector complex consisting of 7 polypeptides.

In eukaryotes, as exemplified by human enzymes, the situation is much more complex and the picture is less clear than in *E. coli*. The functional units of the human replication complex are: (i) the three DNA pols α, δ and ε core enzymes, each containing 4 subunits,[30] (ii) the sliding clamp called proliferating cell nuclear antigen (PCNA) existing as a homotrimeric ring[16,31] and (iii) the clamp loader replication factor C (RF-C) formed by 5 different subunits.[17,32] How the three

DNA pols act together is not known in detail. **Figure 2.4** documents a simplified view of how they might act dynamically at the DNA replication fork. The model of Alberts (1980) proposes that the DNA forms a loop and the two DNA pols dimerise via a putative dimerisation factor, thus rendering the same directionality for the leading and the lagging DNA strands.[29] After the complex events of initiation of DNA replication (for a review and references therein see Ref. 33), the DNA at the origin of replication eventually becomes single-stranded. Starting from this point, the DNA replication fork is formed by recruitment of the diverse DNA replication factors including the DNA pols, the clamp and the clamp loader. The single-stranded DNA is immediately covered by RP-A, which also participates in unwinding of DNA. DNA pol α/primase is then recruited thanks to its interaction with RP-A and Cdc45 (see below for other proteins interacting with DNA pols) and initiates *de novo* synthesis by first synthesizing an RNA fragment of about ten bases, succeeded by a 30-nucleotide DNA fragment. This initiation step and the following events occur during synthesis at the site of each Okazaki fragment on the lagging strand, but only once for the leading strand. After the initiation step, an ATP-dependent DNA pol switch from DNA pol α to DNA pol ε (at the leading strand) and δ (at the

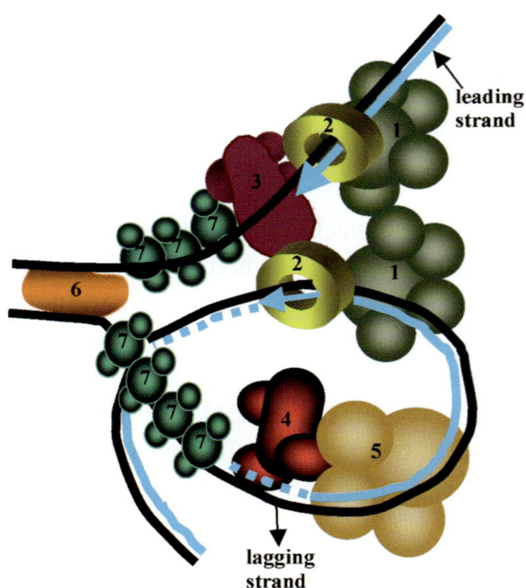

Figure 2.4   Dynamics at the eukaryotic DNA replication fork. DNA forms a loop and dimerisation of the two polymerases via a putative dimerisation factor is proposed, thus rendering the same directionality for the leading and the lagging DNA strands. 1: Replication factor C (RF-C); 2: proliferating cell nuclear antigen (PCNA); 3: DNA pol ε; 4: DNA pol α; 5: DNA pol δ; 6: DNA helicase; 7: replication protein A (RP-A); light blue squares: initiator RNA and DNA. For more details see text. Adapted from Toueille and Hübscher.[34]

lagging strand) occurs, for which hydrolysis of ATP by RF-C is required. PCNA is simultaneously loaded by RF-C at the primer end. DNA pol ε bound to PCNA is then able to synthesize long stretches of DNA. When DNA pol δ encounters the 5′ end of the previous Okazaki fragment, it can perform strand displacement DNA synthesis. The flap structure that is generated by this displacement is eventually a substrate for combined cleavage by flap endonuclease 1 (Fen 1) and the Dna2 endonucleases, for whom the single-stranded binding protein RP-A, PCNA and RF-C can act as modulators. The resulting nick is finally sealed by DNA ligase I (for a review and references therein see Ref. 34).

2.6 DNA Polymerases, Ring-Like Clamps and Clamp Loaders

Proliferating cell nuclear antigen (PCNA) was originally identified as a DNA sliding clamp for replicative DNA pols and as an essential component of the eukaryotic chromosomal DNA replisome.[16] Beyond this function, PCNA can interact with multiple partners, which are involved in several metabolic pathways, such as Okazaki fragment processing, DNA repair (BER, NER, MMR), translesion DNA synthesis (TLS), DNA methylation, chromatin remodelling and cell cycle regulation.[31] PCNA in mammalian cells thus appears to play a key role in controlling a variety of reactions through the coordination and organization of different partners.[35] From these data two major questions emerged: how do these proteins access PCNA in a coordinated manner and how does PCNA temporally and spatially organise their functions? Here we only expand on the functional interactions of PCNA with DNA pols. For other PCNA functions the reader might consult a specific book about PCNA (*Proliferating Cell Nuclear Antigen* (PCNA), Lee H. ed., 2006, ISBN 81-308-0096-9).

PCNA belongs to the so called DNA sliding clamp family, which includes also the *E. coli* DNA pol III β-subunit and the bacteriophage T4 gene45 protein. These proteins perform the essential functions of providing replicative DNA pols (III in *E. coli* and ε/δ in eukaryotes) with the high processivity required to duplicate an entire genome. Crystallographic studies of the β-subunit and PCNA revealed that both these proteins adopt superimposable ring-like structures in solution, with a central hole sufficiently large to accommodate the double helix of DNA, even though they show almost no sequence similarity.[36] In eukaryotes, three identical PCNA monomers, each comprising two similar domains, are joined in a head-to-tail arrangement to form a homotrimer (**Figure 2.5**). This is composed of six repeating domains and exhibits six-fold symmetry. The resulting ring has two non-equivalent surfaces: an outer surface composed of β- sheets and a layer of α-helices rich in basic residues lining the inner side of the hole. Those α-helices are positioned perpendicularly to the phosphate backbone of DNA. Owing to this unique structure, PCNA

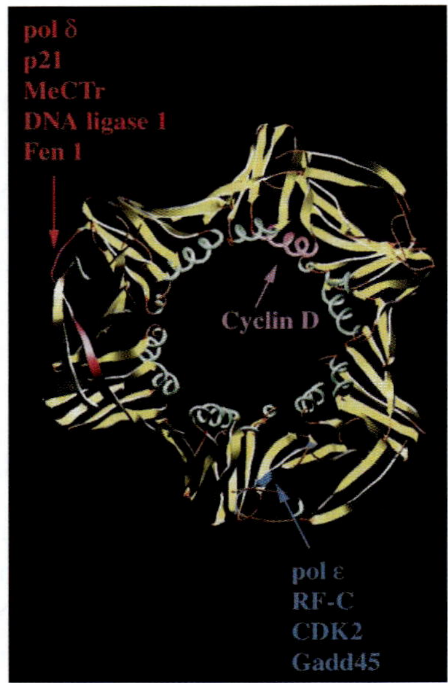

Figure 2.5   Structure of human PCNA. Note that several proteins have overlapping binding sites. For more details see Chapters 5: Global functions of DNA polymerases. Reproduced from Jonsson and Hübscher.[37]

is topologically linked to the double helix by encircling it, but still able to freely slide along the DNA lattice by virtue of the α-helices lining the inner channel. This is believed to happen because both the inner α-helices and the DNA backbone are negatively charged and thus repelling each other. PCNA and its homologs increase the processivity of a DNA pol by engaging in protein-protein interactions with their outer surface and tethering it to the DNA. Thus, those ring-like structures prevent the DNA pol from dissociating while advancing along the template DNA. Hence they were also named sliding clamps.

From **Table 2.7** it is evident that PCNA indeed interacts with 10 different DNA pols from the families B, X and Y. As discussed above, by interacting with the leading strand replicase DNA pol ε and the lagging strand replicase DNA pol δ, PCNA provides processivity to these enzymes. PCNA's interaction with the BER enzymes DNA pols β and λ allows them to perform accurate repair synthesis. The translesion DNA pols ζ, η, κ, ι, and Rev1 all interact with PCNA as well.

For these interactions to take place, posttranslational modifications of both PCNA and DNA pols are necessary. They will be described in more detail in Chapter 5.

The ring-like structure prevents PCNA from binding to and encircling DNA on its own. A protein complex called replication factor C (RF-C) has been found to load

Table 2.7 Functional consequences of the clamp interactions with different DNA polymerases[a]

		Clamp[b]	
Family	**DNA polymerase (DNA pol)**	**PCNA (function for DNA pol)**	**Rad9/Rad1Hus1 (function for DNA pol)**
B	DNA pol α	—	DNA pol α recruits the 9-1-1 complex to DNA
B	DNA pol δ	replication lagging strand	—
B	DNA pol ε	replication leading strand	—
B	DNA pol ζ	translesion DNA synthesis (extender), regulates pol ζ dependent mutagenesis	mutagenesis, recruitment to double-stranded breaks
X	DNA pol β	base excision repair	stimulation of DNA pol β
X	DNA pol λ	base excision repair, DSBR, translesion DNA synthesis	—
X	TdT	DNA damage tolerance	—
Y	DNA pol η	translesion DNA synthesis	—
Y	DNA pol ι	translesion DNA synthesis	—
Y	DNA pol κ	translesion DNA synthesis	—
Y	Rev 1	translesion DNA synthesis	recruitment to double-stranded breaks

a: For detailed reviews of other PCNA interacting proteins see Refs. 16 and 31. b: Posttranslational modifications of the clamps will be discussed in Chapter 5.

PCNA onto DNA. RF-C is a complex of five polypeptides with distinct functions and belongs to the AAA+ machine of ATPases.[17]

The loading mechanism is believed to work as follows[38,39] (**Figure 2.6**): (a) RF-C binds ATP and forms a binary complex with PCNA; (b) this leads to a conformational change in RF-C with the consequence that the PCNA ring is opened; (c) in a next step this binary complex, having a higher affinity to DNA, encircles the DNA whereupon the PCNA ring is closed again and thus encircling the DNA; (d) by hydrolyzing ATP to ADP and P_i, RF-C leaves PCNA and the DNA, allowing the DNA pol to bind to PCNA. It is likely that the DNA pols δ and ε help displacing RF-C while binding to PCNA. After dissociation from PCNA, RF-C binds another ATP molecule and is ready for loading the next PCNA molecule.

2.7 DNA Polymerases, Alternative Clamps and Clamp Loaders

Alternative clamps[40]: Genotoxic stress activates checkpoint-signaling pathways that halt cell cycle progression, regulate DNA repair and trigger apoptosis. Studies

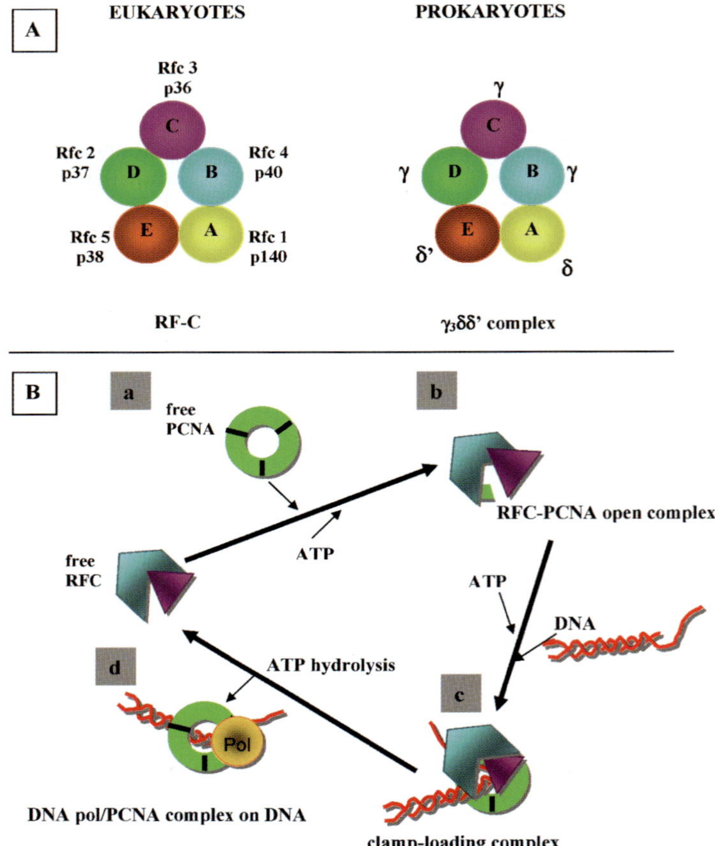

Figure 2.6   PCNA is loaded onto DNA by RF-C. A. Subunit compositions of eukaryotic RF-C and the bacterial γ-complex. Rfc1-5 refer to yeast *S. cerevisiae* while p140 and p36, 37, 38 and 40 refer to the mammalian subunits. B. a: the ATP bound RF-C forms a complex with the ring closed PCNA; b: a conformational change in RF-C opens the PCNA ring; c: this open complex has a higher affinity to DNA. The PCNA ring is closed; d: finally upon hydrolysis of ATP, RF-C dissociates from the complex, the DNA pol binds to PCNA and can thus synthesize DNA. The cycle is then repeated.

in yeast and humans suggest that Rad17, Rad9, Rad1, Hus1 play key roles in checkpoint activation. Three of these proteins Rad9, Hus1, and Rad1 form a heterotrimeric complex (called the 9-1-1 complex), which resembles a PCNA-like sliding clamp, whereas Rad17 is part of a clamp-loading complex that is related to the PCNA clamp loader, RF-C. In response to DNA damage, the 9-1-1 complex is loaded onto DNA in a Rad17-RF-C$_{2-5}$-dependent manner. The crystal structure of the 9-1-1 complex has recently been solved and suggested a toroidal heterotrimeric structure similar to the homotrimeric structure of PCNA.[41,42] The DNA-bound 9-1-1 complex then facilitates ATR-mediated phosphorylation and activation of

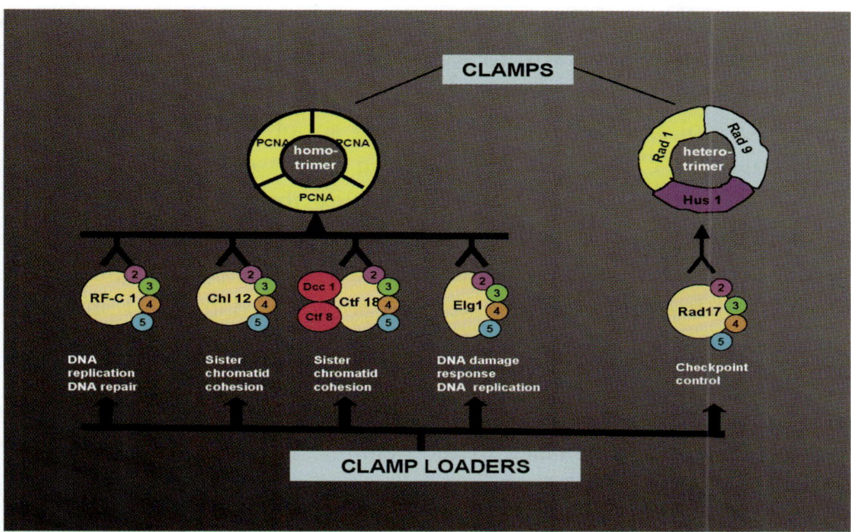

Figure 2.7   Clamps and clamp loaders. Under normal conditions e.g. during replication the RF-C heteropentameric clamp loader (RF-C$_{1-5}$) loads PCNA in an ATP dependent manner. In other cellular metabolic events such as sister chromatid cohesion, DNA damage response and checkpoint control, alternative clamp loaders (consisting of the four small subunits RF-C$_{2-5}$ and an additional protein) can either load PCNA or the alternative clamp, the 9-1-1 complex. For details see text. Reproduced from Toueille and Hübscher.[34]

Chk1, a protein kinase that regulates S-phase progression, G2/M arrest, and replication fork stabilization. In addition, the 9-1-1 complex also directly participates in DNA repair by interacting with and/or stimulating several components of different repair pathways including BER, NER, DSBR and TLS.

The 9-1-1 complex interacts with at least four different DNA pols from the three families B, X and Y (**Table 2.7**). The first description was the interaction of the 9-1-1 complex with the BER enzyme DNA pol β,[43] stimulating its activity, increasing its affinity for the primer-template and enhancing its DNA strand displacement synthesis. The interaction of the 9-1-1 complex with DNA pol ζ regulates its mutagenicity and helps to control this DNA pol to properly extend from a lesion.[44] A similar effect has been described by the interaction of the 9-1-1 complex with Rev1, a deoxycytidyltransferase interacting with many TLS DNA pols.[45] DNA pol α, together with the protein TopBP1 (see below), can directly recruit the 9-1-1 complex to stalled replication forks.[46]

Alternative clamp loaders[34]: In addition to the "classical" RF-C protein involved in DNA replication and the checkpoint clamp-loader Rad17-RF-C$_{2-5}$, the four small subunits of RF-C are part of other alternative clamp-loaders (**Figure 2.7**). First, the sister chromatid cohesion factor, Chl12, exhibits sequence similarity to the large subunit of RF-C. The complex formed by association of Chl12 with the RF-C$_{2-5}$ subunits is structurally similar to RF-C and was shown to be able to load PCNA in an ATP dependent manner onto circular DNA. The Chl12-RF-C$_{2-5}$ complex can stimulate DNA pol δ activity. Second, another set of proteins involved in sister chromatid cohesion, Ctf18, Dcc1 and Ctf8 can also associate with the four small RF-C subunits. Similarly to the previous complex, this complex is also a clamp-loader for PCNA, and can substitute for the replicative RF-C in a DNA pol δ catalyzed DNA replication assay. Furthermore Ctf18 RF-C$_{2-5}$ can unload PCNA in the presence of RP-A and can stimulate DNA pol η. Third, the Elg1 protein shares sequence homology with the large subunit of the clamp loader RF-C and forms a complex with the RF-C$_{2-5}$ subunits.

Genetic data from yeast studies indicated that Elg1 has a role in DNA damage response and in maintenance of the genomic integrity, as well as in DNA replication. Moreover, Elg1-RF-C$_{2-5}$ contributes to genome stability via telomere length regulation through a putative replication-mediated pathway. The telomeric function of Elg1 was shown to be independent of recombination and dependent on an active telomerase and DNA pol α suggesting that Elg1-RFC$_{2-5}$ plays a role in checkpoint pathway and DNA replication.

Finally, the checkpoint clamp-loader Rad17-RF-C$_{2-5}$, when phosphorylated, preferentially associates with the site of ongoing replication and interacts with the leading strand replication enzyme DNA pol ϵ.

2.8 Replicative DNA Polymerases Interacting with Other Proteins

DNA pols, in whatever context they act, have to interact with many other proteins. As an example we introduce some selected proteins that interact with the three replicative DNA pols α, δ and ϵ. First, DNA pol α, the initiating replicase, interacts with the putative replicative DNA helicase Mcm 2-7.[47]. It furthermore binds to the GINS proteins,[48,49], and Cdc45[50] which acts as connectors of DNA pol α to the replication machinery. A physical interaction of DNA pol α with Mcm10 and DNA has been described.[51] This interaction with Mcm10 recruits DNA pol α to chromatin and the interaction with And1/Ctf4 stabilizes DNA pol α at the replication fork.[52] A key role for Ctf4 has been described for coupling the Mcm 2-7 helicase to the initiating DNA pol α.[53] Second, DNA pol ϵ, the leading strand replicase, interacts with Mrc1 (ortholog of metazoan claspin) that links replication and the S phase checkpoint[54] and it can interact with Cdc45.[50] Interaction with Dpb11 (in *S. cerevisiae*)/Cut5 (in *S. pombe*)/TopBP1 (in human) recruits DNA pols α and ϵ to DNA.[55] TopBP1 and DNA pol α together recruit the 9-1-1 complex to DNA when replication forks are stalled after DNA damage.[46] Third, DNA pol δ, the lagging strand replicase, interacts with Werner and Bloom DNA helicases,[56,57] that help to unwind secondary structures in the DNA. Finally fourth, INO80 interacts with all three DNA pol thus regulating replisome function and stability.[58]

2.9 DNA Polymerases and the Single-Stranded DNA Binding Protein Replication Protein A

DNA replication, DNA repair and DNA recombination all expose single-stranded DNA that is extremely prone to genotoxic attacks. Replication protein A (RP-A) protects single-stranded DNA from nucleolytic attack, removes hairpins and blocks DNA reannealing until the above mentioned pathways are terminated.[18,59]

Many DNA repair studies showed that RP-A plays a major role in coordinating the repair mechanisms and is therefore more than a simple "protector" for single-stranded DNA. Essential roles in BER, NER, DSBR and MMR were identified for RP-A. Moreover, RP-A can sense DNA damage and activate checkpoints.[60] Several studies suggested that in this step RP-A is a key player. Indeed, RP-A coated single-stranded DNA is a common structure generated at the sites of DNA damage. The ATR checkpoint kinase is recruited to the sites of single-stranded DNA by binding of its regulatory partner, ATRIP, to the RP-A protein. Furthermore, experiments in yeast suggested that RP-A covered DNA filaments allow the sensing of stalled replication forks via recruitment of Ddc1 (homologue to the human Rad9) and Ddc2 (homologue to the human ATRIP). This recruitment leads to checkpoint activation and to stabilization of the stalled replication fork and confirmed the crucial role of RP-A in the DNA damage-sensing step of the checkpoint response. RP-A contains three subunits with Mr of 70 000 (RP-A1), 32 000 (RP-A2) and 14 000 (RP-A3). An alternative form of RP-A, containing a different 32 kDa subunit called RP-A4, has recently been described. This form may prevent cellular proliferation by inhibition of DNA replication and might play a role in maintaining genetic stability in quiescent cells.[61]

Studies carried out in our laboratories suggest important roles of RP-A in helping DNA pols to act faithfully: (i) RP-A can act as a "fidelity clamp" for DNA pol α[62,63]; (ii) RP-A, together with PCNA and Fen 1 can modulate the strand displacement activity of DNA pol δ in Okazaki fragment processing[63]; (iii) RP-A, together with PCNA, can suppress completely the deoxyribonucleotidyltransferase activity of DNA pol λ, while stimulating its polymerase activity[64]; (iv) RP-A, but not PCNA, selectively prevents the misincorporation of an incorrect nucleotide by DNA pol λ, without affecting the correct incorporation[65]; (v) RP-A and PCNA allowed correct incorporation of C opposite a template 8-oxo-G 1200- fold more efficiently than the incorrect A by DNA pol λ, and 68- fold by DNA pol η, respectively. These findings suggested for the first time a novel accurate mechanism to reduce the deleterious consequences of oxidative damage and, in addition, point to an important role for RP-A and PCNA in determining a functional hierarchy among different DNA pols in lesion bypass[66]; (vi) RP-A and PCNA enhance the fidelity of translesion synthesis by DNA pol λ over 2-hydroxy-adenine (2-OH-A) in a repeated sequence[67]; (vii) RP-A and PCNA act as molecular switches to activate the DNA pol λ-dependent highly efficient and faithful repair of A:8-oxo-G mismatches in human cells, while they inhibit the less faithful DNA pol β thus coordinating the DNA pol selection in 8-oxo-guanine repair.[68]

2.10 Chapter Summary

In this Chapter we summarized that synthesis of DNA in nature needs robust enzymes called DNA pols. The DNA pols are used for all DNA transactions (DNA replication, the different DNA repair events, DNA recombination and physiological recombination in the immune system) in a cell. The DNA pol reaction is complex and its structure resembles a right hand, consisting of fingers, a palm and a thumb. Nature has brought forward seven different DNA pol families and higher eukaryotic cells contain a total of 15 DNA pols and the telomerase. They harbor different, mostly redundant functions. Replicative DNA pols act in large complexes called holoenzymes. They consist, besides the DNA pol core, of a clamp and a clamp loader. In eukaryotes the clamp is called proliferating cell nuclear antigen (PCNA) and the clamp loader replication factor C (RF-C). The fact that PCNA interacts with 10 different DNA pols suggests a central regulatory task for this important protein. Additionally, alternative clamps and several alternative clamp loaders exist. The fact that many of these are involved in checkpoint sensing suggests an important role for them in guiding the appropriate DNA pol under certain conditions to the right place to work. Replicative DNA pols interact with many proteins to build a stable replisome. Finally, the single-stranded DNA binding protein-replication protein A (RP-A) does more than bind and protect DNA. It has additional functions in checkpoint regulation. The investigations of RP-A interacting with different DNA pols have recently revealed key roles for this very important protein in various

Table 2.8   Redundant functions of DNA polymerases

DNA replication	DNA repair	V(D)J recombination
DNA pols α, γ, δ and ε	DNA pols β, η, κ, λ, ζ, θ, υ and Rev1	DNA pols ι, λ and μ
Fidelity: 10^{-4}–10^{-6}	Fidelity: 10^{-2}–10^{-4}	Fidelity: 10^{0}–10^{-2}

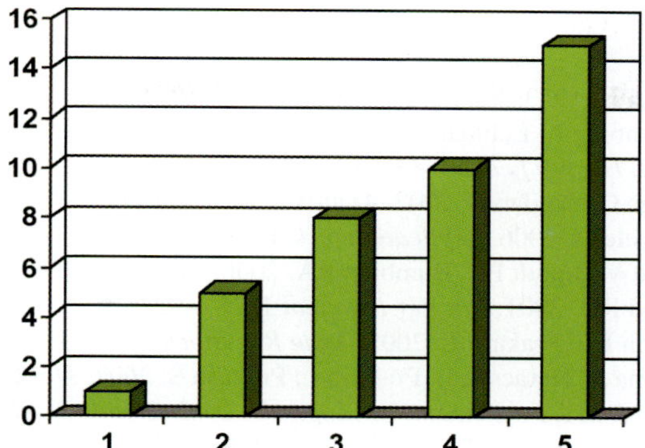

Figure 2.8   The number of different DNA polymerases increases in evolution from viruses to mammals. 1: viruses; 2: prokaryotes; 3: eukaryotes (single cell); 4: eukaryotes (plants); 5: eukaryotes (mammals).

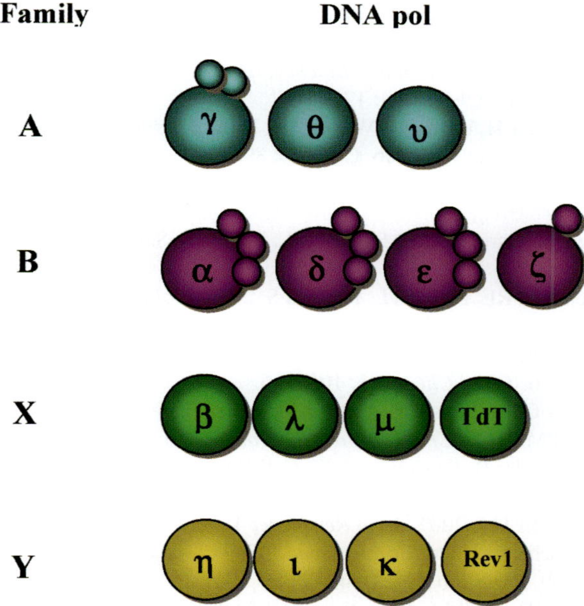

Figure 2.9   Subunit compositions of the 15 mammalian DNA polymerases. For further details see Chapter 4. The small subunit of DNA pol γ is a homodimer.

DNA transactions. **Table 2.8** documents a gradient of fidelity of different mammalian DNA pols in cellular events such as DNA replication, DNA repair and V(D)J recombination while **Figure 2.8** documents that the number of different DNA pols increases in evolution from viruses to mammals. **Figure 2.9** lists the different subunit compositions of mammalian DNA pols.

References

1. Watson JD, Crick FH. 1953. *Nature* 171: 964–7.
2. Bessman MJ, Kornberg A, Lehman IR, Simms ES. 1956. *Biochim Biophys Acta* 21: 197–8.
3. Kornberg A. 1957. *Harvey Lect* 53: 83–112.
4. Hubscher U, Maga G, Spadari S. 2002. *Annu Rev Biochem* 71: 133–63.
5. Hübscher U, Shevelev I. 2006. *Replication fork*. Berlin Heidelberg New York: Springer Verlag.
6. Rossi ML, Purohit V, Brandt PD, Bambara RA. 2006. *Chem Rev* 106: 453–73.
7. Gearhart PJ, Wood RD. 2001. *Nat Rev Immunol* 1: 187–92.
8. Prakash S, Johnson RE, Prakash L. 2005. *Annu Rev Biochem* 74: 317–53.
9. Acharya N, Brahma A, Haracska L, Prakash L, Prakash S. 2007. *Mol Cell Biol* 27: 7266–72.
10. Kelleher C, Teixeira MT, Forstemann K, Lingner J. 2002. *Trends Biochem Sci* 27: 572–9.
11. Wang SW, Toda T, MacCallum R, Harris AL, Norbury C. 2000. *Mol Cell Biol* 20: 3234–44.
12. Kunkel TA, Bebenek K. 2000. *Annu Rev Biochem* 69: 497–529.
13. Kunkel TA. 2004. *J Biol Chem* 279: 16895–8.
14. Shevelev IV, Hubscher U. 2002. *Nat Rev Mol Cell Biol* 3: 364–76.
15. Jiricny J. 2006. *Nat Rev Mol Cell Biol* 7: 335–46.
16. Maga G, Hubscher U. 2003. *J Cell Sci* 116: 3051–60.
17. Davey MJ, Jeruzalmi D, Kuriyan J, O'Donnell M. 2002. *Nat Rev Mol Cell Biol* 3: 826–35.
18. Fanning E, Klimovich V, Nager AR. 2006. *Nucleic Acids Res* 34: 4126–37.
19. Joyce CM, Steitz TA. 1994. *Annu Rev Biochem* 63: 777–822.
20. Steitz TA. 1999. *J. Biol. Chem.* 274: 17395–8.
21. Sawaya MR, Pelletier H, Kumar A, Wilson SH, Kraut J. 1994. *Science* 264: 1930–5.
22. Bailey S, Wing RA, Steitz TA. 2006. *Cell* 126: 893–904.
23. Leulliot N, Cladiere L, Lecointe F, Durand D, Hubscher U, van Tilbeurgh H. 2009. *J Biol Chem* 284: 11992–9.
24. Garcia-Escudero R, Garcia-Diaz M, Salas ML, Blanco L, Salas J. 2003. *J Mol Biol* 326: 1403–12.
25. Haracska L, Johnson RE, Prakash L, Prakash S. 2005. *Mol Cell Biol* 25: 10183–9.
26. McHenry C, Kornberg A. 1977. *J Biol Chem* 252: 6478–84.
27. Johnson A, O'Donnell M. 2005. *Annu Rev Biochem* 74: 283–315.
28. O'Donnell M. 2006. *J Biol Chem* 281: 10653–6.
29. Sinha NK, Morris CF, Alberts BM. 1980. *J Biol Chem* 255: 4290–3.
30. Garg P, Burgers PM. 2005. *Crit Rev Biochem Mol Biol* 40: 115–28.
31. Moldovan GL, Pfander B, Jentsch S. 2007. *Cell* 129: 665–79.
32. Mossi R, Hubscher U. 1998. *Eur J Biochem* 254: 209–16.
33. Sclafani RA, Holzen TM. 2007. *Annu Rev Genet* 41: 237–80.
34. Toueille M, Hubscher U. 2004. *Chromosoma* 113: 113–25.
35. Tsurimoto T. 1999. *Front Biosci* 4: D849–58.
36. Hingorani MM, O'Donnell M. 2000. *Curr Biol* 10: R25–9.
37. Jonsson ZO, Hubscher U. 1997. *Bioessays* 19: 967–75.
38. Bowman GD, O'Donnell M, Kuriyan J. 2004. *Nature* 429: 724–30.
39. Trakselis MA, Bell SD. 2004. *Nature* 429: 708–9.
40. Sancar A, Lindsey-Boltz LA, Unsal-Kaccmaz K, Linn S. 2004. *Annu Rev Biochem* 73: 39–85.
41. Dore AS, Kilkenny ML, Rzechorzek NJ, Pearl LH. 2009. *Mol Cell* 34: 735–45.
42. Xu M, Bai L, Gong Y, Xie W, Hang H, Jiang T. 2009. *J Biol Chem* 284: 20457–61.
43. Toueille M, El-Andaloussi N, Frouin I, Freire R, Funk D, *et al.* 2004. *Nucleic Acids Res* 32: 3316–24.

44. Sabbioneda S, Minesinger BK, Giannattasio M, Plevani P, Muzi-Falconi M, Jinks-Robertson S. 2005. *J Biol Chem* 280: 38657–65.

45. Waters LS, Minesinger BK, Wiltrout ME, D'Souza S, Woodruff RV, Walker GC. 2009. *Microbiol Mol Biol Rev* 73: 134–54.

46. Yan S, Michael WM. 2009. *J Cell Biol* 184: 793–804.

47. Forsburg SL. 2004. *Microbiol Mol Biol Rev* 68: 109–31.

48. De Falco M, Ferrari E, De Felice M, Rossi M, Hubscher U, Pisani FM. 2007. *EMBO Rep* 8: 99–103.

49. Tercero JA, Labib K, Diffley JF. 2000. *EMBO J* 19: 2082–93.

50. Zou L, Stillman B. 2000. *Mol Cell Biol* 20: 3086–96.

51. Warren EM, Huang H, Fanning E, Chazin WJ, Eichman BF. 2009. *J Biol Chem* 284: 24662–72.

52. Zhu W, Ukomadu C, Jha S, Senga T, Dhar SK, *et al.* 2007. *Genes Dev* 21: 2288–99.

53. Gambus A, van Deursen F, Polychronopoulos D, Foltman M, Jones RC, *et al.* 2009. *EMBO J* 28: 2992–3004.

54. Lou H, Komata M, Katou Y, Guan Z, Reis CC, *et al.* 2008. *Mol Cell* 32: 106–17.

55. Masumoto H, Sugino A, Araki H. 2000. *Mol Cell Biol* 20: 2809–17.

56. Kamath-Loeb AS, Johansson E, Burgers PM, Loeb LA. 2000. *Proc Natl Acad Sci USA* 97: 4603–8.

57. Selak N, Bachrati CZ, Shevelev I, Dietschy T, van Loon B, *et al.* 2008. *Nucleic Acids Res* 36: 5166–79.

58. Papamichos-Chronakis M, Peterson CL. 2008. *Nat Struct Mol Biol* 15: 338–45.

59. Wold MS. 1997. *Annu Rev Biochem* 66: 61–92.

60. Zou L, Elledge SJ. 2003. *Science* 300: 1542–8.

61. Mason AC, Haring SJ, Pryor JM, Staloch CA, Gan TF, Wold MS. 2009. *J Biol Chem* 284: 5324–31.

62. Maga G, Frouin I, Spadari S, Hubscher U. 2001. *J Biol Chem* 276: 18235–42.

63. Maga G, Villani G, Tillement V, Stucki M, Locatelli GA, *et al.* 2001. *Proc. Natl. Acad. Sci. USA*, 98: 14298–303.

64. Maga G, Ramadan K, Locatelli GA, Shevelev I, Spadari S, Hübscher U. 2005. *J Biol Chem* 280: 1971–81.

65. Maga G, Shevelev I, Villani G, Spadari S, Hübscher U. 2006. *Nucl Acids Res* 34: 1405–15.

66. Maga G, Villani G, Crespan E, Wimmer U, Ferrari E, *et al.* 2007. *Nature* 447: 606–8.

67. Crespan E, Hubscher U, Maga G. 2007. *Nucleic Acids Res* 35: 5173–81.

68. Maga G, Crespan E, Wimmer U, van Loon B, Amoroso A, *et al.* 2008. *Proc Natl Acad Sci USA* 105: 20689–94.

CHAPTER 3

Structural and Functional Aspects of the Prokaryotic and Archaea DNA Polymerase Families

As every other DNA pol, prokaryotic and archaea DNA pols are classified into several distinct families based on amino acid sequence homology of their catalytic subunits. **Table 3.1** summarizes some of their characteristics. In this chapter we will describe the main functional and structural aspects of DNA pols from *Escherichia coli, Bacillus subtilis, Mycobacteria, Deinococcus radiodurans* and *Archaea*.

3.1 *Escherichia coli*

Escherichia coli served as a prototype of DNA replication studies in prokaryotes. In this chapter *Escherichia coli* DNA pols are classified in distinct families and some of their main biochemical features are described.

3.1.1 *Family A: DNA Polymerase I*

The DNA pol I of the Gram-negative bacteria *Escherichia coli* (*E. coli* DNA pol I) was the first DNA pol discovered.[1] It is an abundant DNA pol (approximately 400 molecules per bacteria) coded by the pol A gene and it is implicated in fundamental cellular processes such as DNA repair and replication.[2] The enzyme is organized in three functional domains: an-N-terminal domain containing the $5' \rightarrow 3'$ exonuclease activity; a central domain bearing the $3' \rightarrow 5'$ exonuclease proofreading activity; and a C-terminal domain associated with the pol activity (for a review see Ref. 3). The first high-resolution structure of a DNA pol was a crystal structure of the Klenow fragment (or large fragment) of *E. coli* DNA pol I.[4] The Klenow fragment of *E. coli* DNA pol I is a protein fragment that is devoid of the $5' \rightarrow 3'$ exonuclease domain but has retained the other two. The crystal structure of the Klenow fragment displayed a DNA pol shaped like a cupped right hand, with

85

Table 3.1   Prokaryotic DNA polymerases: Properties and functions

Name	Family	Gene	Crystal structure	Associated Activities	Main functions
E. coli DNA pol I	A	*pol A*	Yes	3′ exonuclease	DNA repair/ replication
B. subtilis DNA pol I	A	*pol A*	No	5′ exonuclease	DNA repair/ replication
D. radiodurans DNA pol A	A	*DR1707*	No	5′ exonuclease	DNA repair/ replication
E. coli DNA pol II	B	*pol B*	No	3′ exonuclease	DNA replication restart/ TLS[a]
Archaea DNA pol B	B	ND	Yes	3′ exonuclease	DNA replication
E. coli DNA pol III	C	*dna E*	Yes	3′ exonuclease (separate subunit)	DNA replication
B. Subtilis DNA pol III	C	*pol C*	No	3′ exonuclease	DNA replication
B. Subtilis DnaE	C	*Dna E*	No	None	DNA replication
Archaea DNA pol D	D	ND	No	3′ exonuclease (separate subunit)	DNA replication
B. Subtilis DNA pol X	X	*yshC*	No	3′ exonuclease	DNA repair
D. radiodurans DNA pol X	X	*DR0467*	Yes	3′ exonuclease (structure specific)	DNA repair
E. coli DNA pol IV	Y	*din B*	No	None	TLS
E. coli DNA pol V	Y	*umu C*	No	None	TLS
B. subtilis DNA pol Y1	Y	*yqjH*	No	None	TLS
B. subtilis DNA pol Y2	Y	*yqjW*	No	None	TLS
Archaea Dbh	Y	ND	Yes	None	TLS
Archaea Dpo4	Y	ND	Yes	None	TLS

[a]TLS, translesion DNA synthesis

fingers, thumb and palm sub-domains surrounding a deep groove (see Chapter 2). The palm sub-domain contained three essential amino acids residues capable of binding the metals necessary for DNA synthesis (for review see Ref. 5). However the 3′ → 5′ exonuclease site was located many angstroms far from the DNA pol site, indicating that a rearrangement of the pol-DNA complex was necessary to allow the proofreading exonuclease to check for an eventual misincorporation. Subsequent crystallization of DNA pols in a ternary complex with a DNA primer template and an incoming nucleotide, taken in conjunction with pre-steady state fluorescence studies, suggests that kinetics steps along the incorporation pathway differ

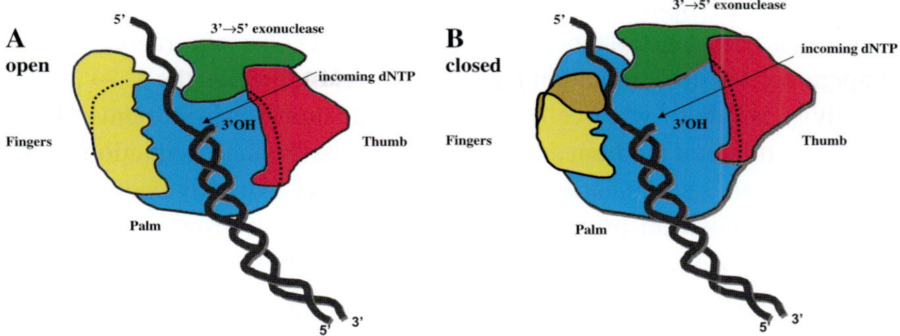

Figure 3.1 **Structure of a DNA polymerase.** Cartoon of an open (A) and closed (B) form of a DNA pol. For details see text.

for Watson-Crick (W-C) and non-W-C base pairs. The ternary structures suggest that the closure of the fingers around substrates bound to the DNA pol active site facilitates incorporation of W-C base pairs (for reviews see Refs. 6 and 7). In the closed conformation the DNA pol active site appears to only accommodate a correct template-nucleotide base pair; subsequently the fingers should reopen to allow binding of a new nucleotide substrate (**Figure 3.1**). In contrast, incorporation of non-W-C base pairs may occur from a partially open state,[8,9] but this remains an open question, as does the possible role of an induced-fit mechanism in regulating pol fidelity.

The purified *E. coli* DNA pol I displays a rather poor processivity (15–20 nt), it avidly binds to nicked and gapped DNA templates and, during polymerization, its $5' \rightarrow 3'$ exonuclease can displace the nick along the duplex DNA, performing the so called "nick translation".[2] These biochemical properties seem to be particularly adapted for an enzyme acting in DNA repair and in lagging-strand DNA replication. In agreement with a central function of *E. coli* DNA pol I in repair, *E. coli* *Pol A* mutants show increased sensitivity to DNA damaging agents.[10] Indeed, *in vitro* evidence suggests that *E. coli* DNA pol I is the main DNA pol acting in base excision repair (BER),[11] while other observations point to an important, although not exclusive, role of this enzyme in nucleotide excision repair (NER).[10,12] A major contribution to the understanding of the biochemistry of DNA replication in *E. coli* came from the isolation by De Lucia and Cairns of the mutant strain *PolA1* having a much reduced level of DNA pol I but capable of growth.[13] However, this is not to say that DNA pol I has no role in replication. In fact it was later shown that the involvement of DNA pol I in replication stems from its $5' \rightarrow 3'$ exonuclease activity that acts to remove the RNA primers of the Okazaki fragments, followed by filling of the resulting gaps by the DNA pol activity of the enzyme. This explains the fact that the *E. coli PolA1* mutant, which lacks DNA pol activity but retains the $5' \rightarrow 3'$ exonuclease, is able to grow, albeit accumulating many Okazaki fragments,[13] while the temperature sensitive mutation *polAex1*, which suppresses the $5' \rightarrow 3'$ exonuclease activity, is lethal[14] when cells are grown in medium rich in nutrients. However, cells grown in minimal medium do survive, albeit at a dramatic reduction in growth rate. When growth is sufficiently retarded, an RNase H-like activity can apparently excise the RNA Okazaki primers. Notably, the cells can grow normally in rich medium simply by providing the $5' \rightarrow 3'$ exonuclease on a plasmid.[15] Recently, *E. coli* DNA pol I was also proposed to be involved in chromosome termination *in vivo*.[16] Clearly, this is a truly important and versatile enzyme, with three enzymatic activities, one pol and two exo activities, all on a single polypeptide chain.

3.1.2 *Family B: DNA Polymerase II*

The *E. coli* DNA pol II was identified independently by three laboratories in extracts of *E. coli Pol A1* mutants in 1970.[2] DNA pol II possesses an active $3' \rightarrow 5'$ proofreading exonuclease but mutations in its encoding gene *E. coli pol B* showed no obvious phenotypes, so that the physiological function of the enzyme has remained unclear for long time. However, recent findings have revealed a role for DNA pol II in the replication restart. First it was found that this DNA pol is induced sevenfold in response to DNA damage[17] and that the *E. coli pol B* gene is in fact the *dinA* gene, one of the genes induced in the course of the so called SOS response.[10] Second, it was discovered that double mutants of DNA pol II and DNA pol V showed increased UV sensitivity compared to the DNA pol V mutant alone.[18]. In the absence of DNA pol II, DNA replication was blocked for roughly 45 minutes, until translesion synthesis (TLS) catalyzed by DNA pol V could take place.[19] The same delay occurred in bacteria proficient for *E. coli* DNA pol II but lacking proteins such as *E. coli* RecF, RecO, RecR or PriA, which are necessary for rescuing replication forks in UV-irradiated *E. coli*.[20] A model shown in Ref. 21 proposes that a lesion in the leading strand uncouples leading and lagging strand synthesis, generating long regions of single-stranded DNA, which are coated by the *E. coli* single-stranded DNA binding protein (SSB). SSB coated DNA is replaced by the RecA nucleoprotein filament followed by induction of the SOS system, which turns on DNA pol II and DNA pol V (see family Y section). A collapsed fork undergoes a RecA nucleoprotein-mediated regression, forming the so called chicken-foot structure[22] in which the uncoupled nascent lagging strand provides a template for synthesis by DNA pol II, which then copies the correct information thus avoiding TLS (**Figure 3.2**).

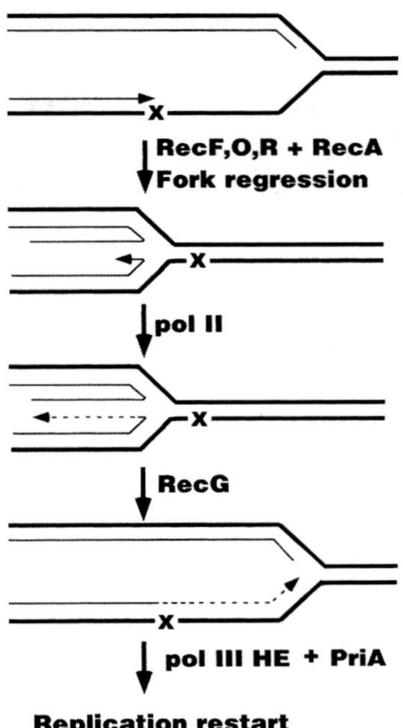

Figure 3.2   Model of error-free replication restart involving *E. coli* DNA polymerase II. Replicative DNA synthesis is blocked by DNA template damage (X). DNA pol III dissociates from the damaged primer-template strand while synthesis continues on the undamaged strand. Regression of the replication fork occurs in the presence of RecF, O, and R proteins, thereby providing an undamaged template strand for DNA pol II to copy that contains the correct coding information at the DNA damage site. RecG is required for progression of the fork past the lesion site and PriA is then involved in reconstituting the replication fork. (Reproduced with permission from Ref. 21.)

The processivity of DNA pol II is stimulated by the DNA pol III subunits β clamp and clamp loader (see Chapter 3.1.3) and by SSB.[23,24] DNA pol II has also been implicated in replication of damaged DNA.[25] No crystal structure of the DNA pol II is available at the present, but the related B family DNA pols RB69 and HSV-1 have been crystallized (see Chapters 2 and 6).

3.1.3 *Family C: DNA Polymerase III Holoenzyme*

The principal function of the DNA pol III holoenzyme is the duplication of the *E. coli* chromosome. However, it can also participate in other DNA metabolic processes such as DNA repair and mutagenesis. DNA pol III holoenzyme acts at the replication fork in combination with other proteins, including SSBs, helicases and topoisomerases. We will attempt to describe here some of the salient features of the *E. coli* DNA pol III holoenzyme but, for more detailed information, we invite the reader to refer to existing reviews and to the references therein.[26-29] DNA pol III was first isolated from *E. coli Pol A* mutants and established as chromosomal replicase by the fact that extracts of temperature sensitive mutants in its encoding gene *dnaE* contained temperature sensitive DNA pol III activity.[2] Initial purification used DNA templates containing nicks and gaps and led to the isolation of what was probably the three subunits DNA pol III core.[30] The subsequent use of more demanding templates, such as primed circular single-stranded DNA, led to the purification of the DNA pol III multi-subunits holoenzyme. There are only 10–20 copies of *E. coli* DNA pol III holoenzyme in the cell but the enzyme catalyzes DNA synthesis at a speed of about 800 nucleotides per second (ntd/s), comparable to the rate of fork movement in *E. coli*.[31]

DNA pol III holoenzyme is composed of ten subunits that are present in various subassemblies (see **Table 2.6** of Chapter 2). The simplest assembly is the DNA pol III core, a 1:1:1 heterotrimer composed of the catalytic subunit α, the $3' \rightarrow 5'$ proofreading exonuclease subunit ε and a small θ subunit of not well established function besides a slight stimulation of ε. DNA pol III core itself is slow and weakly processive, incorporating 20 nt/s and a maximum of 10 bases per binding event; however its association with the β sliding clamp makes the core enzyme very fast (up to 750 nt/s) and highly processive (>50 kb). The β clamp subunit is therefore a processivity factor structurally related to PCNA of eukaryotes (see Chapters 2 and 4) that has to interact with the DNA pol III core and the DNA at the primer terminus. However, because of its closed, ring shaped structure the β clamp cannot assemble on DNA by itself. Another component of the DNA pol III holoenzyme, the γ complex clamp loader, is capable of topologically link the β clamp to primed DNA in a reaction dependent on ATP binding and hydrolysis. The clamp loader is composed of five different subunits in defined stoichiometry: $\gamma_3\delta_1\delta'_1\chi_1\psi_1$. It should be noted, however, that the majority of the clamp loader is in free solution and that the clamp loader associated with the DNA pol III holoenzyme contains a different form of the *dna X* gene encoding for the γ subunit. As a matter of fact, the γ subunit is produced through a ribosomal frameshift in the *dna X* gene that causes termination of translation to produce a 47.5 kDa protein. The full length product of *dna X* is the τ subunit (71.1 kDa) which contains the γ sequence plus a 23.6 kDa terminal region (τ_c). The τ_c region is not essential for clamp loading but is necessary for cell viability, probably via its capacity to organize the replisome at the replication fork (see Ref. 29 for details). It is thought that the *E. coli* replicase contains two DNA pol III cores attached to a $\tau_2\gamma_1\delta\delta'\chi\psi$ clamp loader and this sub-assembly is termed DNA pol III* (or DNA pol III star). The β clamp associates then with the DNA pol III* in an ATP dependent manner to form the final DNA pol III holoenzyme assembly. However, the DNA pol III holoenzyme can also assemble into a particle that contains three DNA pols. The three DNA pols appear capable of simultaneous activity and are fully functional at the replication fork with helicase, primase and sliding clamp. It has been proposed that two pols can function at the leading and lagging strand and that the third can act as a reserve enzyme to overcome certain types of obstacles at the replication fork.[32]

Today many of the subunits of the *E. coli* DNA pol III holoenzyme have been crystallized. The structure of a large fragment of the catalytic α subunit has been recently solved[33] (**Figure 3.3A**).

It consists of the usual fingers, palm and thumb domains, but, unexpectedly, the active site of the enzyme is closely related to those of DNA pols of the X family, such as the eukaryotic DNA pol β (see Chapter 4). The structure also suggests a model for the interaction of the α subunit with the sliding clamp and DNA. The structure of the $3' \rightarrow 5'$ exonuclease proofreading subunit ε bound with HOT, the bacteriophage P1-encoded homolog of the subunit θ, has been also recently published.[34] This structure provides insight into how HOT and, by implication, θ may stabilize the α subunit, thus promoting efficient proofreading during chromosomal replication. Structural analysis of the sliding clamp subunit β showed that the clamp is a head-to-tail dimer, with two protomers that form a closed ring of an interior diameter capable of accommodating duplex DNA[35] (**Figure 3.3B**). Biochemical studies show that one interface of the clamp is opened by the clamp loader to allow passage of a primed DNA to his interior.[36] It may be noted that, although the ring-shaped architecture of the β sliding clamp is similar to correspondent clamps in eukaryotes and archaea, the proteins do not share readily detectable sequence similarities. Recently the structure of the β clamp has been solved in complex with primed DNA.[37] The fact that the clamp recognizes primed DNA has implications for clamp loading and also suggests mechanisms by which different proteins that bind the clamp may switch with one another on β and the DNA that it encircles. Clamp loaders subunits, both prokaryotic and eukaryotic, are members of the AAA+ family.[38] Recent structures of intact *E. coli* $\gamma_3\delta\delta'$ clamp loader revealed five core subunits arranged on a circle[39] (**Figure 3.3C**). ATP binds only to the three γ subunits while most if not all the $\gamma_3\delta\delta'$ subunits interact with the clamp (see Ref. 40 for a review).

Concerning the roles of DNA pol III holoenzyme besides DNA replication *in vivo*,[10] *in vitro* data[41] indicated that the DNA pol III holoenzyme is the *E. coli* DNA pol responsible for the final re-synthesis step of a DNA metabolic process of mutation avoidance capable of repairing mismatched base pairs: the mismatch repair (MMR) system.[10] Interestingly, the DNA pol III holoenzyme subunit β clamp interacts with two essential proteins of the MMR system, MutL and MutS,[42] possibly coordinating MutS and MutL activities with replication. It should also be noted that the β clamp also interacts with the Y family DNA pols IV and V and the current

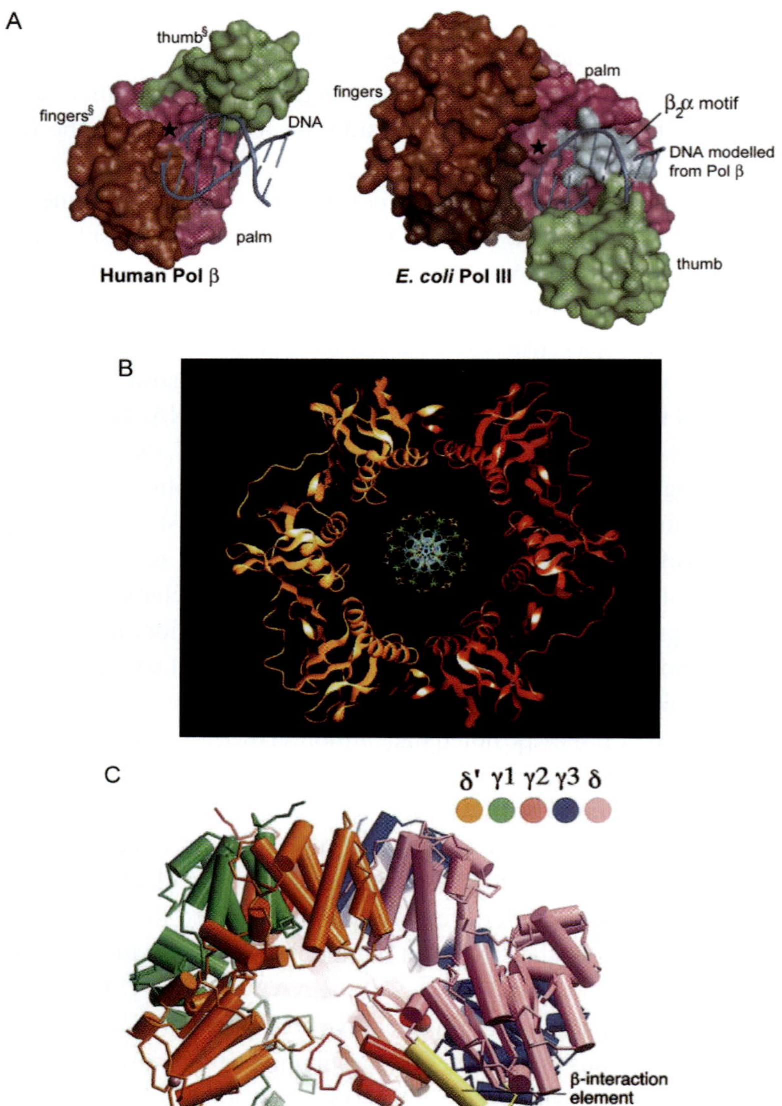

Figure 3.3   Structural view of *E. coli* replicative proteins. **A**. Left: top view of human DNA pol β bound to DNA; right: the same view of DNA polymerase III with the DNA placed into the structure. **B**. Ribbon representation of the polypeptide chain of the clamp β subunit dimer, looking down the 2-fold axis of the ring. The α helices are shown as spirals and the β sheet as flat ribbons. The two monomers are coloured yellow and red. A standard model of B-form DNA is in the middle of the structure. **C**. front view of the γ complex structure (Reproduced with permission from Refs. 33, 35 and 39).

view is that it regulates the exchange of replicative and translesion DNA pols during DNA replication of damaged DNA (see Sec. 3.1.4).

3.1.4　*Family Y: DNA Polymerases IV and V*

DNA damages are a threat for cell survival. Initial work from *E. coli* led to the discovery that these damages can induce physiological responses aimed at rescuing cell growth, and several mechanisms for such rescue have been identified in prokaryotes and eukaryotes.[10] Instrumental in the understanding of these mechanisms was the discovery of the SOS response, a molecular mechanism that, in response to damage, initiates the RecA$^+$ LexA$^+$-regulated de-repression of a number of genes.[10] Among the genes induced, three encode for the DNA polymerases II, IV and V respectively. Two of these DNA pols, namely DNA pols IV and V, belong to the newly discovered Y family of DNA pols and we will briefly summarize some of their properties below. For more details on their functions we address the reader to recent reviews.[10,43].

DNA polymerase IV. DNA pol IV activity was first reported in 1999 and shown to be encoded by the SOS gene *dinB*.[44] DNA pol IV is a relatively abundant DNA pol (about 250 molecules/cell) and its levels increase further 10-fold upon DNA damage.[45] DNA pol IV is devoid of intrinsic $3' \rightarrow 5'$ exonuclease proofreading activity and its fidelity is therefore achieved solely by its capacity to discriminate between correct and incorrect base pair formation. Nevertheless, its fidelity, at least at the incorporation step, is relatively high, the enzyme's misincorporation frequency being in the 10^{-3} to 10^{-4} range.[46] DNA pol IV mediates untargeted mutagenesis *in vivo* and is required for adaptive mutagenesis.[47,48] DNA pol IV has also the capacity to perform *in vitro* DNA synthesis across a wide panel of base modifications, either alone or in presence of the β clamp, but its interaction with the clamp is essential for its translesion synthesis (TLS) *in vivo*.[43]

DNA polymerase V. DNA pol V was identified in 1999 too.[49,50] It is encoded by the *umuDC* locus and is a dimer of the post-translationally modified UmuD$'$ protein in a complex with UmuC.[51] DNA pol V is present at only 15 molecules per cell but, upon DNA damage, its concentration can reach 200 molecules per cell. As in the case of DNA pol IV, DNA pol V lacks intrinsic $3' \rightarrow 5'$ proofreading activity and is also considered a low fidelity DNA pol. Importantly, DNA pol V is responsible for most *in vivo* damage-induced mutagenesis in *E. coli*.[10] *In vitro* DNA pol V was shown capable of replicating a variety of lesions and, notably, TT cis–syn cyclobutane dimer and TT (6–4) photoproducts with specificities similar to those observed *in vivo*.[46] Several groups have investigated the *in vitro* biochemical requirements for DNA pol V-mediated *in vitro* TLS. Despite differences in DNA substrates and lesions used, all these studies pinpointed to the necessity of at least RecA protein and β clamp as necessary cofactors for TLS by DNA pol V (see Ref. 43 for details).

In vitro systems have been established that recapitulate the genetic requirements for TLS of base damage in *E. coli* by DNA pol V.[52–55] The genetic requirements for SOS mutagenesis are minimally UmuD$_2'$C, identified as DNA pol V, and RecA, which *must* be assembled as a filament on single-stranded DNA, often referred to by the term "activated RecA nucleoprotein filament" RecA*. RecA* is ubiquitous in the cell; it is essential for homologous recombination, and for turning on SOS, using its protease function to facilitate autocleavage of the LexA repressor, and in a similar cleavage reaction converting UmuD to UmuD′, which then forms DNA pol V (UmuD$_2'$C). The requirement for RecA* is also unconditional for DNA pol V-catalyzed TLS for reasons that remained a mystery since the late 80s, but no longer. It has recently been shown that RecA* is needed to transfer a single molecule of RecA and ATP from its 3′-filament tip to UmuD$_2'$C so as to activate DNA pol V for TLS.[56] Thus, the mutagenically active form of DNA pol V is not UmuD$_2$C by itself, but is instead a more complex protein-nucleotide assembly that includes a single molecule of RecA and ATP, referred to as a DNA pol V "mutasome" (DNA pol V Mut). DNA pol V Mut is composed of UmuD$_2'$C-RecA-ATP. *In vivo*, a DNA lesion in one strand blocks the progression of the replicative DNA pol III holoenzyme and generates long stretches of single-stranded DNA downstream the lesion as a result of transient uncoupling of the coordinate leading and lagging strand synthesis.[57] Further requirements for TLS include the DNA pol III holoenzyme, associated with the β clamp. When copying damaged DNA, DNA pol III holoenzyme reaches DNA damage and stops at the nucleotide preceding the lesion. When the replication fork stalls at a lesion, the DNA pol V-RecA-ATP complex is loaded onto the β clamp at the primer terminus thereby allowing DNA pol V to synthesize, in a single binding event, a TLS patch long enough to support further extension by the replicative DNA pol III holoenzyme after its interaction with the β clamp has been re-established (**Figure 3.4**). Once DNA pol V Mut has dissociated, it is rapidly inactivated, ensuring that DNA pol V-catalyzed error-prone synthesis will cease once RecA* filaments supporting SOS induction are gone. Left open is the question of where RecA* is located in the cell.[56,58] Perhaps it forms on the region of single-stranded DNA downstream of a blocked replication fork (as drawn in **Figure 3.4**). Or RecA* might form remotely from the blocked replication fork.

Other works have shown that the β clamp is capable of binding simultaneously to DNA pol III and DNA pol IV *in vitro*[59] and plays an important role in the interplay between DNA pol III and both DNA pol II and DNA pol V at a DNA lesion.[60] A very recent paper has reported that DNA pol II and DNA pol IV function with *E. coli*

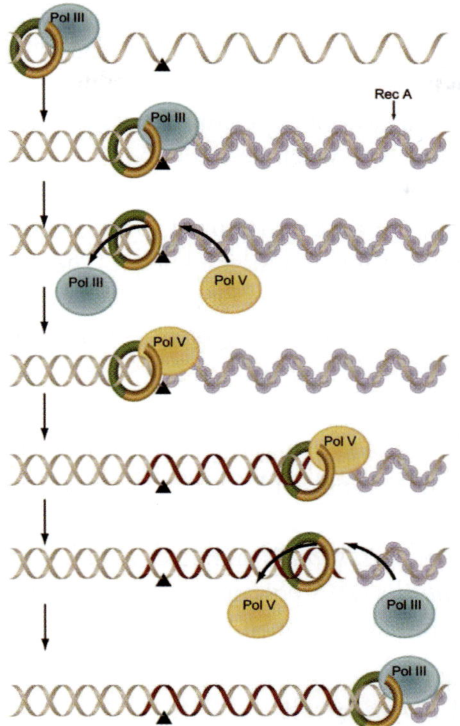

Figure 3.4   Model for TLS in *E. coli*. The replicative DNA pol III associated with the β clamp processivity factor (green and yellow circles), reaches a non-coding region (triangle) and arrests. A region of single-stranded DNA forms downstream of the lesion as a consequence of the transient progression of the replicative helicase and triggers the formation of an extended Rec A filament. DNA pol V binds to the 3′-OH terminus of the primer and forms a stable complex by interacting with both the RecA filament and the β clamp. The "DNA pol V-RecA-β clamp" complex loaded onto the primer terminus allows DNA Pol V to synthesize a patch of DNA of about 20 nucleotides (shown in red). The TLS is then completed and DNA pol III can resume synthesis. (Reproduced with permission from Ref. 54.)

DNA B helicase to regulate its rate of undwinding slowing it to as little as 1 bp/s.[61] Furthermore, DNA pol II and IV freely exchange with the DNA pol III replicase on the β-clamp and function with DNA B helicase to form alternative replisomes. The authors propose that these dynamic actions provide additional time for DNA repair of the lesions before the replication fork reaches them and also enables the appropriate translesion DNA pol to sample each lesion as it is encountered.

The complete DNA pol IV and DNA pol V proteins have not been crystallized yet. However, the crystal structure of the complex between the C-terminal domain of DNA pol IV, comprising the so called "little finger" domain (see Sec. 3.4 Archaea), and the β clamp has been solved.[62] The salient feature of this interaction is that the complex has no access to the template-primer junction (OFF position). However, as far as the β clamp surface is not obstructed, such an "inactive" complex may accommodate a second DNA pol, making a transition to a productive complex at the junction (ON position) (see Ref. 43 for details). Thus, the β clamp may act as a "tool belt"[63] to allow the switch between the *E. coli* replicative DNA pol III and the three DNA pols II, IV and V at DNA damage. Although the DNA pols of the *E. coli* family Y have not been crystallized so far, important structural features of prokaryotic DNA pols of this family are known. They are mainly the result of studies with the DNA pol IV orthologs Dbh and Dpo4 of *Sulfolobus solfataricus* and will be discussed in the Archaea paragraphs (see Section 3.4).

3.2 *Bacillus subtilis*

The comparison between DNA pols in *B. subtilis* and *E. coli* reveals both similarities and differences. A recent comprehensive review summarizes our knowledge about chromosomal replication in *B. subtilis* and other related Gram-positive bacteria.[64] In this chapter the *B. subtilis* DNA pols are classified in distinct families and some of their important features are described.

3.2.1 *Family A: DNA Polymerase I*

The DNA pol I of the Gram-positive bacteria *B. subtilis* was first purified more than four decades ago.[65] It is encoded by the *PolA* gene and studies with *PolA* temperature sensitive mutants showed that, when raised at restrictive temperature, the bacteria accumulated large amounts of newly synthesized short RNA pieces. These observations suggested that DNA pol I is involved in the removal of the RNA attached to the nascent short DNA pieces as well as in filling the gap between these pieces.[66] *B. subtilis PolA* mutant are also sensitive to ultraviolet light, X-rays and MMS, indicating that the DNA pol I has an important role also in DNA repair.[67] Although possessing an active $5' \rightarrow 3'$ exonuclease activity, DNA pol I belongs to a subfamily of DNA pols that have a vestigial and non functional $3' \rightarrow 5'$ exonuclease proofreading site. Since the proofreading activity is expected to oppose TLS catalyzed by a DNA pol by preventing stable incorporation of mispaired bases opposite DNA lesions,[68,69] the role of DNA pol I in TLS mutagenesis was recently investigated. The results of such a study[70] showed that DNA pol I was catalytically required for the UV-induced mutagenesis mediated by the *B. subtilis* Y family DNA pols Y1 and Y2. Furthermore, these data also indicated that the role of DNA pol I in TLS is to extend the mismatches generated by incorporation in front of the lesion by DNA pols Y1 or Y2. These findings illustrate a striking difference between TLS mechanisms in *E. coli* and *B. subtilis*, pointing to a novel function of DNA pol I, i.e. resumption of DNA synthesis following incorporation at a lesion by a Y family DNA pol that appears to be performed by DNA pol III in *E. coli*.[52] However, a similar rescuing function could also be performed by *B. subtilis* DnaE pol (see next subsection).

No crystal structures of the *B. subtilis* DNA pol I are available yet, although a model of its 3D structure was generated by superimposition onto the known *B. stearothermophilus* BF DNA pol I structure.[70,71]

3.2.2 *Family C: DNA Polymerase C and DnaE*

B. subtilis DNA pol III, encoded by the *pol C* gene, is essential for chromosomal replication.[67] The pol C polypeptide carries a DNA pol active site in its C-terminal part and a $3' \rightarrow 5'$ proofreading exonuclease activity in its N-terminal domain.[72,73] This is in contrast to the *E. coli* DNA pol III in which the DNA pol and exonuclease activities are encoded by two separate genes. Purification of the DNA polC from *B. subtilis* cell extracts, however, yielded only the catalytic subunit of the DNA pol and no holoenzyme form similar to the one existing in *E. coli* could be obtained.[74] The DNA pol holoenzyme could be reconstituted from two other Gram-positive pathogens, *Streptococcus pyogenes* and *Staphylococcus aureus*, by expressing the separate subunits in *E. coli*.[75,76] It was found that the interaction between DNA polC and the τ subunits of Gram-positive bacteria were not as strong as the contact between the *E. coli* α and τ subunits, explaining perhaps why the DNA polC holoenzyme could not be isolated from *B. subtilis* extracts. Addition of the clamp loader complex and of the β clamp to DNA pol C endows the DNA pol with a high speed (~700 nucleotides/second) and processivity comparable to the one of *E. coli* DNA pol III holoenzyme.

The *pol C* containing Gram-positive bacteria also encode a second DNA pol, DnaE (formerly referred as DNA pol II) which has homology with the *E. coli* DNA pol III α subunit. Like DNA pol C, DnaE is essential for chromosomal replication. *Streptococcus pyogenes* DnaE can bypass damaged DNA to promote cell survival.[77] DNA polC and DnaE are required for the elongation phase of chromosomal replication *in B. subtilis* and it has been suggested that DNA polC works on the leading strand and DnaE on the lagging strand, while in other Gram-positive bacteria DnaE replicates both strands.[78] Thus *B. subtilis* has two different replicative DNA pols, as in eukaryotes, while the *E. coli* holoenzyme replicase is in the form of an asymmetric dimeric DNA pol, suggesting that the need for specialized leading and lagging strand DNA pols could be evolutionarily conserved.[79] DnaE alone can replicate SSB coated DNA at a speed from 60 to 240 nucleotide/s; addition of β clamp increases processivity but not the intrinsic speed, which remains lower than for DNA pol C. DnaE lacks an intrinsic $3' \rightarrow 5'$ exonuclease proofreading activity and the purified protein can bypass many types of DNA damage that generally block other replicative DNA pols, although it cannot replicate lesions that highly distort DNA.[77,80] The bypass is efficient and highly error prone and can be performed either by extension of misaligned 3′ termini or extension of mispaired termini, depending upon the nature of the lesion.[77,80] DnaE expression is SOS inducible; however DnaE overexpression does not increase the rate of spontaneous mutagenesis *in vivo*, whereas overproduction of error prone Y family DNA pols does.[70,80] Altogether, these findings point at a possible role of DnaE in "assisting" mutagenesis *in vivo* by the specialized Y family DNA pols, at least as far as "bulky" DNA lesions are concerned. It should also be noted that these data do not formally contradict the fact that DnaE might function as a replicative DNA pol, but rather suggest that yet undiscovered factors exist that increase its speed, processivity and fidelity in the cell. Finally, the studies on DnaE also underscore fundamental differences in DNA replication between Gram-positive and Gram-negative bacteria. No crystal structure of these DNA pols is currently available.

3.2.3 *Family X: DNA Polymerase X*

Members of the X family of DNA pols have been identified not only in eukaryotes but also in bacteria and Archaea.[81] The *B. subtilis yshC* gene encodes a monomeric, 64-kDa family X DNA pol (DNA pol X *Bs*) that acts on short gaps bearing a 5 phosphate group at the 5′-end of the downstream DNA.[82] The fact that DNA pol X *Bs* is able to conduct gap filling of a single nucleotide gap, allowing further sealing of the resulting nick by a DNA ligase, points to a putative role of this enzyme in BER during *B. subtilis* life cycle. It should be noted that, differently from eukaryotic family X DNA pols but similarly to the *Deinococcus radiodurans* DNA pol X,[81] DNA pol X *Bs* possesses an intrinsic 3′ → 5′ exonuclease activity that resides in its histidinol phosphatase (PHP) domain and which is specialized in resecting unannealed 3′-termini, including tails, in a gapped DNA substrate.[81,83] Taken together the data from *B. subtilis* and *D. radiodurans* suggest that these two X family pols could participate not only to BER but also in a Non-Homologous End Joining system (NHEJ) under circumstances in which the processing of 3′-flap structures is required.

3.2.4 *Family Y: DNA Polymerases Y1 and Y2*

B. subtilis possesses two Y-family members, DNA pol Y1 and DNA pol Y2, encoded by the *yqjH* and *yqjW* genes, respectively.[70] Both DNA pols lack any associated 3′ → 5′ exonuclease activity.

DNA pol Y1 is expressed constitutively at low levels during growth and is not SOS controlled. DNA pol Y1 seems to be involved in "adaptive" mutagenesis[84] and, when overproduced, promotes untargeted mutagenesis in a process requiring its binding to the β clamp. These properties are reminiscent of the *E. coli* DNA pol IV. DNA pol Y2 is expressed only upon induction of the SOS system and is responsible for most of the UV-induced mutagenesis and its overproduction also leads to untargeted mutagenesis. DNA pol Y2 mediated mutagenesis requires RecA and interaction with β clamp. Based on these properties, DNA pol Y2 appears to be the functional homolog of *E. coli* DNA pol V. However, whereas DNA pol V is the heterotrimeric UmuCD$'_2$ complex, *B. subtilis* DNA pol Y2 seems to act as a single polypeptide. Another notable difference with *E. coli* is that TLS by DNA pols Y1 and Y2 is assisted by the *B. subtilis* DNA pol I.[64]

3.3 Other Bacteria

3.3.1 *Mycobacteria*

The genus Mycobacterium includes the major human pathogens *Mycobacterium tuberculosis* and *Mycobacterium leprae*, important opportunistic pathogens such a *Mycobacterium avium-intracellulare* and a wide variety of non-pathogenic saprophytes. *Mycobacterium tuberculosis* (Mtb) is responsible for the largest number of deaths attributable to a single infectious agent. Control strategies for tuberculosis rely heavily on chemotherapy and the development of drug resistance is a serious threat. Error-prone replication induced by the SOS system can contribute to high rate acquisition of drug resistance. Indeed, induction of the SOS response following DNA damage during infection increased the frequency of both targeted and untargeted mutagenesis in intracellular pathogens.[85] Mtb has two B family DNA pols, known as DnaE1 and DnaE2.[86] Deletion of DnaE1 was found to be lethal, indicating that this is the true replicative DNA pol, while deletion of DnaE2 did not sensibly affect survival.[86] Surprisingly, Mtb homologs of SOS genes *dinP* and *umuC*, belonging to the Y family, were found not up-regulated after DNA damage.[87] A recent work[86] has shown that irradiation of *M. tuberculosis* results in increased mutation frequency in the surviving bacteria. It was confirmed that neither *dinP* nor *umuC* homologs were induced by a variety of DNA damaging agents while the gene coding for the DNA pol DnaE2 was up-regulated through an SOS-type response under *LexA* regulation. The UV-induced mutations in Mtb were characteristic of error prone TLS across UV lesions and were strictly dependent upon the presence of DnaE2. Furthermore the authors infected mice with wild type Mbs and measured *DnaE2* levels in various tissues. Enhanced expression of *DnaE2* was detected for several weeks after infection. They also infected mice with a mutant strain and found that the colonies formed were significantly lower that those counted on wild type. Mice infected with wild type strain and then treated with rifampicin yielded an increased fraction of antibiotic-resistant colonies than did animal infected with the DnaE2 mutated strain. Taken together, these data indicate that DnaE2 and not a member of the Y family of error-prone DNA pols, is the primary mediator of survival through inducible mutagenesis and can contribute directly to the emergency of drug resistance *in vivo*. In this respect it is perhaps noteworthy that Mycobacteria seem to lack the highly conserved mismatch repair system.[88] A review on error prone mechanisms of replication, including mycobacteria and archaea has been published.[89]

3.3.2 *Deinococcus radiodurans*

The bacterium *Deinococcus radiodurans* (*D. radiodurans*) is an extremophile organism that survives extreme desiccation and exceptionally high doses of ionizing radiations. Sequence analysis of *D. radiodurans* genome indicated the presence of three DNA pols.[90,91] *D. radiodurans* encodes a single α subunit that is most similar to the one of the protobacterial replicative DNA pols and a separate $3' \rightarrow 5'$ exonuclease that is encoded by a gene ortholog to *E. coli* dnaQ. It also encodes for a *PolA* gene having 35% sequence identity with the *E. coli* DNA pol I. Finally *D. radiodurans* encodes for a DNA pol of the X family (DNA pol X_{DR}). Interestingly no orthologs of *E. coli* DNA pol II or DNA pol V have been found in *D. radiodurans*. While very little is known on the *D. radiodurans* replicative DNA pol, DNA pol A, which has been shown to be necessary for DNA damage survival of the bacterium.[92] Recently it was also found that *D. radiodurans* DNA pol A is necessary for reassembly of shattered chromosomes by a process called "extended synthesis dependent strand annealing" (ESDSA), requiring significant synthesis of new DNA.[93] *D. radiodurans* DNA pol A harbors both $5' \rightarrow 3'$ exonuclease and DNA pol domains. The domain containing the DNA pol activity (DNA polA*) has now been cloned, amplified and purified.[94] The obtained enzyme is able to efficiently catalyze DNA-dependent DNA synthesis requiring Mg^{2+} as divalent metal ion. Additionally, the enzyme shows strand displacement activity. It was further found that DNA pol function of DNA polA* is modulated by the presence of Mn^{2+}, which facilitates replication by the DNA pol of certain DNA damages that occur trough radiation such as oxidative damages. These observations parallel reports that *D. radiodurans* accumulates intracellular Mn^{2+} in cases of irradiation.[95] The DNA pol X of *D. radiodurans* (DNA polX$_{DR}$) has been recently cloned and characterized.[81,96] DNA polX$_{DR}$ is an abundant protein (2000–4000 molecules per cell) and possesses both a DNA pol (DNA pol Xc) and a Histidinol Phosphate (PHP) domain. In addition, it possesses a structure-modulated $3' \rightarrow 5'$ exonuclease activity that resides in the DNA pol Xc domain, but its structure specificity requires also the PHP domain. Mutation of two conserved glycines in the DNA pol Xc domain leads to a specific loss of the structure-modulated exonuclease activity but not of the exonuclease activity in general. Both DNA pol and exonuclease activities are stimulated by Mn^{2+} and DNA polX$_{DR}$ knockout cells show a delay in double-stranded break repair and increased sensitivity to γ-radiation. Noticeably, the wild type protein, an active-site mutant and the two domains were expressed separately in $\Delta PolX_{DR}$ cells. The wild type protein could restore the radiation resistance, whereas the mutant proteins showed a significant negative effect on survival of γ-irradiated cells. Taken together these results suggest that both DNA polA* and DNA polX$_{DR}$ play important roles in double-strand break repair in *D. radiodurans*. DNA polX$_{DR}$ has recently been crystallized. The present crystal structure of DNA polX$_{DR}$ solved at $2.46 \, \text{Å}$ resolution reveals that the enzyme has a novel extended conformation in strong contrast to the closed right hand conformation commonly observed for DNA pols. This extended conformation is stabilized by the C-terminal PHP domain, whose putative nuclease active site is obstructed by its interaction with the DNA pol domain. The overall conformation and the presence of non standard residues in the active site of the DNA pol X domain makes DNA polX$_{DR}$ the founding member of a novel class of DNA pols involved in DNA repair but whose detailed mode of action still remains enigmatic.[97]

3.4 Archaea

Archaea have been considered for a long time close to bacteria but genetic analysis based on genome sequencing has now established that they constitute a distinct domain of life, being a complex mixture of bacterial, eukaryotic and unique features. In this section the *Archaea* DNA pols are classified into families and some of their important characteristics are described.

3.4.1 *Family B: DNA Polymerase B*

Archaea family B DNA pol (*a*DNA pol B) has counterparts in bacteria (e.g. *E. coli* DNA pol II). The biochemical properties of *a*DNA pol B from different species vary but all possess a potent $3' \rightarrow 5'$ exonuclease proofreading activity (for reviews on archaeal replication see Refs. 98 and 99). Many *a*DNA pol B proteins also contain from one to three inteins,[100] which are located in highly conserved regions within the DNA pol catalytic subunit. Another unique feature found in *a*DNA pol

B from *M. thermautotrophicus* and not yet identified in any other DNA pol, is that the catalytic activity is split into two polypetides and the association of both is needed to create an active enzyme.[101] A function of *a*DNA pol B in replication can be inferred by the presence of a proofreading activity and by its stimulation by cognate PCNA and RF-C. Due to the lack of archaeal genetic model it is not known if both *a*DNA pols B and D are necessary for viability nor is known the eventual division of labor at a replication fork. Nevertheless, recent biochemical work from *Pyrococcus abyssi* led to the suggestion that *a*DNA pol B replicates the leading strand while *a*DNA pol *D* acts on the lagging strand.[102] One interesting feature of both *euryarchaeal* and *crenarchaeal* family B DNA pols is their ability to sense uracil ahead of the DNA pol in the template strand, leading to a stall of the DNA pol four nucleotides before the residue. This property is conferred by a small pocket that lies on the N-terminal domain of the DNA pol.[103] The stalling of the *a*DNA pol B presumably signals to the repair machinery to facilitate removal of the uracil base. The crystal structures of three eukaryal B-type DNA pols, two from *Thermococcus* and one from *Desulphurococcus* strains, have recently been solved (for a review see Ref. 104 and references therein). Each of these enzymes has the classic finger-palm-thumb domains, in addition of distinct $3' \rightarrow 5'$ and N-terminal domains, characteristic of other A or B family DNA pols, such as the *E. coli* Klenow enzyme or the bacteriophage RB69 gp43 protein. However, a comparison of the structures of these three archaeal enzymes with the related RB69 gp43DNA pol shows that the archaeal proteins have a more compact design thus providing an insight into how the archeal enzymes are adapted to growth at extreme temperatures.[105–107] A recent structure of the *a*DNA pol B of the archaeon *Sulfolobus solfataricus* indicates the unusual presence of two extra α helices in the N-terminal domain, interacting with the fingers helices to form an extended fingers subdomain.[108] This work also describes how the N-terminal subdomain pocket of archaeal DNA pols could allow specific recognition of deaminated bases such as uracil and hypoxantine in addition to typical DNA bases.

3.4.2 *Family D: DNA Polymerase D*

The archaeal family D DNA pol (*a*DNA pol D) is restricted to euryarchaeal species. The *a*DNA pol D enzyme is a heterodimer composed of a large DP2 and a small DP1 subunits.[109] DP2 is the catalytic subunit while DP1 serves as an accessory factor. The interaction of the two subunits has been reported to be necessary for optimal DNA pol and $3' \rightarrow 5'$ exonuclease activity.[109,110] Accordingly DP1 has been shown to possess an intrinsic proofreading activity.[111] The DP1subunit shows homology to small, non catalytic subunits, of eukaryal DNA pols α (p 70 subunit), DNA pol δ (Cdc27p) and DNA pol ε (p 55 subunit).[112] Similarly to *a*DNA pol B, *a*DNA pol D is also stimulated by PCNA and RF-C.[113] However, the DP1 subunit also directly interacts with RadB, an homolog of the eukaryotic proteins Rad51/Dmc1 in *Pyrococcus furiosus* (*P. furiosus*) suggesting that *a*DNA pol D may participate to recombination and /or repair in addition to its role in replication.[114] The capacity of *a*DNA pol B and *a*DNA pol A of the hyperthermophilic archaea *Pyrococcus abyssi* (*P. abyssi*) (*Pab* DNA pol B and *Pab* DNA pol A) to replicate endogenous frequent DNA lesions such abasic site has been recently investigated. Both DNA pols could incorporate in front of the AP site (mainly dAMP) but only *Pab* DNA pol B could extend, albeit inefficiently, after the lesion.[115] However, both $3' \rightarrow 5'$ exonuclease deficient forms of these DNA pols could perform TLS synthesis, reiterating the importance of the proofreading activity in preventing bypass of lesions not severely distorting the DNA. Consistent with the existing translesional systems and the lack of specialized family Y DNA pols in *P. abyssi*, it can be speculated that one or both *Pab* DNA pols could be involved in damage tolerance. No crystal structure of any of the family D enzymes is currently available.

3.4.3 *Family Y: DNA Polymerases Dbh and Dpo4*

As all other DNA pols of the Y family, archaeal DNA pols Dbh and Dpo4 lack any associated $3' \rightarrow 5'$ exonuclease activity. According to a general property of Y family DNA pols, Dbh was found capable of replicating across an abasic site,[116] while Dpo4 can bypass non-distorting DNA lesions, such as abasic sites,[117] the major oxydative lesion 7,8-dihydro-8-oxo-deoxyguanosine[118] and also "bulkier" lesions such as thymine dimer,[119] the cisplatin intrastrand lesion[120] or the benzo(a)pyrene-dG adduct.[121] One of the first DNA pols of the Y family to be crystallized was the catalytic fragment of the *DinB* (Pol IV) homolog (Dbh) DNA pol from *Sulfolobus solfataricus*.[116] The enzyme was shown to be non processive and able to bypass an abasic site. It was found that the structure of the catalytic site (palm domain) was nearly identical to those of most other DNA pols families. However it appeared that two novel features distinguished the Dbh structure from other DNA pol structures. First, the fingers domain appeared to be in a close conformation, even if not bound to any substrate. Second, Dbh is a minimal enzyme with little contact predicted between the protein and its substrates. These features provided structural insights on the low fidelity, processivity and translesion capacity of the enzyme. Crystallization of another *DinB* from *Sulfolobus solfataricus*, Dpo4, extended our understanding of the molecular basis for the Y-family's low fidelity DNA synthesis on undamaged DNA templates and their ability to traverse normally replication-blocking lesions in DNA.[122] Dpo4 was crystallized in ternary complexes with DNA and incoming

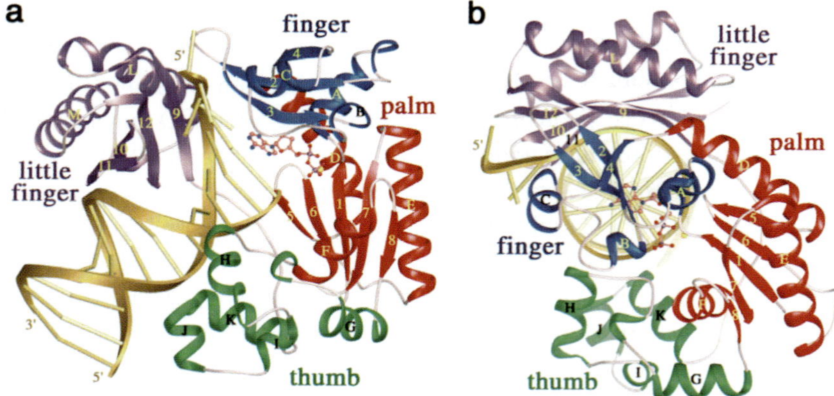

Figure 3.5   Crystal Structure of the Dpo4 Ternary Complex. Both the protein and DNA backbones are shown in ribbon diagrams, DNA bases as rods, and the incoming nucleotide and the Ca^{2+} ion in a ball-and-stick model. The Dpo4 structural domains are shown in red (palm), green (thumb), blue (finger), and purple (little finger); the DNA is in gold. Secondary structures are named in alphabetic or numeric order for helices and strands, respectively, following their order in the primary sequence. (a) A view looking into the active site with the palm, finger, thumb, and little finger domains well separated. (b) A view down the DNA-helical axis. (Reproduced with the permission from Ref. 122.)

nucleotide, either correct or incorrect. The properties of making limited and non specific contacts with replicating base pairs observed with Dbh were confirmed. However, Dpo4 also contained an additional fourth domain at the C terminus which was not included in the Dbh catalytic core structure. This domain was tethered to the thumb domain in sequence but physically located next to the finger domain and interacts with DNA in the major grove. It was thereafter referred as "little finger" (LF) domain (**Figure 3.5**). Finally, Dpo4 was also captured in the crystal translocating two templated bases to the active site at once, suggesting a possible mechanism for bypassing of thymine dimers. Although originating from two closely related strains, Dbh and Dpo4 pols exhibit different properties *in vitro*. For instance both can bypass a variety of DNA lesions (see above and Ref. 123) but Dbh does it less efficiently. When replicating undamaged DNA, Dpo4 is prone to make base pair substitutions, whereas Dbh predominantly makes deletions.

To investigate the role of the LF domain in the fidelity and lesion bypass ability of Dph and Dpo4, chimeras of the pols have been generated in which their LF domains have been interchanged.[123] It was found that, by replacing the LF domain of Dbh with that of Dpo4, the enzymatic properties of the chimeric enzyme were now more Dpo4 like. Conversely, the Dpo4-LF-Dbh chimera was more Dbh like. Therefore this study indicated that the unique but variable LF domain of the two *Sulfolobus solfataricus* DNA pols played a major role in determining the enzymatic properties of each individual enzyme. Many DNA pols are coordinated by sliding clamps (β-clamp/PCNA) in TLS. The sliding clamp of *Sulfolobus solfataricus* P2 is a heterotrimer of three distinct subunits (PCNA 1, 2 and 3). The PCNA heterotrimer, but not individual subunits, stimulates the activity of DNA pol, DNA ligase I and of the flap endonuclease (FEN1). However, distinct PCNA subunits contact DNA pol, DNA ligase or FEN1, imposing a defined structure at the replication fork.[124] Additionally, unique subunit-specific interactions between components of the clamp loader, RF-C, suggest a model for clamp loading of PCNA. Very recently, the crystal structure of Dpo4 in complex with the heterodimeric *Sulfolobus solfataricus* PCNA1-PCNA2 has been reported.[125] Dpo4 exhibits an extended conformation that differs from the Dpo4 structures in apo-or DNA-bound form and appears to bind, via its C-terminal PIP-box to the conserved PIP binding site of PCNA1. Based on the results in Ref. 125 and from the known properties of the Y-family DNA pols, a model is proposed suggesting that the conformational flexibility of Dpo4 would turn it off when replicative more efficient DNA pols work on undamaged DNA and turns it on onto DNA templates when DNA synthesis is stalled at lesions. For a general view on the structures/properties of the Y-family DNA pols the reader can refer to several reviews.[126–128]

3.5 Chapter Summary

In this chapter we have summarized the main functional and structural aspects of the prokaryotic and archaea DNA pols families (**Table 3.1**). From *E. coli* to *Archaea* these enzymes display both common and different features. All the crystallized proteins show the usual right hand structure, although notably differences exist in the relative importance of the fingers palm and thumb structures within each family. One example is the Y family TLS DNA pols, as exemplified by the Dpo4 enzyme, where the fingers are small and stubby compared to the structures of DNA pols of other families and a fourth domain, called "little finger" exists. In addition Dpo4 can also capture two templated bases in the active site at once, thus providing a structural explanation for the capacity of the Y family DNA pols to replicate across bi-functional lesions. All the replicative DNA pols, belonging mainly to the family C, appear to be multi-subunit enzymes possessing an associated $3' \rightarrow 5'$ proofreading exonuclease activity (with the exception of *B. subtilis* DnaE). The $3' \rightarrow 5'$ exonuclease is also associated with DNA pols of the B and D families and can even be DNA structure specific, as in the case of the family X *D. radiodurans* DNA pol X. Sliding clamps, charged at the DNA primer-template junction by clamp loader, confer high processivity to replicative DNA pols. However, sliding clamps can also participate in TLS by families Y or B DNA pols. Finally, family A DNA pols also contain a $5' \rightarrow 3'$ exonuclease activity that seems to be implicated in the repair processes and removal of Okazaki fragments during replication.

References

1. Bessman MJ, Kornberg A, Lehman IR, Simms ES. 1956. *Biochim Biophys Acta* 21: 197–8.
2. Kornberg A, Baker T. 1992. *DNA Replication, 2nd edition.* New York: W.H.Freeman.
3. Patel PH, Suzuki M, Adman E, Shinkai A, Loeb LA. 2001. *J Mol Biol* 308: 823–37.
4. Ollis DL, Brick P, Hamlin R, Xuong NG, Steitz TA. 1985. *Nature* 313: 762–6.
5. Steitz TA. 1998. *Nature* 391: 231–2.
6. Doublie S, Sawaya MR, Ellenberger T. 1999. *Structure* 7: R31–5.
7. Kunkel TA, Bebenek K. 2000. *Annu Rev Biochem* 69: 497–529.
8. Florian J, Goodman MF, Warshel A. 2005. *Proc Natl Acad Sci U S A* 102: 6819–24. Epub 2005 Apr 29.
9. Tsai YC, Johnson KA. 2006. *Biochemistry* 45: 9675–87.
10. Friedberg EC WG, Siede W, Wood RD, Schultz RA, Ellenberger T. 2006. *DNA Repair and Mutagenesis, 2nd edition.* Washington: ASM Press.
11. Dianov G, Lindahl T. 1994. *Curr Biol* 4: 1069–76.
12. Cooper PK, Hanawalt PC. 1972. *Proc Natl Acad Sci U S A* 69: 1156–60.
13. De Lucia P, Cairns J. 1969. *Nature* 224: 1164–6.
14. Konrad EB, Lehman IR. 1974. *Proc Natl Acad Sci U S A* 71: 2048–51.
15. Joyce CM, Grindley ND. 1984. *J Bacteriol* 158: 636–43.
16. Markovitz A. 2005. *Mol Microbiol* 55: 1867–82.
17. Bonner CA, Randall SK, Rayssiguier C, Radman M, Eritja R, *et al.* 1988. *J Biol Chem* 263: 18946–52.
18. Rangarajan S, Woodgate R, Goodman MF. 1999. *Proc Natl Acad Sci U S A* 96: 9224–9.
19. Sommer S, Boudsocq F, Devoret R, Bailone A. 1998. *Mol Microbiol* 28: 281–91.
20. Courcelle J, Carswell-Crumpton C, Hanawalt PC. 1997. *Proc Natl Acad Sci U S A* 94: 3714–9.
21. Goodman MF. 2002. *Annu Rev Biochem* 71: 17–50.
22. Postow L, Ullsperger C, Keller RW, Bustamante C, Vologodskii AV, Cozzarelli NR. 2001. *J Biol Chem* 276: 2790–6.
23. Bonner CA, Stukenberg PT, Rajagopalan M, Eritja R, O'Donnell M, *et al.* 1992. *J Biol Chem* 267: 11431–8.
24. Hughes AJ, Jr., Bryan SK, Chen H, Moses RE, McHenry CS. 1991. *J Biol Chem* 266: 4568–73.
25. Napolitano R, Janel-Bintz R, Wagner J, Fuchs RP. 2000. *Embo J* 19: 6259–65.
26. McHenry CS. 1988. *Annu Rev Biochem* 57: 519–50.
27. Marians KJ. 1992. *Annu Rev Biochem* 61: 673–719.
28. Kelman Z, O'Donnell M. 1995. *Annu Rev Biochem* 64: 171–200.
29. Johnson A, O'Donnell M. 2005. *Annu Rev Biochem* 74: 283–315.
30. Kornberg T, Gefter ML. 1972. *J Biol Chem* 247: 5369–75.
31. Studwell PS, O'Donnell M. 1990. *J Biol Chem* 265: 1171–8.
32. McInerney P, Johnson A, Katz F, O'Donnell M. 2007. *Mol Cell* 27: 527–38.
33. Lamers MH, Georgescu RE, Lee SG, O'Donnell M, Kuriyan J. 2006. *Cell* 126: 881–92.
34. Kirby TW, Harvey S, DeRose EF, Chalov S, Chikova AK, *et al.* 2006. *J Biol Chem* 281: 38466–71.
35. Kong XP, Onrust R, O'Donnell M, Kuriyan J. 1992. *Cell* 69: 425–37.
36. Stewart J, Hingorani MM, Kelman Z, O'Donnell M. 2001. *J Biol Chem* 276: 19182–9.
37. Georgescu RE, Kim SS, Yurieva O, Kuriyan J, Kong XP, O'Donnell M. 2008. *Cell* 132: 43–54.
38. Neuwald AF, Aravind L, Spouge JL, Koonin EV. 1999. *Genome Res* 9: 27–43.
39. Jeruzalmi D, O'Donnell M, Kuriyan J. 2001. *Cell* 106: 429–41.
40. O'Donnell M, Kuriyan J. 2006. *Curr Opin Struct Biol* 16: 35–41.
41. Lahue RS, Au KG, Modrich P. 1989. *Science* 245: 160–4.

42. Lopez de Saro FJ, Marinus MG, Modrich P, O'Donnell M. 2006. *J Biol Chem* 281: 14340–9.
43. Fuchs RP, Fujii S, Wagner J. 2004. *Adv Protein Chem* 69: 229–64.
44. Wagner J, Gruz P, Kim SR, Yamada M, Matsui K, *et al.* 1999. *Mol Cell* 4: 281–6.
45. Kim SR, Matsui K, Yamada M, Gruz P, Nohmi T. 2001. *Mol Genet Genomics* 266: 207–15.
46. Tang M, Pham P, Shen X, Taylor JS, O'Donnell M, *et al.* 2000. *Nature* 404: 1014–8.
47. Wagner J, Nohmi T. 2000. *J Bacteriol* 182: 4587–95.
48. McKenzie GJ, Lee PL, Lombardo MJ, Hastings PJ, Rosenberg SM. 2001. *Mol Cell* 7: 571–9.
49. Tang M, Shen X, Frank EG, O'Donnell M, Woodgate R, Goodman MF. 1999. *Proc Natl Acad Sci U S A* 96: 8919–24.
50. Reuven NB, Arad G, Maor-Shoshani A, Livneh Z. 1999. *J Biol Chem* 274: 31763–6.
51. Woodgate R, Rajagopalan M, Lu C, Echols H. 1989. *Proc Natl Acad Sci U S A* 86: 7301–5.
52. Fujii S, Fuchs RP. 2004. *Embo J* 23: 4342–52.
53. Tang M, Bruck I, Eritja R, Turner J, Frank EG, *et al.* 1998. *Proc Natl Acad Sci U S A* 95: 9755–60.
54. Friedberg EC, Lehmann AR, Fuchs RP. 2005. *Mol Cell* 18: 499–505.
55. Schlacher K, Goodman MF. 2007. *Nat Rev Mol Cell Biol* 8: 587–94.
56. Jiang Q, Karata K, Woodgate R, Cox MM, Goodman MF. 2009. *Nature* 460: 359–63.
57. Pages V, Fuchs RP. 2003. *Science* 300: 1300–3.
58. Fujii S, Fuchs RP. 2009. *Proc Natl Acad Sci U S A* 106: 14825–30.
59. Indiani C, McInerney P, Georgescu R, Goodman MF, O'Donnell M. 2005. *Mol Cell* 19: 805–15.
60. Fujii S, Fuchs RP. 2007. *J Mol Biol* 372: 883–93.
61. Indiani C, Langston LD, Yurieva O, Goodman MF, O'Donnell M. 2009. *Proc Natl Acad Sci U S A* 106: 6031–8.
62. Bunting KA, Roe SM, Pearl LH. 2003. *Embo J* 22: 5883–92.
63. Pages V, Fuchs RP. 2002. *Oncogene* 21: 8957–66.
64. Noirot P, Polard P, Noirot-Gros MF. 2007. In *Bacillus Cellular and Molecular Biology.*, ed. PL Graumann (ed), pp. 1–42. Norfolk, U.K.: Caister Academic Press.
65. Okazaki T, Kornberg A. 1964. *J Biol Chem* 239: 259–68.
66. Tamanoi F, Okazaki T, Okazaki R. 1977. *Biochem Biophys Res Commun* 77: 290–7.
67. Gass KB, Cozzarelli NR. 1973. *J Biol Chem* 248: 7688–700.
68. Villani G, Boiteux S, Radman M. 1978. *Proc Natl Acad Sci U S A* 75: 3037–41.
69. Khare V, Eckert KA. 2002. *Mutat Res* 510: 45–54.
70. Duigou S, Ehrlich SD, Noirot P, Noirot-Gros MF. 2005. *Mol Microbiol* 57: 678–90.
71. Kiefer JR, Mao C, Hansen CJ, Basehore SL, Hogrefe HH, *et al.* 1997. *Structure* 5: 95–108.
72. Sanjanwala B, Ganesan AT. 1991. *Mol Gen Genet* 226: 467–72.
73. Barnes MH, Hammond RA, Kennedy CC, Mack SL, Brown NC. 1992. *Gene* 111: 43–9.
74. Low RL, Rashbaum SA, Cozzarelli NR. 1976. *J Biol Chem* 251: 1311–25.
75. Bruck I, O'Donnell M. 2000. *J Biol Chem* 275: 28971–83.
76. Bruck I, Georgescu RE, O'Donnell M. 2005. *J Biol Chem* 280: 18152–62.
77. Bruck I, Goodman MF, O'Donnell M. 2003. *J Biol Chem* 278: 44361–8.
78. Dervyn E, Suski C, Daniel R, Bruand C, Chapuis J, *et al.* 2001. *Science* 294: 1716–9.
79. McHenry CS. 2003. *Mol Microbiol* 49: 1157–65.
80. Le Chatelier E, Becherel OJ, d'Alencon E, Canceill D, Ehrlich SD, *et al.* 2004. *J Biol Chem* 279: 1757–67.
81. Blasius M, Shevelev I, Jolivet E, Sommer S, Hubscher U. 2006. *Mol Microbiol* 60: 165–76.
82. Banos B, Lazaro JM, Villar L, Salas M, de Vega M. 2008. *J Mol Biol* 384: 1019–28.
83. Banos B, Lazaro JM, Villar L, Salas M, de Vega M. 2008. *Nucleic Acids Res* 36: 5736–49.
84. Sung HM, Yeamans G, Ross CA, Yasbin RE. 2003. *J Bacteriol* 185: 2153–60.

85. Schlosser-Silverman E, Elgrably-Weiss M, Rosenshine I, Kohen R, Altuvia S. 2000. *J Bacteriol* 182: 5225–30.

86. Boshoff HI, Reed MB, Barry CE, 3rd, Mizrahi V. 2003. *Cell* 113: 183–93.

87. Brooks PC, Movahedzadeh F, Davis EO. 2001. *J Bacteriol* 183: 4459–67.

88. Springer B, Sander P, Sedlacek L, Hardt WD, Mizrahi V, *et al.* 2004. *Mol Microbiol* 53: 1601–9.

89. Tippin B, Pham P, Goodman MF. 2004. *Trends Microbiol* 12: 288–95.

90. Makarova KS, Aravind L, Wolf YI, Tatusov RL, Minton KW, *et al.* 2001. *Microbiol Mol Biol Rev* 65: 44–79.

91. Blasius M, Sommer S, Hubscher U. 2008. *Crit Rev Biochem Mol Biol* 43: 221–38.

92. Gutman PD, Fuchs P, Ouyang L, Minton KW. 1993. *J Bacteriol* 175: 3581–90.

93. Zahradka K, Slade D, Bailone A, Sommer S, Averbeck D, *et al.* 2006. *Nature* 443: 569–73.

94. Heinz K, Marx A. 2007. *J Biol Chem* 282: 10908–14.

95. Daly MJ, Gaidamakova EK, Matrosova VY, Vasilenko A, Zhai M, *et al.* 2004. *Science* 306: 1025–8.

96. Lecointe F, Shevelev IV, Bailone A, Sommer S, Hubscher U. 2004. *Mol Microbiol* 53: 1721–30.

97. Leulliot N, Cladiere L, Lecointe F, Durand D, Hubscher U, van Tilbeurgh H. 2009. *J Biol Chem* 284: 11992-9. Epub 2009 Feb 26.

98. Grabowski B, Kelman Z. 2003. *Annu Rev Microbiol* 57: 487–516.

99. Barry ER, Bell SD. 2006. *Microbiol Mol Biol Rev* 70: 876–87.

100. MacNeill SA. 2009. *Biochem Soc Trans* 37: 108–13.

101. Kelman Z, Pietrokovski S, Hurwitz J. 1999. *J Biol Chem* 274: 28751–61.

102. Henneke G, Flament D, Hubscher U, Querellou J, Raffin JP. 2005. *J Mol Biol* 350: 53–64.

103. Fogg MJ, Pearl LH, Connolly BA. 2002. *Nat Struct Biol* 9: 922–7.

104. MacNeill SA. 2001. *Mol Microbiol* 40: 520–9.

105. Hopfner KP, Eichinger A, Engh RA, Laue F, Ankenbauer W, *et al.* 1999. *Proc Natl Acad Sci U S A* 96: 3600–5.

106. Zhao Y, Jeruzalmi D, Moarefi I, Leighton L, Lasken R, Kuriyan J. 1999. *Structure* 7: 1189–99.

107. Rodriguez AC, Park HW, Mao C, Beese LS. 2000. *J Mol Biol* 299: 447–62.

108. Savino C, Federici L, Johnson KA, Vallone B, Nastopoulos V, *et al.* 2004. *Structure* 12: 2001–8.

109. Cann IK, Komori K, Toh H, Kanai S, Ishino Y. 1998. *Proc Natl Acad Sci U S A* 95: 14250–5.

110. Uemori T, Sato Y, Kato I, Doi H, Ishino Y. 1997. *Genes Cells* 2: 499–512.

111. Jokela M, Eskelinen A, Pospiech H, Rouvinen J, Syvaoja JE. 2004. *Nucleic Acids Res* 32: 2430–40.

112. Makiniemi M, Pospiech H, Kilpelainen S, Jokela M, Vihinen M, Syvaoja JE. 1999. *Trends Biochem Sci* 24: 14–6.

113. Cann IK, Ishino S, Hayashi I, Komori K, Toh H, *et al.* 1999. *J Bacteriol* 181: 6591–9.

114. Hayashi I, Morikawa K, Ishino Y. 1999. *Nucleic Acids Res* 27: 4695–702.

115. Palud A, Villani G, L'Haridon S, Querellou J, Raffin JP, Henneke G. 2008. *Mol Microbiol* 70: 746–61.

116. Zhou BL, Pata JD, Steitz TA. 2001. *Mol Cell* 8: 427–37.

117. Kokoska RJ, McCulloch SD, Kunkel TA. 2003. *J Biol Chem* 278: 50537–45.

118. Eoff RL, Irimia A, Angel KC, Egli M, Guengerich FP. 2007. *J Biol Chem* 282: 19831–43.

119. Boudsocq F, Iwai S, Hanaoka F, Woodgate R. 2001. *Nucleic Acids Res* 29: 4607–16.

120. Brown JA, Newmister SA, Fiala KA, Suo Z. 2008. *Nucleic Acids Res* 36: 3867–78.

121. Bauer J, Xing G, Yagi H, Sayer JM, Jerina DM, Ling H. 2007. *Proc Natl Acad Sci U S A* 104: 14905–10.

122. Ling H, Boudsocq F, Woodgate R, Yang W. 2001. *Cell* 107: 91–102.

123. Boudsocq F, Kokoska RJ, Plosky BS, Vaisman A, Ling H, *et al.* 2004. *J Biol Chem* 279: 32932–40.
124. Dionne I, Nookala RK, Jackson SP, Doherty AJ, Bell SD. 2003. *Mol Cell* 11: 275–82.
125. Xing G, Kirouac K, Shin YJ, Bell SD, Ling H. 2009. *Mol Microbiol* 71: 678–91.
126. Beard WA, Wilson SH. 2001. *Structure* 9: 759–64.
127. Yang W. 2005. *FEBS Lett* 579: 868–72.
128. Lehmann AR. 2006. *Mol Cell* 24: 493–5.

CHAPTER 4

Structural and Functional Aspects
of the Eukaryotic DNA Polymerase Families

4.1 The High Number of Specialized Pathways in Eukaryotic Cells Requires a Plethora of Specialized DNA Synthesizing Enzymes

Based on sequence homology and structural similarities, DNA polymerases (pols) have been grouped in seven different families: A, B, C, D, X, Y and reverse transcriptase (RT).[1] The evolution of DNA pols from prokaryotic organisms to metazoan, can be correlated to the increasing levels of complexity, from single-cell organisms to the highly organized hierarchical development of tissues and organs during embryogenesis occurring in vertebrates and plants. Very likely the replicative DNA pols of prokaryotes and eukaryotes evolved from two independent lineages.[2] Eukaryotic-type B family enzymes probably evolved from ancient RNA-dependent RNA polymerases and are present in both Archaea and eukaryotes. Prokaryotic-type C family enzymes, instead, evolved later in the history of life, likely from a template-independent nucleotidyl transferase such as poly(A) polymerase. However, regardless of their phylogenetic relationships, the universally conserved function of DNA pols among all DNA-based life forms (i.e. all cellular-based organisms and several viruses) is their ability to copy the genetic information stored in the double-helix, both for its duplication (DNA replication) or for the maintenance of its integrity (DNA repair). As a consequence, a minimal set of replicative and repair DNA pols is present in all eukaryotic forms (**Table 4.1**).

The transition from prokaryotic to eukaryotic cells, marked by the inner compartmentalization of cellular functions through the development of membrane-limited subcellular compartments (nucleus) or the endosymbiotic acquisition of organelles (mitochondria, plastids), brought up the need of additional specialized functions. For example, the development of a cellular cycle, made of temporally and spatially defined different phases, required the confinement of the DNA replication and repair machineries within the appropriate phases as well as their integration in the cell cycle regulatory circuits.[3–8]

Table 4.1   DNA polymerases in diverse eukaryotic organisms

DNA polymerase					
Saccaromyces cerevisiae	*Arabidopsis thaliana*	*Homo sapiens*	Subunits[a]	Family	Main Function
DNA pol α/ primase	DNA pol α/ primase	DNA pol α/ primase	p180 p68 p55(Pri2) p48(Pri1)	B	DNA replication (initiation)
DNA pol δ	DNA pol δ	DNA pol δ	p125 p66 p50 p12[b]	B	DNA replication (lagging strand)
DNA pol ε	DNA pol ε	DNA pol ε	p261 p59 p17 p12	B	DNA replication (leading strand)
DNA pol ζ	DNA pol ζ	DNA pol ζ	p350(Rev3L) p24(Rev7)	B	TLS[c] (extender)
DNA pol γ	DNA pol γ-like	DNA pol γ	p140 p55(dimer)	A	Mitochondria/ plastid DNA replication
—	DNA pol θ	DNA pol θ	p250	A	DNA repair
—	—	DNA pol υ	p100	A	DNA repair
DNA pol IV	DNA pol λ	DNA pol λ	p70	X	DNA repair/ V(D)J recombination
—	—	DNA pol β	p39	X	DNA repair
—	—	DNA pol μ	p55[d]	X	DNA repair/ V(D)J recombination
—	—	TdT	p58	X	DNA repair/ V(D)J recombination
DNA pol η	DNA pol η	DNA pol η	p79	Y	TLS (inserter)
Rev1	Rev1	Rev1	p138	Y	TLS (inserter)
—	—	DNA pol ι	p80	Y	TLS (inserter)
—	DNA pol κ	DNA pol κ	p98	Y	TLS (inserter/ extender)

a: The molecular weights refer to the human proteins. b: the orthologue of p12 has not been found yet in *S. cerevisiae*. c: TLS, translesion DNA synthesis. d: p55 forms a homodimer.

The organelles of endosymbiotic origins such as mitochondria and plastids,[9] possess their own DNA and replicative machineries, distinct from those responsible for the replication of nuclear DNA, even though encoded by the nuclear genome.[10,11] Thus, as a consequence of both their larger genomes and their more complex regulation eukaryotic cells possess a higher number of replicative and repair DNA pols (including organelle-specific enzymes) than prokaryotes, as a consequence of both their larger genomes and their more complex regulation (**Table 4.1**). An example of this diversification is the structure of the replisome. As detailed in Chapter 3, at the prokaryotic DNA replication fork a dimer of DNA pol III holoenzyme performs both leading and lagging strand DNA synthesis, whereas RNA priming is made by the primase activity. The eukaryotic replication fork, on the other hand, employs three different DNA pols[12–17]: DNA pol α/primase initiates DNA synthesis on both strands, DNA pol ε synthesizes the leading strand, whereas DNA pol δ works at the lagging strand. RNA priming is catalyzed by a heterodimeric DNA primase which is an intrinsic component of the DNA pol α holoenzyme (See also Chapter 2, **Figure 2.4**). Such a division of labor at the eukaryotic DNA replication fork allows for several layers of regulation, being each DNA pol a sensor and a transducer of signals for the cell cycle regulatory circuits.

The transition from unicellular eukaryotes (such as yeast) to metazoan is reflected by a further specialization of DNA pol functions. The requirement for tissue- or even cell-specific events, hierarchically organized within a rigid temporal program during embryogenesis, as well as the division of labor among different tissues and organs in the adult organism, determined the need of highly regulated and specialized enzymes. As a consequence, all the four DNA pol families typical of eukaryotic cells (A, B, X and Y), were expanded in metazoan through a series of duplication/diversification events, leading to the acquisition of new members with specialized functions (**Table 4.1**). An example of such expansion is the V(D)J recombination process, an essential component of the immune system. In vertebrate cells, at least three DNA pols from the family X, DNA pol λ, DNA pol μ and Terminal-deoxynucleotidyl Transferase (TdT), are involved in V(D)J recombination at specific steps of the immunoglobulin repertoire development.[18–20] In all other eukaryotes, including multicellular organisms such as plants, with tissues and organs but without a vertebrate-like immune system, only one member of family X (the ortholog of DNA pol λ), is present (**Table 4.1**). This suggests that in vertebrate cells, the X family of DNA pols underwent a large expansion as a consequence of immune system development.

4.2 Eukaryotic DNA Polymerase Structure: The "Right Hand" of the Cell

4.2.1 *Common Features*

Albeit DNA pols from different families are quite dissimilar in terms of their primary sequence, they fold into a conformation resembling a human right hand

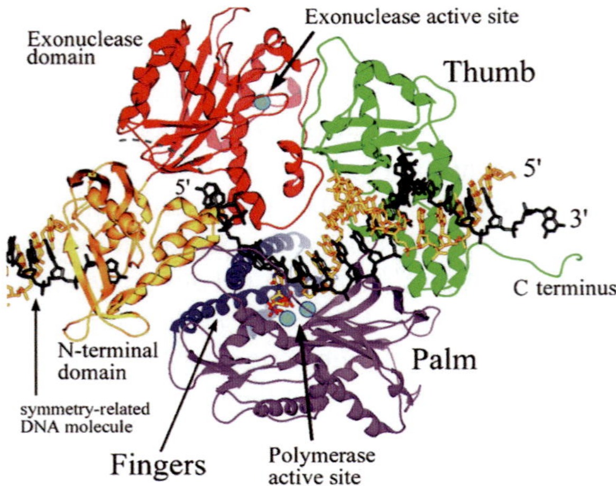

Figure 4.1　Structure of the B family DNA polymerase from bacteriophage RB69. The different domains are highlighted in different colors. The right-hand analogy includes the fingers, palm and thumb subdomain, while the exonuclease domain is positioned on top of the palm with its catalytic site facing the polymerase active site. (Reproduced from Ref. 27 with permission.)

composed of three distinct domains designated as palm, thumb, and fingers[1,21] (see also Chapter 2). The palm subdomain, which contains the catalytically important residues, can be superimposed among members of the A, B, Y and RT families. This subdomain consists of four- to six-stranded β-sheets flanked by two α-helices. Comparison with the structure of the bacteriophage RB69 DNA pol (**Figure 4.1**) showed that the most conserved residues are located 10 Å from the active site, into three regions that emanate from the palm, the fingers, and the thumb and converge at the catalytic site, forming a continuous conserved surface.

Unlike the palm, the other two subdomains, thumb and fingers, are unrelated among the different families, but they function similarly using analogous secondary structural elements. The fingers are involved in correctly positioning the template and the incoming complementary dNTP, and the thumb is important in DNA binding and processivity. All DNA pols catalyze the $5' \rightarrow 3'$ addition of nucleotides in a template-dependent manner, following an universally conserved reaction mechanism.[22] The phosphoryl transfer reaction is catalyzed by a two-metal-ion mechanism. Two Mg^{2+} ions form a pentacoordinated transition state with the phosphate groups of the incoming nucleotide, through interaction with conserved carboxylate residues in region I and region II. Another common feature of all DNA pols is the obligatory sequential ordered reaction mechanism, whereby the DNA substrate binds first, followed by the dNTP. The sequential association of each substrate with the enzyme is accompanied by a rearrangement of the relative positions of regions within the structural domains. Of particular significance is the concerted movement of the finger subdomains that rotate toward the palm to switch from an "open" to a "closed" conformation, forming the binding pocket for the incoming dNTP. This transition sets the enzyme in its catalytically competent state, properly aligning the active site's carboxylate residues and the α-phosphate of the incoming dNTP for nucleophilic attack and phosphodiester bond formation. In most DNA pols, the rate of the open-to-close transition is slower than the rate of chemical bond formation, thus representing the limiting step of the overall DNA pol reaction.

4.2.2 *Specific Features of the Different Families*

Family B. As listed in **Table 4.1**, the eukaryotic replicative enzymes DNA pols α, δ and ε consist each of multiple subunits, with the largest one being the one responsible for the catalytic activity. Both from sequence and structural analysis, it appears that the catalytic subunits of eukaryotic DNA pols are composed of a central domain that is evolutionarily very conserved.[1] On the basis of the DNA pol α sequence, six highly similar regions termed I–VI have been identified, whose relative position along the primary sequence is conserved. From the N- to the C-terminus the order is: IV, II, VI, III, I, and V. The highly conserved catalytic site of family B DNA pols is located in the palm, and contains three invariant aspartic residues, two in region I (-**YGDTDS**- motif) and one in region II (-**D**xxSLYPS- motif). The -SLYPS- motif is also important for deoxyribonucleoside triphosphate (dNTP) binding. These motifs are absolutely conserved in DNA pol α and δ subfamilies, whereas DNA pol ε has considerably diverged from the consensus, so that the region I motif became -ELDTDG- and region II has become -**D**xxAMYPN-. The large subunits of DNA pol α, δ, and ε are representative of three distinct subfamilies within the B-type DNA pols. DNA pol δ is the most conserved, with 90% identity between human and mouse and 49% between human and yeast. In comparison, the identity between human and yeast DNA pol α and DNA pol ε is 35% and 39%, respectively. The N-terminal part of DNA pol δ (amino acid residues 1–305), albeit generally poorly conserved, contains three regions of high homology: a nuclear targeting signal (NTS) and the so-called NT-1 and NT-2 regions. The C-terminal part (amino acids 850–1105) is more conserved, with three nearly identical regions termed CT-1 to 3 and a zinc-finger domain (ZnF2), which is 89% identical between human and yeast DNA pol δ. C-terminal zinc-finger domains are also present in α- and ε-like DNA pols. Both the N- and C-terminal parts of DNA pol ε show about 25% identity between human and yeast. However, the C-terminal region of DNA pol ε contains about 1000 extra amino acids found only in the DNA pol ε subfamily members, characterized by an highly acidic region (residues 1918–1948) and a zinc-finger domain (residues 2125–2222).

The DNA pols δ and ε catalytic subunits both contain a $3' \rightarrow 5'$ proofreading exonuclease domain at their N terminus. Recently, the structure of the catalytic subunit of yeast DNA pol δ has been solved[23] and found similar to the other known crystal structures of exonuclease proficient family B members[24–28] (**Figure 4.1**). In these structures, the exonuclease domain is folded around a central β-sheet that contains the active site and, together with the palm domain, creates a ring-shaped structure with a central hole, where the template/primer duplex DNA is positioned.

The catalytic mechanism of the exonuclease activity involves two metal ion phosphoryl transfer, analogous to the one responsible for polymerization. The $3' \rightarrow 5'$ exonuclease activity allows the removal of misincorporated nucleotides, ensuring the high fidelity of DNA synthesis required for faithful genome replication. During DNA synthesis, DNA pols δ and ε repetitively shuttle between a polymerizing and an editing mode. A mismatched base pair prevents the fingers from rotating toward the palm to bind the incoming dNTP. This leaves the 3′ mismatched end available for binding to the exonuclease active site, which removes the wrong nucleotide. During the switch between polymerizing and editing modes, the DNA moves toward the exonuclease active site with a rotation in the double-helix axis. This movement is aided by the tip of the thumb subdomain, which holds contact with the DNA during the movement, guiding it on a path between the two sites. Thus, the balance between polymerase and exonuclease is regulated by a competition for the 3′ end of the primer between the respective active sites. Recently, the low-resolution structure of the DNA pol ε holoenzyme from *S. cerevisiae* has been solved by cryo-electron microscopy.[29] This study revealed an extended shape for the DNA pol ε holoenzyme, where the double stranded region of the DNA template is accomodated in a large cleft residing in the "tail" constituted by the three small subunits, whereas the single-stranded region and the primer/template junction is bent while protruding into the active site. The tail seems to be connected to the catalytic subunit through a flexible link, which might help to keep the DNA in the right orientation.

Family X. The eukaryotic family X is composed of four members: DNA pols λ, β, μ and TdT.[1,30] They all possess a C-terminal β-like domain, comprising the catalytic site. In addition, all members except DNA pol β contain an extended N-terminus, with BRCT-, NLS- and Proline-rich domains. Crystal structures are available for all members of this family, giving plenty of information on their structure-function relationships.[31] The first X-ray crystal structure of the ternary complex of a DNA pol with its DNA substrate and an incoming nucleotide bound in the active site of a family X DNA pol was solved for mammalian DNA pol β (**Figure 4.2**).[32]

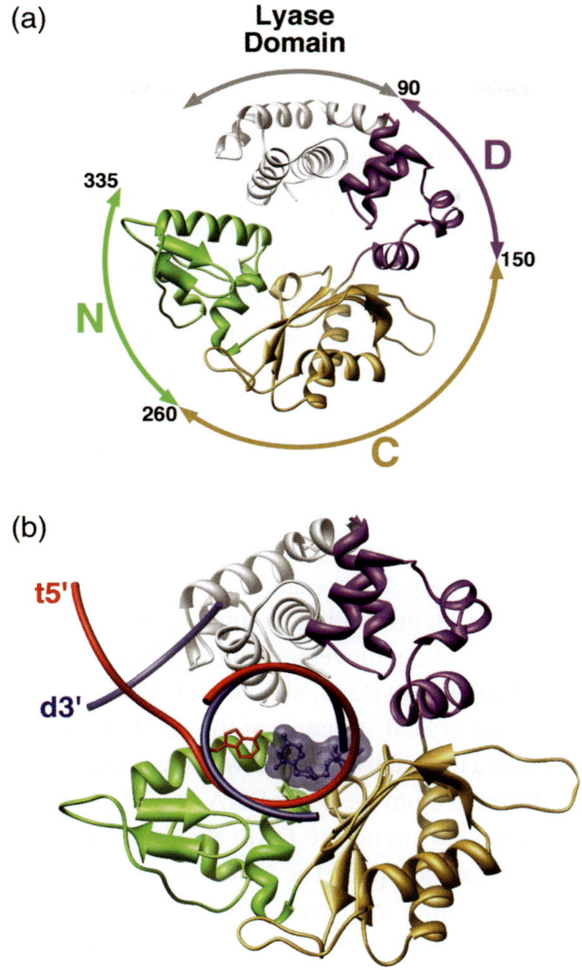

Figure 4.2   Structure of mammalian DNA polymerase β. a. Shown are the DNA pol (colored) and the N-terminal lyase (grey) domains. D, C and N refers to the polymerase subdomains. Numbers indicate the amino-acidic residues for the subdomain boundaries. **b.** Top view of DNA pol β bound to a one-nucleotide gapped DNA and an incoming dNTP. Red is the template strand, solid blue is the primer strand, shaded blue is the dNTP. The template (t) and downstream (d) ends are indicated (t5′ and d3′). (Reproduced from Ref. 32 with permission.)

Although the topology of the palm subdomains of family X DNA pols is not homologous to those of other families, the geometry of the active site including the position of the substrates (metals, dNTP, and DNA) can be functionally aligned. Three conserved aspartic acid residues (D190, D192, and D256) in the palm subdomain of DNA pol β define the active site and coordinate two Mg^{2+} ions necessary for catalysis. One metal ion is coordinated by the oxygen atoms from all three phosphates of the incoming nucleotide, and by two aspartate residues. The other metal ion interacts with the α-phosphate, the three aspartate residues, and the 3′-OH of the terminal deoxyribose sugar of the primer strand. The catalytic mechanism proceeds via in-line displacement of a proton from the 3′-OH which is transferred to the nearby D256 residue of DNA pol β. Nucleophilic attack of the α-phosphate generates a bipyramidal pentacoordinated transition state. This transition state species leads to inversion of the α-phosphate stereochemistry and release of the pyrophosphate group. The active site then returns to the pre-catalytic state, ready for the next round of catalysis. The chemistry step in DNA pol β is dependent upon large-scale subdomain motions. The incoming nucleotide triggers movement of the carboxyl-terminal thumb subdomain of DNA pol β from its "open" conformation to a "close" one, so that the α-helix N of the thumb subdomain can directly interact with the nascent base pair. The side chain of R283 from α-helix N shifts to interact with the DNA minor groove. Such minor groove interactions help the polymerase to assess the proper geometry of the incoming nucleotide and the primer terminus. In addition, the residue R258 in the open conformation bonds with catalytic aspartate D192, preventing its interaction with the metal ions. Upon binding of the correct nucleotide, movement of the aromatic residue F272 disrupts the interaction between D192 and R258, allowing D192 to re-orient and coordinate the active site metals. The crystal structure of the catalytic domain of DNA pol λ complexed with a gapped DNA substrate[33] showed that the architecture of the active site is substantially different from the other DNA pols.[34] The interactions between DNA pol λ and the template strand are fewer than for family B DNA pols, and involve only the 3′ terminal base pair. The N-terminal 8 kDa lyase domain makes contacts with the downstream 5′-phosphate. Crystal structures of the DNA pol λ ternary complex with DNA and dNTP, showed that the active site is in a "closed" conformation even prior to dNTP binding. Binding of the nucleotide shifts the template strand and causes the repositioning of a β-hairpin in the palm subdomain, a loop in the thumb subdomain and of three key residues: Y505, N513 and R517. These side chains establish hydrogen bonds with the 3′-terminal base pair and with the incoming nucleotide. Thus, DNA pol λ takes much less interactions with both the template strand and the incoming nucleotide than family A or B DNA pols. Structures of DNA pol λ in complex with a mismatched (G:G) primer/template,[35] showed that the templating G adopted a *syn* conformation, with little or no distortion of either the amino acid sidechains or the DNA backbone. These data indicated that DNA pol λ is able to easily accommodate even a bulky purine:purine mismatch in its active site, providing a rationale for its TLS ability. DNA pol λ shows a very high rate of −1 frameshifts generation. A structure has been solved of DNA pol λ in complex with a frameshift intermediate.[36] The DNA contained an extrahelical template nucleotide upstream the active site; however, even this highly distorted primer did not cause significant rearrangement of DNA pol λ active site. Thus, DNA pol λ seems to have specialized for the elongation of aberrant primer ends, such those arising from the presence of lesions, or of primer/template pairs with minimal homology, such as those generated during NHEJ.

Family X comprises the only known teminal deoxynucleotidyl transferases of mammalian cells: TdT, DNA pol λ and DNA pol μ.[20] While DNA pols λ and μ are also capable of template-dependent DNA polymerization, TdT is a strictly template-independent polymerase. This feature can be explained at the structural level[31] by the presence of an extended loop between the β-strands 3 and 4, called Loop I, which occupies the site where normally the primer/template junction is bound, occluding the space for the template strand, so that only a single-stranded DNA can be bound. Loop I is also present in DNA pol μ, and is of a similar length of that of TdT, but has higher conformational flexibility, explaining the limited template-dependent capability of DNA pol μ. Interestingly the equivalent regions in DNA pols β and λ are much shorter and thus less able to interfere with binding of the template strand. A second important determinant to template-independent activity is a histidine residue present in the active sites of DNA pol μ (H329) and TdT (H342) but absent in DNA pols β and λ.

Family Y. The Y-family DNA pols η, ι and κ share five conserved motifs (motifs I–V). Motifs I (I/V-D-M/L/F) and III (A/L-SIDE-V/A-F/Y) are responsible for catalysis and coordination of the metal ions. Motif II (Y-x-A-R/K) is involved, together with motif IV, in dNTP binding. Finally, motifs IV and V contain hairpin-helix-hairpin domains. Crystal structures[37−39] of the polymerase domain of archaeal (Dbh and Dpo4) and eukaryotic Y-family members (REV1, DNA pol η, ι, and κ) have been solved (**Figure 4.3**). They show that the first 250–350 N-terminal residues including the five conserved motifs constitute the catalytic core of these DNA pols, folding in the thumb, palm, and fingers subdomains found in all known DNA and RNA pols. Despite a lack of apparent sequence homology, the palm domain of the A-, B-, and Y-family DNA pols, as well as RT's, are highly conserved. In addition to the palm, DNA pol η contains a small α-helical "wrist" domain. The thumb and fingers are smaller in Y-family polymerases than in other families. The thumb domain is constituted by a small and stubby α-helical structure, whereas the fingers are shorter than in family B DNA pols and composed of a mixed α + β structure. Intriguingly, the fingers of Y-family DNA pols lack the O α-helix, which is present in high fidelity DNA pols and has been implicated in selectivity against mismatched nucleotides. At the C-terminus, Y-family DNA pols possess a unique domain of

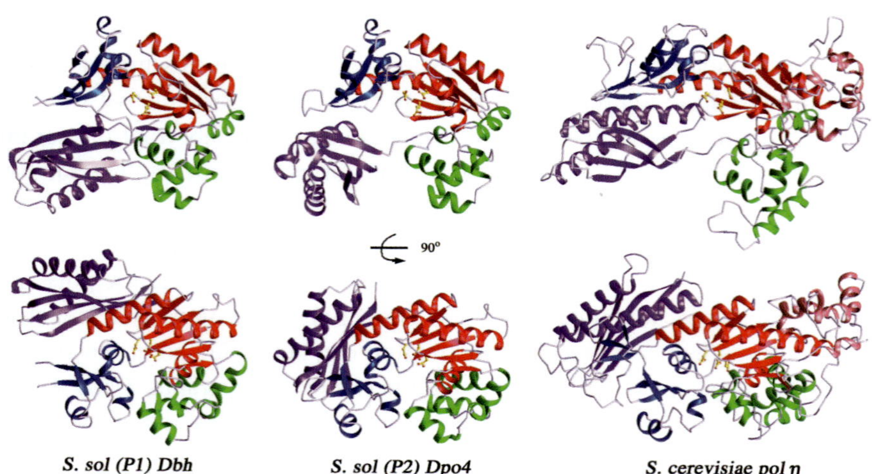

Figure 4.3   Structures of archaeal and eukaryotic family Y DNA polymerases. The different domains are highlighted in different colors. Red, palm; blue, fingers; green, thumb; purple, little finger. (Reproduced from Ref. 37 with permission.)

about 100 residues, called the little finger (LF) domain. A common characteristic of these enzymes is the presence of a more spacious active site, with little contacts between the enzyme and the DNA template or the incoming dNTP. This may explain their ability to accommodate bulky DNA lesions which are normally excluded by the tighter active sites of high-fidelity DNA pols.

4.3   Eukaryotic DNA Polymerases Accessory Subunits

Among eukaryotic DNA pols, the three replicative enzymes DNA pols α, δ, ε, the translesion DNA pol ζ of family B and the mitochondrial replicase DNA pol γ from family A are heteromultimeric enzymes consisting of a large (>100 kDa) subunit and up to three smaller subunits (**Table 4.1** and **Figure 2.9** in Chapter 2). The large subunit contains the DNA polymerase activity and, in the case of DNA pols γ, δ, and ε, also the $3' \rightarrow 5'$ proofreading exonuclease. Here we summarize some structural features of the small subunits.

DNA primase subunits (p55, p48). DNA pol α is the only family B enzyme showing an additional catalytic activity associated to two of its small subunits. The p55/p48 subunits of the DNA pol α holoenzyme, in fact, are responsible for its associated DNA-dependent RNA pol activity, commonly referred to as DNA primase. Eukaryotes and Archaea often possess a DNA primase physically associated with a DNA pol, whereas other organisms using a DNA priming mechanism such

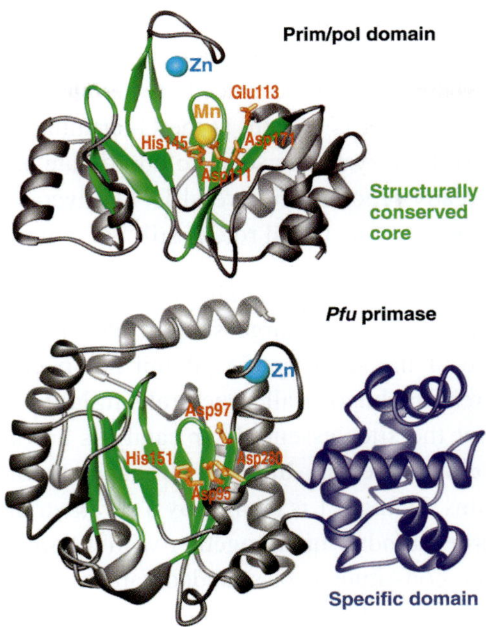

Figure 4.4   Structures of archaeal DNA primases. Comparison of the primase-polymerase domain of pRN1 (top) with the archaeal Pfu primase (bottom). Catalytic domains are in silver. Equivalent amino acids are in orange. Pfu species-specific domain is in blue. (Reproduced from Ref. 40 with permission.)

as prokaryotes and some DNA viruses, have an independent DNA primase or a multifunctional helicase/primase protein (see also Chapters 3 and 6). The structure of the archaeal pRN1 and Pfu DNA pol-primases have been solved,[40] providing support for a common evolutionary ancestor of eukaryotic and archaeal primases (**Figure 4.4**).

Interestingly, the DNA pol α-associated primase is both structurally and evolutionarily distinct from family B DNA pols.[41] In fact, the primases appear related to enzymes cleaving a DNA strand using an internal tyrosine residue in the protein to accept the 5′ end of the cleaved DNA, such as DNA topoisomerases. One hypothesis is that an active site tyrosine might have originally provided to the ancestral protein the hydroxyl group required for the priming reaction, whereas a more sophisticated *de novo* polymerase activity was acquired by the priming protein later in evolution, as a consequence of the development of more complex replisomes. The conserved catalytic core of the heterodimeric archaeo-eukaryotic primase (AEP) family consists of two units packed against each other, bearing six conserved strands (β1-6) and four helices (α1-4). The N-terminal unit contains an AEP-specific α-helix/β-sheet repetitive unit, whereas the C-terminal unit contains an RNA-recognition motif related to the one previously reported in the palm domain of DNA replication and repair DNA pols. The active sites residues (a -DxD- motif in β3, a histidine in β5 and an aspartate in β6) are lying in the space between the two units. In addition, a conserved four-cysteine cluster is present at the C-terminus of the large DNA primase subunit[42]. These four cysteines seem to coordinate an [Fe$_2$-S$_2$] iron-sulfur cluster of the type already reported in several DNA glycosylases, endonucleases and helicases, whose precise functional role is still not completely clear.

DNA polymerase γ small subunit (p55). The mitochondrial replicase DNA pol γ is a heterotrimer composed of one large subunit and a dimer of the small p55 subunit. The structure of the small (p55) subunit has been recently solved.[43,44] The protein shows a remarkable overall structural similarity with the glycyl-tRNA synthetase protein, but the sites essential for catalysis are not conserved. In the solved structure, p55 crystallized as a dimer, with each monomer consisting of three distinguishable domains. Domain 1 is formed by a seven-stranded β sheet, whereas domain 2 contains three strands which, together with the same structural elements of the facing monomer, constitute a six-stranded parallel β sheet which defines the monomer-monomer interface. Domain 3 contains a five-stranded β sheet. Another prominent feature is a four-helix bundle, formed by α-helices D and E in the domain 2 of each monomer. The dimer interface, comprising domains 1 and 2, contains the binding sites for two metal ions, whereas domain 3 contains the region interacting with the DNA pol γ catalytic subunit. Electron microscopy studies revealed that the DNA pol γ holoenzyme is a trimer composed of a large and two small subunits, where one small subunit is involved in tethering the catalytic subunit to the second small subunit, which is largely exposed to the solvent.

The B subunits superfamily. The second largest subunits of the family B heterotetrameric enzymes DNA pols α, δ and ε are generally designated as B subunits. Multiple sequence analysis of the B subunits of different species revealed extensive sequence similarities, identifying 12 conserved boxes (I–XII).[45] Boxes I-III define an oligonucleotide binding (OB)-fold domain which is separated by a short proline-rich region from the C-terminal calcineurin-like phosphoesterase domain spanning boxes IV to XII. Secondary structure predictions also suggest that all the B subunits share a common fold. The recently solved structure of the N-terminal part of the DNA pol ε B subunit,[46] revealed a predominant α-helical structure with significant similarity to the α-helical C-terminal domains of the ATPase associated with various cellular activity (AAA+) proteins, a superfamily of ATP-hydrolizing enzymes. The AAA+ C-terminal domain forms an α-helical bundle which forms the lid of the ATP binding site and is involved in the functional regulation of the ATPase activity. However, no catalytic activities have been found for any of the

B subunits so far. Phylogenetic analysis of the B subunits superfamily members suggested two distinct lineages: the first comprises DNA pol α p70 and DNA pol ε p59, the second comprises DNA pol δ p66 and also some archaeal proteins.[45] One possibility would be that the B subunits originated from a single ancestor before the archaea–eukaryotes separation, with two subsequent gene-duplication events leading to the separate lineages. Alternatively, two B subunits might have existed already prior to the separation between the two kingdoms, and one of them could have been lost during archaea evolution, while a duplication in eukaryotes gave rise to the DNA pol α and ε subunits.

4.4 Eukaryotic DNA Polymerase Fidelity: Structural and Functional Aspects

Fidelity is the ability of a DNA pol to discriminate between a complementary base pair and a mismatch.[47] The current models predict that fidelity arises from the combined effects the size of the nascent base pair and the plasticity (or tightness) of the active site, which affect the nature of different kinetic checkpoints along the reaction pathway.[48–50] Overall, both kinetic and structural data are in agreement with the so called "induced fit" model, where fidelity is the result of a complex series of small-scale conformational changes occurring upon substrate binding, during stabilization of the ternary complex (pre-chemistry) and in phosphodiester bond formation (chemistry). These steps involve interaction of the enzyme with different chemical entities including water molecules, salts and metal ions. The microscopic rates of these small-scale motions can vary among different enzymes, thus resulting in different free energy profiles and, ultimately, in different combinations of rate limiting steps for correct vs. incorrect incorporation.[51]

This "unified theory" of DNA pol fidelity can also help to explain why eukaryotic DNA pols exhibit an extraordinary wide range of fidelity values. As summarized in **Table 4.2**, the error rates for single-base substitutions due to DNA pols can vary from 10^{-1} to $>10^{-6}$, depending on the polymerase, the mispair, and the local sequence context. Such a wide variation of fidelity values correlates with the existing structural differences among the DNA pols of the different families (see Section 4.2), which, in turn, have evolved to fulfil specific functions. As shown in **Table 4.2**, replicative and repair DNA pols show high and moderate fidelity, respectively. This reflects the need for replicative enzymes to catalyze hundreds of thousands of incorporation events, whereas the activity of repair enzymes is limited to very short stretches of DNA (typically 1–10 nucleotide incorporation events). Indeed, for longer DNA tracts to be synthesized (for example such those arising in nucleotide excision repair, mismatch repair or recombination) replicative DNA

Table 4.2   Biochemical properties of eukaryotic DNA polymerases

| DNA pol | Template[a] | Metal | Catalytic efficiency (k_{cat}/K_m, $M^{-1}s^{-1}$) | | Error rate ($\times 10^{-5}$)[a] | | TSL Lesion (Effect) |
			3′OH	dNTPs	exo[−b]	exo[+b]	
exo[+b]							
α	gap	Mg^{2+}	4.3×10^6	8.4×10^5	10	n.a	**T-T dimer** (blocking) **8-oxo-G** (miscoding) **AP site** (blocking)
δ	p/t	Mg^{2+}	3.3×10^{8c}	0.37×10^{6c}	10	0.1	**T-T dimer** (blocking) **8-oxo-G** (miscoding) **AP site** (miscoding)
ε	p/t	Mg^{2+}	2.1×10^8	0.5×10^6	20	0.5	**T-T dimer** (blocking) **8-oxo-G** (miscoding) **AP site** (n.a.)
ζ	mismatch	Mg^{2+}	n.a.[g]	1.4×10^5	100	n.a.	**T-T dimer** (blocking miscoding) **8-oxo-G** (blocking)
γ	DNA/RNA hybrid	Mg^{2+}	4.5×10^9	5×10^7	10	0.1	**T-T dimer** (n.a.) **8-oxo-G** (miscoding) **AP site** (blocking)
θ	p/t	Mg^{2+}	$2.3–4.2 \times 10^6$	0.4×10^6	1000[d]	n.a	**T-T dimer** (blocking) **8-oxo-G** (n.a.)
υ	p/t	Mg^{2+}	n.a.	$1.6–5.1 \times 10^6$	350[e]	n.a	**5S Thymine glycole** (coding)

(Continued)

Table 4.2   (*Continued*)

| DNA pol | Template[a] | Metal | Catalytic efficiency (k_{cat}/K_m, $M^{-1}s^{-1}$) | | Error rate ($\times 10^{-5}$)[a] | | TSL Lesion (Effect) |
			3′OH	dNTPs	exo^{-b}	exo^{+b}	
β	gap	Mg^{2+}/Mn^{2+}	5×10^8	1.7×10^6	10	n.a	**T-T dimer** (miscoding) **8-oxo-G** (miscoding) **AP site** (miscoding)
λ	gap	Mn^{++}/Mg^{2+}	5×10^5	1.2–2.7×10^6	100	n.a	**T-T dimer** (n.a.) **8-oxo-G** (miscoding) **AP site** (miscoding)
μ	gap	Mn^{++}/Mg^{2+}	n.a.	6.4×10^4	10	n.a	**T-T dimer** (coding^f) **8-oxo-G** (miscoding) **AP site** (miscoding)
η	p/t	Mg^{2+}	6×10^6	4.3×10^5	3500	n.a	**T-T dimer** (coding) **8-oxo-G** (miscoding) **AP site** (blocking)
ι	p/t	Mg^{2+}	1.3×10^7	1.2×10^5	72000^e	n.a	**T-T dimer** (blocking) **8-oxo-G** (miscoding) **AP site** (blocking)
κ	p/t	Mg^{2+}	n.a.	1.4×10^4	1000	n.a	**T-T dimer** (blocking) **8-oxo-G** (miscoding) **AP site** (blocking)

a: p/t, primer-template. Errors are base substitution rates. b: For the enzymes with intrinsic $3' \rightarrow 5'$ exonuclease activity, the indicated exo$^-$ values are those for the correspondent exo$^-$ mutant proteins, where available. c: In the presence of its auxiliary factor PCNA. d: For the preferred G:T/T:T mismatches. e: For the preferred T:G mismatch. f: Coding, neither miscoding nor blocking. Template information is retained. g: n.a., not available pols are usually employed. The high fidelity of replicative DNA pols of the A and B families is further enhanced by the presence of intrinsic $3' \rightarrow 5'$ exonuclease or "proofreading" activities (**Table 4.2**). Exonucleases catalyze the excision of nucleoside monophosphates (dNMPs) from 3'- or 5'-DNA termini.[52] To be qualified as a "proofreader" an exonuclease needs to satisfy precise criteria: it has to act unidirectionally starting from the 3' end of a DNA strand; it needs to be non-processive in its DNA degradation; it has to act on single-stranded DNA as the optimal substrate; it must show preferential excision on a mispaired primer terminus. A non-processive or distributive manner of nucleotide hydrolysis is required in order to allow for the immediate subsequent elongation of primers during error correction.

At the opposite extreme of the spectrum of fidelity variation strands the recently discovered DNA pol family Y.[39] While the differences in fidelity between replicative and repair enzymes are moderate, the eukaryotic TLS enzymes DNA pol η, ι and κ, can display up to 10^5–10^6-fold increased error rates with respect to replicative enzymes (**Table 4.2**). The high "infidelity" of TLS DNA pols is a direct consequence of their ability to support DNA synthesis past lesions that cannot be negotiated by the high-fidelity replicative DNA pols. Structural data offered the possibility to understand the molecular basis for the great tolerance towards various kinds of DNA damage showed by TLS DNA pols.[53,54] As already mentioned in Section 4.2, the thumb and fingers subdomains are smaller in Y-family DNA pols than in the other families. The thumb domain is constituted by a small and stubby α-helical structure, whereas the fingers are shorter than in family B DNA pols and composed of a mixed α + β structure. Intriguingly, the fingers of Y-family DNA pols lack the O α-helix, which is present in high fidelity DNA pols and has been implicated in selectivity against mismatched nucleotides. In addition, a common characteristic of these enzymes is the presence of a more spacious active site, with very little contacts between the enzyme and either the DNA template and the incoming dNTP. This may explain their ability to accommodate even bulky DNA lesions which are normally excluded by the tighter active sites of high-fidelity DNA pols. Family Y DNA pol ι incorporates G 10-fold better than the correct A opposite a template T.[55] The molecular basis for this relaxed nucleotide selectivity lies in the fact that DNA pol ι favors Hoogsteen base pairing, rather than the classical Watson-Crick.[56] In the ternary structure of DNA pol ι,[37] the templating A is switched to its *syn*-conformation through interaction with L62 and Q59 in the fingers subdomain, making hydrogen bonds between its N7, N6 atoms and the N3 and O4 of the incoming dTTP in *anti*-conformation. Hoogsteen base pair can be formed also between *syn*-guanine and *anti*-thymine, thus explaining the nucleotide incorporation properties of DNA pol ι.

In conclusion, comparison of replicative DNA pols with the specialized TLS enzymes of the Y-family suggests that the latter are more tolerant towards lesions in the DNA template thanks to a more open active site, in combination with the ability to use non-Watson-Crick base pairing. These structural features, in turn, determine the very low ability of these enzymes to discriminate against mismatched nucleotides, thus explaining their low fidelity.

4.5 Biochemical and Functional Properties of the Different Eukaryotic DNA Polymerases

4.5.1 *Family A DNA Polymerases*

The eukaryotic members of family A, whose prototypic enzyme is *E. coli* DNA pol I, are prokaryotic-related DNA pols specialized in the replication of the small circular DNA of organelles (mitochondria, plastids), termed DNA pol γ-like enzymes. However, in vertebrate cells, this family comprises two additional enzymes whose functions are still not completely known, DNA pol θ and υ.

DNA polymerase γ. The DNA pol γ-like enzymes are conserved in all eukaryotic cells.[57] They are named after the animal cell mitochondrial DNA replicase, DNA pol γ,[58] but related enzymes are also present in plants for the replication of plastid DNA[59–62] and in Apicomplex protozoa for the duplication of plastid-like elements.[63,64] While replication and repair of the nuclear genomes in eukaryotic cells is accomplished by at least 14 different DNA pols (**Table 4.1**), the duplication and maintenance of the genetic information stored within the small organelles genomes solely relies on DNA pol γ-like enzymes,[65] with the notable exception of the trypanosome *T. brucei* kinetoplasts, which possess several plastidic DNA pols unrelated to DNA pol γ.[66] In animal cells mitochondrial (mt) DNA represents 1% of the total DNA content. Accordingly, DNA pol γ accounts for about 1–5% of the total cellular DNA pol activity. Homology modeling allowed the reconstruction of the three-dimensional structure of the p140 catalytic subunit of the human enzyme[65] (**Figure 4.5**). In Chapter 8 we will discuss the processive human mitochondrial DNA synthesis and the disease related DNA pol mutations as elaborated by the structure of the DNA pol γ.

DNA pol γ is an heterotrimer composed of a single large subunit and a dimeric small subunit.[67,68] According to cDNA sequence,[69] the predicted molecular weights for the large catalytic subunit range from 115 kDa (*S. pombe*) to 143 kDa (*S. cerevisiae*), with the human protein being 140 kDa. Primary sequence alignments of the large subunits from different organisms, reveal the DNA pol family A highly

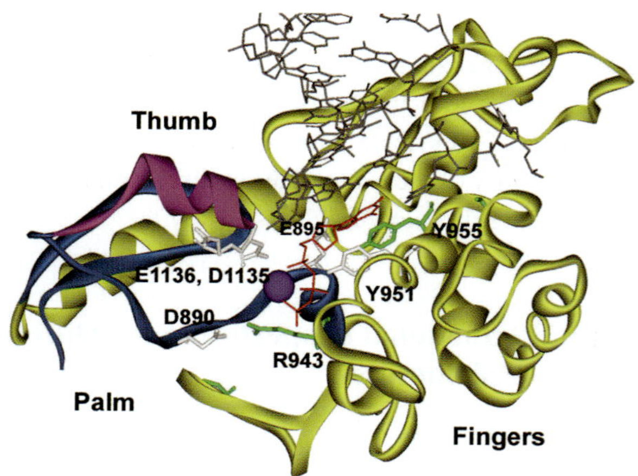

Figure 4.5   Structural model of the human DNA polymerase γ large subunit. The thumb, fingers and palm subdomains are in magenta, yellow, and blue, respectively. Primer and template DNA strands are gray, and the incoming ddGTP is red. The purple sphere is an Mg^{2+} ion. (Reproduced from Ref. 65 with permission.)

conserved motifs A, B and C, as well as the exonuclease consensus motifs I, II and III. In addition, six DNA pol γ-specific elements (1–6) were recognized. Elements 5 and 6 are localized at the C-terminus of the DNA pol domain, flanking motif C, whereas elements 1–4 are in the long linker region between the exonuclease and the polymerase domains. This region spans 337 aa in yeast and 482 aa in humans, being twice as big as the corresponding linker regions of the other family A DNA pols. Mutations in the elements 1–4 have been shown to affect DNA binding and processivity, as well as the interaction with the small subunit. The 55 kDa small subunit of human DNA pol γ (p55) forms a stable complex with the p140 large subunit in solution. As mentioned in Section 4.2.2, it is likely that the functional DNA pol γ holoenzyme is composed of a dimer of p55 subunits associated to one p140 subunit. Electron microscopy analysis of the DNA pol γ holoenzyme, showed that one p55 subunit interacts more closely with the p140 subunit, while the other small subunit is more exposed to the solvent.[70]

The biochemical properties of DNA pol γ are summarized in **Table 4.2**. Human DNA pol γ has a $3' \to 5'$ proofreading exonuclease activity[71] and can utilize a variety of nucleic acid substrates, including homopolymeric RNA/DNA hybrids such as poly(rA)/oligo(dT), the typical substrate of viral RT's.[72–74] Dissociation studies showed a strong interaction between the catalytic and accessory subunits of *Drosophila DNA* pol γ. The accessory subunit was important to maintain the structural integrity and catalytic efficiency of the holoenzyme.[58,65,75] In fact, the *Drosophila* catalytic subunit alone, showed very low catalytic activity (2% of that of the holoenzyme), especially at near physiological salt concentrations. Intrinsic processivity of the catalytic subunit is low (<30 nt), however, biochemical studies showed that the p55 subunit stimulates the processivity of DNA pol γ up to >1000 nt. Thanks to its 3′ → 5′ exonuclease activity, DNA pol γ is intrinsically a high-fidelity enzyme, as expected for an enzyme responsible for DNA replication.[71] Again, the small subunit was shown to stimulate the proofreading activity of DNA pol γ. As noted above, DNA pol γ is responsible also for the repair of mtDNA damages. Since mitochondria are the sites of oxygen metabolism, mtDNA is subjected to high levels of endogeneous oxidative damage. The main repair pathway operating the removal of oxidized bases in the DNA is the base excision repair (BER). When acting on nuclear DNA, BER relies on the activity of two specialized family X DNA pols: DNA pol β and λ. These enzymes in fact possess intrinsic 5′-deoxyribose phosphate (dRP) lyase activity, which is essential for BER. Remarkably, DNA pol γ also has dRPlyase activity,[76] which is consistent with its role as the main mtDNA BER DNA pol.[77,78]

DNA polymerase θ. DNA pol θ is encoded by the human POLQ gene on chromosome 3, with an expressed mRNA of 8.5 kb encoding for a protein of 2592 amino acids. Human POLQ gene is homologous to the *D. melanogaster* mus308 protein product, a putative polymerase–helicase involved in repair of interstrand crosslinks. Flies carrying a mutation in this gene showed sensitivity to DNA crosslinking agents, elevated frequency of chromosomal aberration and altered DNA metabolism.[79] The C-terminal region of DNA pol θ contains the canonical pol motifs A, B, and C found in the family A type of DNA pols. The N-terminal region contains an ATPase-helicase domain.[80] Full-length human DNA pol θ was expressed in a baculovirus system and purified as a recombinant protein. Before recombinant DNA pol θ was available, a DNA pol was purified from HeLa cells, based on its ability to replicate a depurinated DNA template and showing all the characteristics of a family A enzyme. Initial classification, based on its cross-reactivity with polyclonal antibodies raised against the C-terminal half of DNA pol θ, identified it as human DNA pol θ.[81] Contrary to the recombinant enzyme,[82] the natural enzyme possessed a fidelity comparable to replicative enzymes, thanks to a 3′ → 5′ exonuclease activity and showed efficient bypass of an AP site. Interestingly, SDS-PAGE and gel filtration analysis under native conditions suggested that the native enzyme was a heterotrimer composed of three subunits of 80, 90 and 100 kDa, respectively. The 100-kDa subunit was responsible for the DNA pol catalytic activity, as revealed by *in situ* activity gel analysis. Whether any of the two other subunits were contributing the observed 3′ → 5′ exonuclease activity is presently unknown.

The main biochemical properties of DNA pol θ are summarized in **Table 4.2.** The enzyme showed polymerase activity on nicked double-stranded DNA and on a singly primed DNA template. In addition, DNA pol θ exhibited a single-stranded DNA-dependent ATPase activity.[80] DNA pol θ is very efficient in bypassing an AP site,[83] inserting A with 22% of the efficiency of a normal template, and continuing extension past the lesion as efficiently as with a normally paired base. Consistently with these findings, DNA pol θ was found to extend quite efficiently A:G, A:T and A:C mismatches.[84] DNA pol θ was unable to incorporate nucleotides opposite a cyclobutane pyrimidine dimer or a (6-4) photoproduct, but when combined with DNA pol ι, an enzyme that can insert a base opposite UV-induced lesions, it could perform extension from the resulting primers allowing complete bypass of a (6-4) photoproduct.[84] According to its putative role as an extender in TLS, recombinant DNA pol θ showed error rates 10- to 100-fold higher than other family A members when copying undamaged DNA, depending on the sequence context.[82] Its average base substition rate is comparable to the very inaccurate family Y DNA pol ι.

While a role for DNA pol θ in TLS was inferred from *in vitro* data, experiments with knockout cells revealed a possible role for this enzyme in BER.[85] In fact, targeted disruption of the POLQ gene in chicken DT40 cells resulted in hypersensitivity to oxidative base damage, but not to UV or cisplatin treatments. This phenotype was synergistically increased by concomitant inactivation of the major BER enzyme, DNA pol β. However, it is not clear how the observed biochemical properties of DNA pol θ might be consistent with a role in BER.

More closely resembling its *in vitro* properties is the role ascribed to DNA pol θ in somatic hypermutations of Ig genes.[86] Knockout mice have revealed that a deficiency in family Y DNA pol η caused an 80% reduction of mutations at A/T basepairs, suggesting that DNA pol η is the major enzyme generating A/T mutations during somatic hypermutation. The absence of DNA pol θ also resulted in measurable (20%) decreases of both A/T and C/G mutations. DNA Pol θ/Pol η double null mice, however, did not show a further decrease of A/T mutations as compared with DNA pol η knockout mice, suggesting that these two enzymes function in the same genetic pathway in the generation of these mutations. The current hypothesis is that the mismatch-extender DNA pol θ might be responsible for the extension of mispaired termini opposite an A or T generated by DNA pol η. The fact that inactivating DNA pol θ causes only a 20% reduction of mutations implies that in its absence another general mismatch extender can take up its role. Currently, at least three other DNA pols have been shown to efficiently extend mispaired termini: family B DNA pol ζ, family X DNA pol λ and family Y DNA pol κ,[1] thus providing a likely backup for the inactivation of DNA pol θ.

DNA polymerase υ. DNA pol υ is the most recently discovered human DNA pol.[87] The human gene comprises 24 exons within 160 kilobases of genomic DNA and encodes a protein of 900 amino acid residues. The C-terminal half of the protein shows 33% identity (48% similarity) with *E. coli* DNA pol I and 29% identity (43% similarity) with human DNA pol θ. Numerous alternatively spliced transcripts for DNA pol υ have been found in human cell lines and expression analysis by northern blotting and *in situ* hybridization showed highest expression of the full-length protein in human and mouse testis. The main biochemical properties of DNA pol υ are summarized in **Table 4.2**. DNA pol υ shows moderate processivity (100 nucleotides) and strand displacement activity. Recombinant DNA pol υ proved to be a low fidelity enzyme incorporating T opposite template G with a frequency higher than most other DNA pols.[88] DNA pol υ is particularly efficient and accurate in translesion synthesis past a 5*S*-thymine glycol,[89] a common product of oxidative damage to DNA which blocks DNA synthesis by most DNA pols.

4.5.2 *Family B DNA Polymerases*

Family B DNA pols are multisubunit enzymes (**Table 4.1**) which, with the exception of DNA pol ζ, are responsible for the replication of nuclear DNA in all eukaryotic organisms.

DNA polymerase α-*primase*. DNA pol α consists of four subunits (**Table 4.1**) whose primary structures are conserved in all eukaryotes, consistent with the primary role of this enzyme in the initiation of leading strand DNA replication and in the repeated priming of Okazaki fragments during lagging-strand DNA replication.[12,90] These functions rely on the unique properties of the p55 and p48 subunits which constitute a dimeric DNA-dependent RNA polymerase called DNA primase.[91,92] Thus, the DNA pol α holoenzyme possesses two distinct yet functionally interacting active sites: one in the large subunit responsible for DNA synthesis and one in the dimeric primase, responsible for RNA synthesis.

The biochemical properties of DNA pol α-primase are summarized in **Table 4.2**. DNA pol α has been among the first discovered eukaryotic DNA pols (See Chapter 1).[93,94] Even though no detailed pre-steady state enzymological analysis is available for this enzyme, steady state kinetic studies combined with site directed mutagenesis, allowed the identification of important residues for catalysis and fidelity, as well as to the definition of a plausible model for the functional interaction between the polymerase and primase activities.[95–97]

Site-directed mutagenesis helped to define a group of amino acid residues important for catalysis. Asp1002 and Asp1003 of Motif I are thought to coordinate the catalytic metal ion, with the participation of the Thr1003 side chain. In the -DFNSLYPII- Motif II sequence, Asp860 and Tyr865 are involved in direct interaction with the incoming dNTP, whereas Ser867 is involved in primer binding. Finally, Lys950 of Motif III has been implicated in binding the α-phosphate of the incoming dNTP.

Kinetic studies with the primase subunits revealed that neither of the purified subunits displayed primase activity alone.[98] The p55 subunit stabilized the binding by p48 of the initiating ribonucleotide (usually adenine) to form the initiation complex. After initiation, the p48 subunit alone was able to extend the RNA primer and its Arg304 residue was inferred to be important for both nucleotide binding and primer elongation. The p55 subunit increased both the affinity for the incoming nucleotide and the catalytic rate of p48. Immunoprecipitation experiments indicated that only the p55 subunit directly contacted the large (p180) subunit of DNA pol α. Finally, it was shown that the DNA pol α-primase holoenzyme does not dissociate from the primer during the transition from RNA to DNA synthesis, suggesting that the switch from the primase to the polymerase active site occurs by an intramolecular mechanism.

The second largest subunit (p70) has no detectable enzymatic activity but is essential for cell viability in *S. cerevisiae* and has been proposed to function at the initiation of S phase and to participate in intra-S phase checkpoint regulation.[99] Consistent with this hypothesis, p70 has been shown to interact with the origin recognition complex proteins Orc1 and Orc2. In addition, p180 and p70 were shown in yeast to bind the telomer capping proteins Stn1 and Cdc13, respectively, suggesting a role of DNA pol α-primase in coupling telomere elongation with completion of DNA replication.[100] DNA pol α is also subjected to a complex regulation during the cell cycle, both at the transcriptional and post-transcriptional level (see Section 4.6).

DNA polymerase δ. DNA pol δ is a heterotetrameric enzyme (**Table 4.1**), formed by one large, catalytic subunit (p125 in humans) and three smaller subunits (p66, p50 and p12) devoid of catalytic activities.[1] DNA pol δ is the most conserved DNA pol of the B family.[101] Sequence identity within the large subunit is 90% among mammalian species and 49% between mammalian and yeast genes. Genetic experiments showed that the exchange of domains within the catalytic subunit between *S. cerevisiae* and *S. pombe* resulted in active chimeric enzymes, whereas the intact catalytic subunit of either *S. cerevisiae* (POL3) or mouse was unable to complement the disruption of the correspondent gene (Pol3) in *S. pombe*, suggesting that species-specificity is mediated by different domains within the large subunit.[102] In human cells the mRNA for the large subunit has been shown to increase at the G1/S border and the protein itself has been shown to be phosphorylated in

S-phase, suggesting a cell cycle-dependent regulation both at the transcriptional and post-transcriptional levels.[103] On the contrary, in *S. pombe* neither mRNA, nor protein levels were found to change during the cell cycle for the large DNA pol δ subunit.[104]

The second subunit of human DNA pol δ (p66) showed a 33% identity with the corresponding *S. pombe* gene (Cdc27).[105] Remarkably, the corresponding *S. cerevisiae* gene (POL32) showed very limited sequence similarity with either the *S. pombe* Cdc27 and the human p66 gene.[106] POL32 was not essential, but mutants in this gene showed hypersensitivity to DNA damaging agents and synthetic lethality when combined with conditional mutations in either the catalytic subunit (POL3) or the p50 subunit (POL31).[107]

The third subunit of human DNA pol δ (p50) showed 25-28% sequence identity to the corresponding genes of *S. cerevisiae* (POL31) and *S. pombe* (Cdc1). The POL31 gene is allelic to HYS2, an essential gene whose conditional mutants were found to show hypersensitivity to the DNA replication inhibitor hydroxyurea.[107]

The fourth subunit of human DNA pol δ (p12) shows 25% identity with the corresponding *S. pombe* gene (Cdm1).[108,109] So far, no homologous genes have been found in *S. cerevisiae*, whose DNA pol δ is still considered to be a three-subunit enzyme. The Cdm1 gene is not essential in *S. pombe*, but interfering-RNA mediated inhibition of the expression of the mammalian homolog p12 in murine cells, caused a significant decrease in proliferation upon growth factor stimulation, suggesting a role of p12 in DNA replication.

The precise quaternary structure of DNA pol δ along with the exact functional roles of the different subunits are still not fully understood. In fact, different sub-complexes of DNA pol δ have been isolated and shown to be active.[109-111] An important characteristic of DNA pol δ is its functional stimulation by the auxiliary protein PCNA. Again, this stimulation appears not to be restricted to a unique form of DNA pol δ.[112-118] Reconstitution of human DNA pol δ subcomplexes have shown that the trimeric form lacking p12 exhibits less than 10% of the activity of the tetrameric holoenzyme. However, an alternative trimeric complex lacking p66 appeared to be fully active in the presence of PCNA. Similarly, a dimeric form composed of p125 and p50 has been shown to be active and fully responsive to PCNA. All three small subunits have been shown to interact with PCNA, whereas a direct interaction between the p125 catalytic subunit and PCNA is still controversial.[106,112,114,119-123] In addition, p50 binds tightly to p125, p12 binds both p50 and p125, whereas p66 interacts with p50. Interestingly, a damaged-induced degradation of p12 has been shown to lead to a trimeric form of DNA pol δ with altered catalytic properties,[124,125] suggesting that this small subunit might play a role in the functional regulation of the DNA pol δ holoenzyme.

The biochemical properties of DNA pol δ are summarized in **Table 4.2**. The main distinguishing feature of DNA pol δ is its dependence on the auxiliary factor PCNA for processive DNA synthesis. The catalytic activity of the DNA pol δ holoenzyne in the absence of PCNA is almost undetectable on primer/template substrates containing long stretches of single strand DNA such as poly(dA)$_{200}$/oligo(dT)$_{20}$ or singly primed M13 DNA. This prevented a full characterization of this enzyme, until more sensitive assays were developed, employing 5′-radioactively labeled oligonucleotide primer/template substrates, enabling the detection of single nucleotide incorporation events. These assays, coupled to pre-steady state kinetic analysis techniques, allowed detailed studies of the biochemical properties of DNA pol δ.[118,126–135] The best characterized form is the dimeric p125–p50 subcomplex from calf thymus. This enzyme, similarly to most DNA pols, showed a steady state rate-limiting step following product formation determined by its dissociation rate from the DNA substrate, and a pre-steady state rate-limiting step (probably a conformational change) preceding the chemical step. The main effect of PCNA was to decrease the dissociation rate of DNA pol δ from the primer/template, consistent with its role as a processivity clamp (See Section 4.5 this Chapter and also Chapter 2).

Robust genetic and biochemical evidences support a main role of DNA pol δ as the lagging strand replicase.[12,13,17,136] In addition, DNA pol δ has been shown to participate in several repair pathways including BER, nucleotide excision repair (NER), mismatch repair (MMR) and homologous recombination (HR)[137–148] and cell cycle checkpoints.[149–151] Consistently, all these pathways also require the presence of PCNA.[152] DNA pol δ possesses the ability to perform strand displacement-coupled DNA synthesis, being able to invade a double-stranded DNA tract located downstream with respect to the direction of polymerization, such as in the case of gapped DNA structures, and to displace several nucleotides.[153] This activity is very important for the proposed role of DNA pol δ as the lagging strand replicase. According to the current model for Okazaki fragment synthesis, the action of DNA pol δ is not only critical for the extension of the newly synthesized Okazaki fragment, but also for the displacement of an RNA/DNA segment of about 30 nucleotides on the pre-existing downstream Okazaki fragment, in order to create an intermediate flap structure which is the target for the subsequent action of the Dna2 endonuclease and the flap endonuclease 1 (Fen-1).[154–157] This process has the advantage of removing the entire RNA/DNA hybrid fragment synthesized by the DNA pol α/primase, potentially containing nucleotide misincorporations due to the lack of a proofreading exonuclease activity of DNA pol α/primase. This results in a more accurate copy synthesized by DNA pol δ. The intrinsic strand displacement activity of DNA pol δ, in conjunction with Fen-1, PCNA and replication protein

A (RP-A), has also been proposed to be essential for S-phase specific long-patch BER pathway.[147]

DNA polymerase ε. DNA pol ε is a multisubunit enzyme, which, similarly to DNA pol α-primase and DNA pol δ, is conserved in all eukaryotic organisms.[158–166] In all eukaryotic organisms studied so far, DNA pol ε is an heterotetramer composed of one large catalytic subunit and four small accessory subunits in a 1:1:1:1 stoichiometry.[167–172] The N-terminal half fo the large subunit contains all the canonical polymerase and exonuclease boxes. In addition, the DNA pol ε large subunit has an extended C-terminal domain, with very little homology to any other DNA pol, and which is likely involved in DNA pol ε specific protein-protein interactions. The second largest subunit (p59 in human cells) interacts with the C-terminus of the DNA pol ε large subunit and is devoid of enzymatic activity. However it likely fulfills important roles in the determination of the correct (i.e. catalytically active) quaternary structure of the DNA pol ε holoenzyme, as suggested by the strong mutator phenotype of yeast strains harboring mutations in the corresponding gene. Similarly, the p17 and p12 subunits lack any detectable enzymatic activity and also interact with the C-terminal part of the large subunit. In addition, they interact with each other through an histone-fold motif, homologous to those found in histone H2A and H2B.

As expected for a protein involved in many cellular processes, DNA pol ε has been shown to physically interact with several partners, such as the checkpoint protein TopBP1, the cell cycle regulator Mdm2, the sister chromatid cohesion protein Trf4, the replication and repair proteins PCNA and DNA ligase I.[173–176] In addition, DNA pol ε was shown to interact with RNA polymerase II.

The biochemical properties of DNA pol ε are summarized in **Table 4.2**.[177] DNA pol ε synthesizes DNA in a processive manner, even though its processivity can be stimulated by PCNA (see Section 4.5). It also possesses a $3' \rightarrow 5'$ proofreading exonuclease activity, which is responsible for its high fidelity of DNA synthesis. Due to these properties, along with the compelling evidences for a role in DNA replication,[178–180] since its discovery DNA pol ε has contended to DNA pol δ the role of the leading strand replicase. Recent genetic evidence in yeast seem to have convincingly demonstrated that DNA pol ε is indeed acting at the leading strand, whereas DNA pol δ shares with DNA pol α-primase the task of replicating the discontinuous lagging strand.[17,181] As is the case of DNA pol δ, DNA pol ε is also involved in many biochemical pathways besides DNA replication, such as BER, NER and HR.[172] A role for DNA pol ε has also been shown in cell cycle control, mainly based on genetic evidences in yeast.[182–185] For example, mutants in the C-terminal part of the catalytic subunit of *S. cerevisiae* (POL2), resulted in hydroxyurea sensitivity, due to a defect in S-phase checkpoint activation. It was shown that this effect was dependent on the interaction of DNA pol ε with the essential helicase SGS1. Also the second largest subunit (DBP2 in *S. cerevisiae*) has been linked to cell cycle regulation through a phosphorylation/ dephosphorylation cycle, which was dependent on the main yeast S-phase cyclin dependent kinase CDC28 and on the CDC14 phosphatase.[186] Genetic evidences also point to an important role of *S. cerevisiae* DNA pol ε in gene silencing.[187] C-terminal POL2 mutants as well as mutations in the third subunit (DPB3), were shown to impact on gene silencing through alterations of the chromatin status. It is likely that proper holoenzyme assembly, rather than the catalytic activity, is essential for this function of DNA pol ε. Another link between DNA pol ε and chromatin structure comes from the discovery that its third subunit in human cells (p17) is also a component of the chromatin remodeling complex CHRAC.[170]

DNA polymerase ζ. Most informations about DNA pol ζ come from the budding yeast *S. cerevisiae*.[188] *S. cerevisiae* DNA pol ζ consists of two subunits: the catalytic polymerase subunit encoded by the REV3 gene and an accessory subunity encoded by the REV7 gene. Homologs of both REV3 and REV7 have been also identified in most eukaryotes **(Table 4.1)**.[189–195] Two transcripts have been identified for the human large subunit (Rev3L). One form encodes a 3130 amino acid protein and the other a 3052 amino acid protein, due to alternative splicing and translation from an internal initiation codon. The predicted size of mammalian Rev3L is 350 kDa, making it the largest eukaryotic DNA pol known so far, twice the size of *S. cerevisiae* REV3 (173 kDa). Yeast REV3 and mammalian Rev3L proteins share three regions of high sequence similarity: (i) the N-terminal region, showing 36% identity between human and yeast, with several motifs also present in DNA pol δ; (ii) the region involved in Rev7 binding, which is 29% identical between human and yeast; (iii) the DNA pol domain, showing 39% identity and bearing the six conserved B-family DNA pol motifs and two zinc-finger motifs. Rev3L is expressed in many different tissues and cell lines. The human Rev7 cDNA encodes a 211 amino acid protein with a predicted molecular weight of 24 kDa. It shows 23% sequence identity to the corresponding *S. cerevisiae* REV7. Similar to its yeast counterpart, human Rev7 not only interacts with Rev3L, but also with several additional proteins such as Cdc20 and Mad2, suggesting that this protein has additional roles besides acting as a component of the DNA pol ζ holoenzyme. For example, human Rev7 was shown to bind the transcription factor ELK1 and a phosphorylated form of the MAP kinase JNK. JNK phosphorylates ELK1, leading to the activation of several cell proliferation genes. Additional studies suggest that Rev7/Mad2 complex can inhibit the anaphase promoting complex through interactions with Cdc20. Genetic evidences in yeast and generation of knockdown mammalian cell lines by interfering RNA supported a role of DNA pol ζ in spontaneous mutagenesis and recombination-dependent DNA repair. Consistent with an essential role, DNA pol ζ $-/-$ mice showed severe embryonic lethality.

The biochemical properties of DNA pol ζ are summarized in **Table 4.2.** Contrary to DNA pol δ or ϵ, DNA pol ζ lacks a $3' \rightarrow 5'$ exonuclease activity, and has relatively low fidelity and processivity (approx 2-3 nt). The best characterized *in vitro* activity of DNA pol ζ is its role as a general extender DNA pol during TLS.[37] DNA pol ζ has a remarkable ability in elongating aberrant primer/template termini arising from incorporation of nucleotides opposite several DNA lesions. Its efficiency in extending mismatched primer termini is >1000 fold higher than that of DNA pol α. Biochemical studies with *in vitro* reconstituted systems showed that a combination of DNA pol ζ with either a specialized TLS DNA pol such as DNA pol ι or even with the replicative DNA pol δ, can lead to efficient bypass of several blocking lesions such as cis-syn thymidine dimers or abasic sites.[196–198] These observations led to the two-polymerase model for TLS, whereby efficient bypass of certain DNA lesions can be achieved through the coordinated action of a specialized inserter DNA pol, responsible for the incorporation of a nucleotide opposite the lesion, and a general extender DNA pol, able to elongate the resulting aberrant primer terminus. DNA pol ζ, thus appears to be the principal extender enzyme in TLS. The *in vitro* lesion bypass activity of DNA pol ζ has been shown to be stimulated by PCNA,[199] even though no PCNA binding motifs exist in either Rev3L or Rev7 and no direct physical interaction between DNA pol ζ and PCNA has been shown.

4.5.3 *Family X DNA Polymerases*

Family X consists of specialized small DNA pols whose primary function is to fill gaps of one to a few nucleotides during DNA repair.[30] However, while in lower eukaryotes as well as in plants only one X-family DNA pol is present, vertebrates have four members (TdT, DNA pols λ, μ and β) with different specific functions in a variety of processes, such as: DNA repair, V(D)J recombination and translesion synthesis.[1]

DNA polymerase λ. DNA pol λ is the product of the gene POLL, which is localized on chromosome 10 in humans and chromosome 19 in mice and comprises nine exons spanning a genomic segment of 8 kb. The human DNA pol λ contains 575 amino acid residues, and the murine 573, accounting for a polypeptide of 67–70 kDa.[200,201] The structure of the C-terminal part of pol λ shows the typical folding with palm, finger, thumb and 8 kDa subdomains. The first 230 N-terminal amino acids of DNA pol λ contain a BRCT and a proline/serine rich domain.[202] The C-terminal part of human DNA pol λ shows 32–33% sequence identity to the corresponding region of human DNA pol β. DNA pol λ is expressed at the highest level in testis, ovary, and fetal liver and in small amount in other tissues.

The main biochemical properties of DNA pol λ are summarized in **Table 4.2**. DNA pol λ is endowed with multiple catalytic activities: in addition to template dependent DNA pol activity, it possesses terminal deoxynucleotidyl transferase, dRPlyase and polynucleotide synthetase activities. Morever, it efficiently synthesizes DNA from an RNA primer.[203,204] The terminal deoxynucleotidyl transferase activity of DNA pol λ prefers single-stranded DNA and is not active on 3′-recessed DNA ends such those in primer/template junctions. It is, at least in part, sequence specific and preferentially incorporates pyrimidine nucleotides.[205] DNA pol λ has been purified from calf thymus tissue and the natural enzyme was shown to efficiently bypass AP sites.[206] This is also true of the recombinant enzyme, which also introduces −1 deletions by a primer misalignment mechanism.[207] Thanks to its dRP-lyase activity, DNA pol λ can efficiently participate in BER of uracil-containing DNA in an *in vitro* reconstituted reaction. A remarkable feature of DNA pol λ is its preference for Mn^{2+}, over Mg^{2+}.[208] *In vitro* experiments suggested that DNA pol λ participates in double strand break DNA repair (DSBR) via NHEJ.[209,210] In addition, recent biochemical evidences support a prominent role of DNA pol λ in the repair of oxidative DNA lesion such as 2-hydroxy-adenine and 8-oxo-guanine in cooperation with the auxiliary proteins RP-A and PCNA[211–213] (**Figure 4.6**).

Knock out mice for DNA pol $λ^{-/-}$ are fertile, and homozygous breeding has been performed up to the third generation without any noticeable problem. However, detailed analysis of immunoglobulin locus rearrangements showed that immunoglobulin heavy chain junctions from DNA pol λ-deficient animals have shorter length with normal N-additions, thus indicating a role of this enzyme during heavy chain rearrangement.[19,214] DNA pol $λ^{-/-}$ mouse embryonic fibroblasts were more sensitive to oxidative DNA damage and this phenotype was stronger when combined with inactivation of the related enzyme DNA pol β, suggesting backup functions for these two proteins in the repair of DNA oxidative lesions.[215,216]

DNA pol λ is unique in possessing all the enzymatic activities (DNA polymerization, terminal transferase, dRPlyase) which are individually present in the other X family members. Indeed, beside vertebrates all other eukaryotic organisms possess only one family X DNA pol, which, based on sequence homology, is the ortholog of DNA pol λ, which thus appears to be the most conserved DNA pol from this family throughout evolution.[217,218] Biochemical characterization of DNA pol λ from *Oriza sativa* and *Arabidopsis thaliana*, has revealed a remarkable functional similarity of the plant enzymes with their animal cell counterpart,

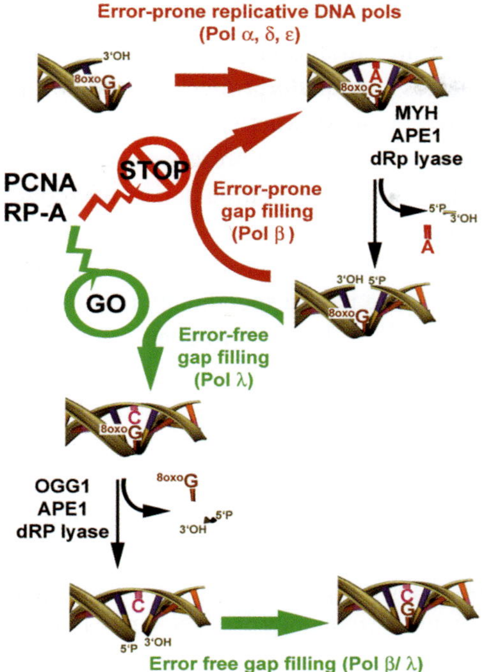

Figure 4.6   A model for the coordinated action of DNA pol λ, PCNA and RP-A in the repair of A:8-oxo-G mismatches. After the removal by the glycosylase MYH of the mismatched A incorporated opposite the 8-oxo-G lesion, the resulting gap has to be filled by a specialized DNA pol, able to incorporate the correct C opposite the lesion. PCNA and RP-A restrict the access to this gapped intermediate by DNA pol β, which would result in a futile cycle (red arrow) favoring the action of the more faithful DNA pol λ (green arrow). Once a "correct" C:8-oxo-G basepair is formed, this can be recognized by the OGG1-dependent BER pathway, leading to the replacement of the oxidized guanine with an undamaged one.

including the ability to participate in BER and to bypass abasic sites and 8-oxo-guanine. Remarkably, plant DNA pol λ also shows terminal transferase activity. In animal cells, this enzymatic activity has always been linked to the V(D)J recombination process during immunoglobulin gene rearrangement, but its presence in plant cells argues for a more fundamental and evolutionarily conserved function. Human DNA pol λ is also regulated at the post-transcriptional level during the cell cycle (see Section 4.6).

DNA polymerase μ. Human DNA pol μ is encoded by the POLM gene located on chromosome 7, containing 11 exons over a 12 kb genomic fragment, encoding a protein of 494 amino acids (55 kDa).[219] Among the pol X family it has the strongest homology to TdT with 42% amino acid identity. The human DNA pol μ possesses both DNA template dependent and independent (terminal transferase) activity and prefers Mn^{2+} over Mg^{2+} as a divalent cation. The main biochemical properties of DNA pol μ are summarized in **Table 4.2**. The DNA pol μ terminal transferase activity requires a template/primer junction for optimal efficiency, contrary to TdT and DNA pol λ, which are active on single-stranded DNA, and preferentially incorporates pyrimidines.[220] In the presence of Mn^{2+}, DNA pol μ behaves as a strong mutator, lacking base discrimination during nucleotide insertion on a DNA template-primer structure.[221,222] DNA pol μ is expressed predominantly in peripheral lymphoid tissues, although basal levels were detected in most other tissues analysed. DNA pol $\mu^{-/-}$ mice have a specific alteration in the $IgM^{-/-}$ to $IgM^{+/+}$ transition in bone marrow.[223] Ig light chain gene rearrangement was impaired at the levels of Vκ–Jκ and Vλ–Jλ junctions. These alterations led to a profound defect in the peripheral B cell compartment which caused 40% reduction in the splenic B cell fractions. DNA pol μ appears, therefore, as a key factor contributing to Ig gene rearrangement. DNA pol μ can efficiently extend a primer terminus located opposite a lesion containing a single N-2-acetyl-aminofluorene (AAF) adduct.[224] It can also efficiently extend mismatched bases by annealing the primer end with a microhomology region in the template 3 or 4 nt downstream resulting in frameshift DNA synthesis with a high frequency.[222] DNA pol μ efficiently bypasses also small DNA lesions such as 8-oxoguanine, AP-site and $1,N^6$-ethenoadenine as well as bulky DNA lesions other than AAF, such as $+$ and $-$ trans-anti-benzo[α]pyrene-N^2-dG, cis-Pt and T-T dimer, through a deletion mechanism.[225,226] Deletion resulted from primer realignment before TLS. Bypass of T-T dimers is achieved mainly in a error-free manner with AA incorporation. A novel bypass mechanism over AP-lesions by DNA pol μ was recently described as a template-dependent sequence independent nucleotidyl transferase activity. By this mechanism DNA pol μ would use its terminal transferase activity to extend primers whose 3′terminal nucleotides are located opposite an AP site, in a manner that does not depend on the sequence of the template.[220] Biochemical data as well as the physical interaction between DNA pol μ, and the end-joining factors Ku and XRCC4-LigIV also support a role of DNA pol μ in NHEJ.[227]

DNA polymerase β. The human gene encoding DNA pol β spans 33 kb and contains 14 exons. DNA pol β is composed of a single 39 kDa polypeptide containing 335 amino acid residues consisting of two domains connected by a protease-sensitive hinge region. The N-terminal 8 kDa domain contains the dRPlyase activity and binds single-stranded DNA, whereas the C-terminal 31 kDa domain carries the polymerase active site.[20] The main biochemical properties of DNA pol β[32,228] are summarized in **Table 4.2**. DNA pol β prefers gapped DNA substrates bearing a 5′-phosphate on the downstream strand in the gap.[229] In the absence of a downstream strand, DNA binding affinity of DNA pol β is strongly reduced. In contrast, high-affinity DNA binding does not require the primer strand. This suggests that DNA pol β binds short DNA gaps first through interaction with the 5′-phosphate on the downstream portion of the gap.[230] Only when such a first contact is established, an availabe 3′-OH end upstream of the gap will be bound and extended. This is consistent with the processive gap-filling DNA synthesis observed with short (<5 nucleotide) DNA gaps. A prominent role of DNA pol β in DNA repair is known since more than 30 years ago.[231,232] In particular, DNA pol β is the main enzyme involved in the gap-filling step of short-patch BER.[233] This fundamental function of DNA pol β is very conserved, since the eukaryotic enzyme can substitute for the *E. coli* BER enzyme DNA pol I. It appears that DNA pol β also acts in meiosis, DSBR, neurogenesis and TLS.[234] Experimental evidences for these additional functions come from different studies. First, DNA pol β has been implicated in meiotic events associated with synapsis and recombination.[235] Second, DNA pol β was found to bypass *in vitro* d(GpG)-cisplatin adduct placed on codon 13 of the human ras gene in a highly mutagenic way.[236] This correlates with the observation that upregulation of DNA pol β in certain cell types might lead to genomic instability and cancer.[234,237,238] Finally, mice carrying a targeted disruption of the DNA pol β gene had growth retardation and died of a respiratory failure immediately after birth.[239] Apoptotic cell death was observed in the developing central and peripheral nervous system in newly generated post-mitotic neurons of DNA pol β knock-out mice, indicating an essential role of this enzyme in neurogenesis. Indeed, DNA pol β is the only DNA pol expressed in post-mitotic neurons, likely due to its essential role in the repair of endogeneous DNA lesions, such as oxidized bases and uracil.

Terminal deoxynucleotidyl Transferase (TdT). TdT belongs to the pol family X, together with DNA pols β, λ and μ.[1] TdT and DNA pol μ are the closest derivatives with 41% identity at the amino acid level and are derived from a common ancestor, which presumably was a template-dependent polymerase.[20] TdT seems to be a recent acquisition through evolution. This is in agreement with the established role of TdT in the V(D)J recombination leading to the generation of the diversity of the antigen-binding region of immunoglobulins and T-cell receptors. This process has only evolved in higher organisms endowed with an immune system, such as vertebrates, and thus has appeared late in evolution. TdT is a nuclear enzyme found predominantly in developing B and T cells in the bone marrow and thymus. It is also found in fetal human B- and T-cell precursors. Human, bovine, mouse and rat TdT genes consist of 13 exons and their expression is restricted to B- and T-cell precursors through a complex regulatory mechanism.[240,241] Three alternative splice variants for TdT have been found in humans and other mammalian organisms, which are termed TdTL1, L2 and S, where L stands for Long (529 aa) and S for Short (509 aa) protein isoform.[242–244] In normal human pre-B and T cells, only TdTS or TdTL2 are expressed. However, this ordered pattern of expression is altered in transformed cells, so that TdTL1 can be detected in transformed B and T cell lines. This expression pattern is also different in evolution, so that in mice TdTL1 is expressed, whereas TdTL2 is not present. This evolutionary constraint to express only one of the two possible long isoforms, can be related to the fact that, whereas the short isoform can perform terminal nucleotide addition, the TdTL1 and 2 forms have a $3' \rightarrow 5'$ exonuclease activity, which is also important for V(D)J recombination. Thus, it appears that only one exonuclease-containing TdT is required during B- and T-cell development, in combination with TdTS. Several conserved domains can be recognized in the TdT sequence.[245] From the N- to the C-terminus they are: a nuclear localisation signal (NLS), a BRCA1 C-terminal (BRCT) domain, exonuclease (Exo) motifs I and II (Exo I and II), a DNA-binding domain (HhH), the Pol X domain and the Exo III box. Three cAMP kinase consensus sites for threonine phosphorylation have been also found, but it is not known whether TdT is phosphorylated *in vivo*.

The main biochemical properties of TdT are summarized in **Table 4.2**. TdT can add nucleotides in a non-templated manner to the 3'-OH end of almost any DNA, either single-stranded or double-stranded with both 3'-recessed or protruding ends. Also blunt end substrates can be utilized, but with lower efficiency. The nature of the nucleotide to be incorporated and the sequence of the 3' end of the DNA substrate can also influence the efficiency of the reaction. Finally, TdT activity is profoundly influenced by the metal activator: Co^{2+} seems to be preferred *in vitro*, but Mn^{2+} and Mg^{2+} are likely utilised *in vivo*.

TdT has been shown to interact with several proteins: the DNA–dependent protein kinase (DNA-PK), the DNA repair and recombination protein Pso4, the replication and repair factor PCNA and the transcription factor p65 homologous protein TdIF1.[242] However, the significance of many of these interactions is still not clear. The best studied is the interaction with the heterotrimeric DNA-PK, because of its importance in the V(D)J recombination pathway. TdT binds to all three DNA-PK subunits (Ku70, Ku86 and DNA-PK catalytic subunit) and this interaction can regulate TdT activity by limiting both the length and composition of nucleotide additions.[246,247]

4.5.4 *Family Y DNA Polymerases*

As outlined in Section 4.3, replicative DNA pols are endowed with a high degree of accuracy during DNA synthesis, but at the cost of being unable to bypass lesions that significantly distort the DNA double helix. TLS is carried out by specialized

DNA pols, particularly of the Y family, able to synthesize DNA past lesions that block replicative polymerases, but at the cost of a much higher error rate.[37] As mentioned in Section 4.2.2, Y family DNA pols share a common basic structural arrangement. Nevertheless, a large degree of functional divergence has occurred, rendering the different members of this family highly specialized for the bypass of specific lesions.

DNA polymerase η. DNA pol η in yeast is encoded by the RAD30 gene, whose mutation causes hypersensitivy to UV-induced mutagenesis. Similarly, mutation in the human gene for DNA pol η is responsible for the so called variant form of the disease Xeroderma pigmentosum (XP-V), characterized by high-level of UV-induced mutagenesis and skin cancer incidence.[248–250] The strong phenotype caused by DNA pol η deficiency as well as biochemical studies on the purified enzyme, suggested that its main role *in vivo* is to replicate through *cis-syn* TT dimers, a strongly blocking DNA lesion commonly induced by UV light, in an error free manner, inserting an adenine opposite each "T" in the dimer.[251–253] Beside cys-syn TT dimers, DNA pol η seems to be involved in error free bypass of other major DNA lesions.[254–260] It bypasses the 6-4 TC or CC photoproducts, by preferentially inserting a guanine opposite the $3'$C of the dimer. DNA pol η is also fairly proficient in error-free bypass of an 8-oxo-guanine lesion, inserting the correct "C" in about 75% of the cases. On the other hand, DNA pol η is strongly inhibited by O^6-methyl guanine and abasic sites and by bulky lesions such as benzo[a]pyrene 7,8-diol 9,10-epoxide, butadiene epoxide and the acrolein-derived adduct γ-hydroxy-1,N2-propano-deoxyguanosine. The inhibition of DNA pol η by lesions disrupting the Watson-Crick base pairing geometry can be accounted for by the particular active site architecture of DNA pol η. The ability of DNA pol η of bypassing cyclobutane pyrimidine dimers derives from its particularly wide active site, able to accommodate both bases of the dimer. As a consequence, most of the structural constraints present in the active site of replicative DNA pols and responsible for the induced-fit fidelity mechanism (see Section 4.3) are absent in DNA pol η.[38] Thus, its ability to discriminate between correct and incorrect base pairs might strictly rely on the intrinsic stability of Watson-Crick hydrogen bonding. In agreement with this hypothesis, kinetic and thermodynamic analysis showed that DNA pol η binds the correct nucleotide only three- to five-fold better than the incorrect one,[261–263] corresponding to a free energy difference ($\Delta\Delta G$) of approx. 1.0 kcal/mol, close to the one provided by Watson-Crick hydrogen bonding of base pairs.

The known biochemical properties of DNA pol η are summarized in **Table 4.2**. Immunolocalization studies in cells overexpressing pol η at low levels, showed that this enzyme is strictly associated to the sites of ongoing DNA replication (the so called replication factories) in the nucleus during S-phase.[264] Distinct domains in the C-terminal part of the protein have been shown to be important for its subnuclear localization, including a bipartite nuclear localization signal. The recruitment of DNA pol η to replication factories and its ability to "switch" with the replicative enzymes in case the replication fork is stalled at a lesion, is thought to be mediated by its interaction with the essential replication and repair protein PCNA.[265] DNA pol η can weakly interact with PCNA through the conserved PCNA-interacting protein (PIP) domain I/L-x-x-F-F, localized at the extreme C-terminus (aa 624–628) of the protein.[266] The addition of a single ubiquitin chain at Lys 164 of PCNA stabilizes this interaction.[267] A novel ubiquitin-binding motif has been characterized in DNA pol η and called UBZ.[268] The UBZ domain of DNA pol η lies close to the PIP-domain (aa 548–577) and is characterized by a mononuclear zinc-finger domain of the C2H2 type. This domain has been shown to bind ubiquitin, to increase DNA pol η affinity for the ubiquitinated form of PCNA and to be important for DNA pol η recruitment to replication factories.[269] DNA pol η itself is ubiquitinated, even though by enzymes different from those responsible for PCNA ubiquitination. Ubiquitinated DNA pol η cannot bind to ubiquitinated PCNA, suggesting a possible regulatory function for this modification. However, the physiological significance of DNA pol η ubiquitination is still under investigation. DNA pol η seems to be involved in homologous recombination.[270–272]

Targeted disruption of the DNA pol η gene in chicken DT40 cells, a B-cell derived population which can perform immunoglobulin gene rearrangements through a recombination-mediated gene conversion mechanism, resulted in a 2- to 6-fold reduction of these gene convertion events. Gain-of-function experiments proved that this defective phenotype could be rescued by reintroducing a catalytically active form of DNA pol η. However, immunoglobulin gene rearrangements in human cells do not require gene conversion and XP-V patients do not show immune system deficiencies.

In vitro reconstituted systems with purified enzymes or XP-V cell extracts suggested that DNA pol η might be involved in the resynthesis step starting from D-loop recombination intermediates. However, an *in vivo* role of DNA pol η in recombination has yet to be proven.

DNA polymerase ι. Human DNA pol ι is encoded by the POLI gene located on chromosome 18 and accounts for a protein of 715 amino acids.[273] DNA pol ι has a low processivity and lacks an intrinsic $3' \rightarrow 5'$ exonuclease. It can insert nucleotides opposite lesions present in DNA but is very inefficient in extending the resulting template/primers.[274,275] DNA pol ι physically interacts with and is stimulated by the auxiliary protein PCNA.[276] It also physically interacts with DNA pol η and co-localizes with DNA pol η following UV irradiation. The fidelity of DNA pol ι is high opposite template A, with error frequencies of 10^{-4} to 10^{-5}. Opposite templates G and C, DNA pol ι has a relatively low fidelity, with error frequencies ranging from 10^{-1} to 10^{-2}. However DNA pol ι has an unprecedented low fidelity opposite template T, with error frequencies $\geq 10^{-1}$, so that DNA pol ι preferentially inserts the incorrect G 10-fold more efficiently than the correct A. The known biochemical properties of DNA pol ι are summarized in **Table 4.2**. As reported in paragraph 4.3, DNA pol ι, in contrast to all known DNA pols, uses Hoogsteen base pairing instead of the regular Watson-Crick one.[37,56] This unusual property of DNA pol ι is related to its function in promoting replication through purine adducts. Only the template purines can form Hoogsteen hydrogen bond with incoming pyrimidines, thus Hoogsteen base pairing can serve as a mechanism to replicate through adducts of purines, such as those covalently attached to the N^3 of adenine or the N^2 group of guanine, where the change from *anti* to *syn* conformation would place the adduct into the major groove, causing less steric hindrance.

DNA polymerase κ. Human DNA pol κ is encoded by the POLK gene on chromosome 5 and contains 870 amino acid residues.[37] Like the other Y-family enzymes, DNA pol κ lacks an intrinsic $3' \rightarrow 5'$ proofreading exonuclease. DNA pol κ has an higher processivity than DNA pol η or ι, and has higher fidelity.[277] In addition, DNA pol κ can efficiently extend mismatched pimer termini, but often generating -1 frameshift deletions through a slippage mechanism.[278] The known biochemical properties of DNA pol κ are summarized in **Table 4.2**. Like pol η, pol κ can copy DNA containing lesions that substantially distort helix geometry, but with a different lesion bypass specificity.[279,280] For example, DNA pol κ does not bypass a cys-syn TT dimer, but it does bypass other lesions, including a benzo[a]pyrene diol epoxide (BPDE) adduct on the N2 of guanine, inserting a C. This has led to the suggestion that pol κ may participate in the bypass of lesions generated by polycyclic aromatic hydrocarbons in a manner that avoids PAH-induced mutations. Indeed, it was shown that the expression of the mouse DNA pol κ gene is under the control of the arylhydrocarbon receptor, which is responsible for the conversion of benzo[a]pyrene into BPDE in mammalian cells. DNA pol κ was shown to be able to efficiently elongate aberrant primer termini resulting from incorporation of a nucleotide opposite 8-oxo-guanine and O^6-methyl guanine by DNA pol δ, thus suggesting a role of this enzyme as an extender DNA pol in TLS.[281]

REV1. The *S. cerevisiae* REV1 gene encodes a polypeptide of 985 amino acid residues.[282] The human Rev1 cDNA consists of 4255 bp and codes for a protein of 1251 amino acid residues with a calculated molecular weight of 138 kDa.[283]

The human Rev1 gene is localized on chromosome 2 and is ubiquitously expressed in various human tissues. In *S. cerevisiae* REV1 depends on the functions of the REV3 and REV7 genes. The yeast and human Rev3 and Rev7 proteins are subunits of family B DNA pol ζ, that appears to carry out the extension step during TLS.[188,194,195,197,284–288] Yeast and human Rev1 proteins have been classified as cytidylate-transferases that insert dCMP opposite template G, A and AP sites. However, further biochemical analysis showed that *S. cerevisiae* Rev1 specifically inserts a C residue opposite template G, and it is approximately 25-, 40-, and 400-fold less efficient at inserting a C residue opposite an AP site, an O^6-methylguanine, and an 8-oxoguanine lesion. Also, Rev1 misincorporates G, A, and T residues opposite template G with a very low efficiency (10^{-3}–10^{-4}). Thus, yeast and human Rev1 can be considered as G template-specific DNA pols, since their preferred substrates are a templating G and an incoming dCTP nucleotide. The known biochemical properties of Rev1 are summarized in **Table 4.2**. Rev1 interacts with Rev7 and with the Y-family DNA pols, DNA pol ι, DNA pol η and DNA pol κ, through the same C-terminal region. Furthermore, Rev7 competes directly with DNA pol κ for binding to the Rev1. However, the polymerase activities of Rev1 and DNA pol κ are unaffected by their physical interaction when extending normal primer-termini or when bypassing a single thymine glycol lesion.

4.6 Interaction with Auxiliary Factors

DNA pols always work within multiprotein complexes. Isolation of such complexes has shown that DNA pols associate with and are modulated by a particular class of proteins called "auxiliary proteins" (see also Chapter 2). The function of these ancillary factors is either to enhance or regulate intrinsic properties of DNA pols, so that their action is best suited to a particular function, or to confer even novel properties to these enzymes.[289] The two main DNA pol auxiliary proteins are: proliferating cell nuclear antigen (PCNA) and replication protein A (RP-A).

Proliferating Cell Nuclear Antigen. PCNA was originally characterized as a processivity factor for the replicative DNA pol δ. Subsequent studies have revealed several additional multiple protein partners for PCNA, involved in several metabolic pathways including Okazaki fragment processing, DNA repair, TLS, DNA methylation, chromatin remodeling and cell cycle regulation.[152] The proteins interacting with PCNA contain a common binding motif (PCNA Interacting Protein- or PIP-box),[290] with the consensus amino-acid sequence: Q-xx-(h)-x-x-(a)-(a), (where "h" represents residues with moderately hydrophobic side chains, e.g., L, I, M; "a"

Table 4.3   Effects of PCNA on the catalytic properties of eukaryotic DNA polymerases

DNA pol	Effect of PCNA
DNA pol δ	Increases processivity, decreases fidelity of TLS
DNA pol ε	Increases primer binding
DNA pol β	Stimulates TLS
DNA pol ζ	Stimulates TLS[a]
DNA pol η	Increases the affinity for the nucleotide substrate, stimulates TLS
DNA pol ι	Increases the affinity for the nucleotide substrate, stimulates TLS
DNA pol κ	Increases the affinity for the nucleotide substrate, stimulates TLS
DNA pol λ	Increases processivity, increases bypass fidelity

[a]No physical interaction has been shown so far.

represents residues with highly hydrophobic, aromatic side chains, e.g., F, Y; and "x" is any residue). An additional PIP-box, termed KA-box, has been identified, which is also present in several PCNA interacting proteins and whose consensus is: K-A-(A/L/I)-(A/L/Q)-x-x-(L/V).[291] Among the PCNA interacting proteins are several DNA pols, and PCNA proved to affect their catalytic activity in different ways (**Table 4.3**). The first DNA pol shown to interact with PCNA was the major replicative enzyme DNA pol δ.[112] A structural model for this interaction was built using the solved three-dimensional structures of the B family RB69 bacteriophage DNA pol and its PCNA homolog gp45[292] (**Figure 4.7**).

PCNA was found to stimulate the replicative DNA pol ε[293] and the TLS enzymes DNA pols η, ι, κ and λ.[265,266,276,294,295] In the case of DNA pols δ and λ, PCNA acts as a processivity factor, allowing synthesis of long stretches of DNA by the polymerase preventing its dissociation from the template strand. In the case of DNA pol ε, PCNA acts by increasing the rate of association of the enzyme to the primer-template, thus increasing the affinity for the DNA template, rather than the processivity. Finally, PCNA was shown to stimulate the TLS enzymes DNA pol η, ι and κ by increasing their affinities for the incoming nucleotide.[37] PCNA was found to promote TLS by DNA pol δ over thymidine dimers, abasic sites and 8-oxo-G lesions.[131,134,296] Interestingly, PCNA was found to specifically increase the rate of dCTP incorporation opposite the 8-oxo-G lesion by DNA pol λ and TLS activity over an abasic site by DNA pol β.

Replication Protein A. RP-A is a heterotrimer consisting of three subunits (p70, p32 and p14) and is the main eukaryotic single-stranded DNA binding protein. RP-A fulfills multiple roles in the cell, during DNA replication initiation and elongation, DNA repair and DNA recombination.[297] RP-A is also able to influence the catalytic activities of DNA pols. It has been shown that RP-A modulates the DNA synthetic

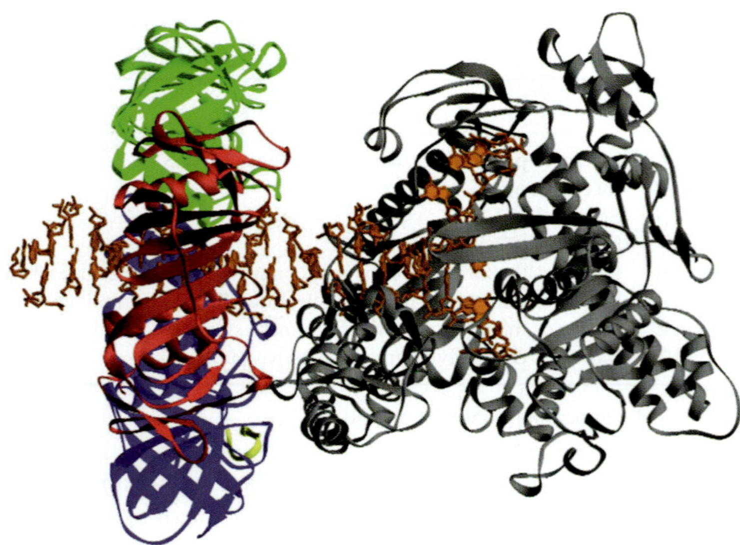

Figure 4.7   Interaction between RB69 DNA polymerase and the gp45 processivity factor. The DNA double helix is in orange. The RB69 DNA pol is in grey, with its C-terminal side facing the gp45 protein. Each monomer of the trimeric gp45 is rendered in blue, red and green, respectively. (Reproduced from Ref. 292 with permission.)

activity of DNA pol α[298] by increasing the stability of DNA pol α binding to the primer and by reducing its overall misincorporation rate. RP-A was found to exert similar effects on DNA pol λ.[299] In addition, RP-A was able, together with PCNA, to significantly increase the fidelity of 8-oxo-G and 2-OH-A bypass by DNA pol λ.[211,212] A recent study[300] also showed that PCNA and RP-A cooperate to efficiently suppress the tendency by DNA pol δ to generate large deletions when replicating templates containing direct repeats.

4.7 Eukaryotic DNA Polymerases Are Tightly Regulated in the Cell Cycle

As it is evident from the properties described so far, the multitude of DNA pols present in eukaryotic cells often participate in different, yet interconnected, metabolic pathways.

Moreover, frequently their action must be restricted both spatially and temporally within the cell cycle (**Figure 4.8**), in order to avoid unscheduled activity, that would otherwise lead to deleterious effects. Finally, a relevant degree of functional redundancy exists among the different eukaryotic DNA pols. All these facts imply

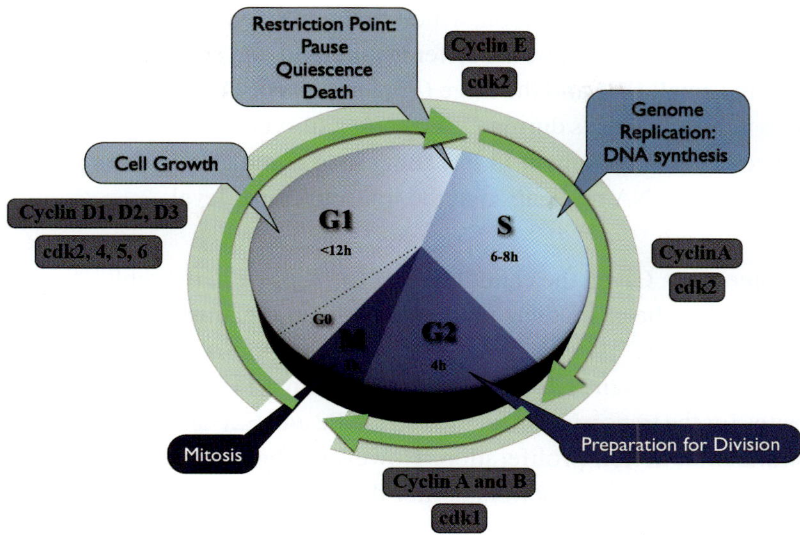

Figure 4.8   The eukaryotic cell cycle and its main regulatory proteins, the cyclin/CDKs. G1, G2: gap phases; S: DNA synthesis (replication) phase; M: mitosis. Green arrows indicate the order of events from G1 to M phase. Boxes indicate the main events taking place at each phase, as well as the main cyclin/CDK complexes which trigger that specific phase transition.

the existence of finely tuned mechanisms of regulation to allow the selection of the right enzyme for any given situation. Beside transcriptional regulation through the alternative use of specific transcription factors, a major regulatory mechanism is through phosphorylation. Phosphorylation of DNA pols can regulate their stability, intracellular localization and functional interactions with other protein partners, as illustrated below for representative DNA pols belonging to the different families.

DNA poymerasel γ. DNA pol γ transcriptional regulation has been studied for both the catalytic and the small subunit.[57,58,65,301] Both subunits of DNA pol γ are encoded by nuclear DNA. In *D. melanogaster* the mRNA levels of the catalytic and small subunit are differentially regulated during development. The mRNA level for the large subunit decreases as development proceeds and it is relatively low when mtDNA replication starts to take place in embryos. On the contrary, the mRNA levels of the small subunit show an opposite trend, reaching their peak when mtDNA replication is activated. The genes for both subunits are located within a gene cluster which contains three promoter regions (P1–P3). The P1 promoter directs the expression of the small subunit gene, and contains a genetic element called DNA replication-related element (DRE), bound by its cognate transcription factor DREF and known to regulate the expression of genes encoding components of the nuclear DNA replication machinery. The expression of the DNA pol γ large subunit is regulated by the P3 region, which contains a weaker promoter and does not have any DRE sequence. Thus, the different strength and structure of the promoter regions for the small (P1) and the large (P3) subunits of DNA pol γ can explain the differences in mRNA levels during *Drosophila* embryonic development. Moreover, the presence of a DRE element in P1, suggests a common regulatory mechanism for nuclear and mtDNA replication at the transcriptional level.

DNA polymerase α. Given the central role of DNA pol α in the replication process, its regulation has been studied in detail both at the transcriptional and post-transcriptional levels.[99,104,302–322] The promoter region of the DNA pol α large subunit gene from human cells was shown to contain a 248 bp region containing the binding sites for the transcription factors Ap1, AP2 and E2F and to be inducible upon serum stimulation of cell proliferation. However, when cells are actively cycling, the levels of DNA pol α transcripts do not fluctuate during the different phases of the cell cycle. The catalytic and the B (the second largest subunit) subunits of the DNA pol α holoenzyme have been shown to be phosphorylated in a cell cycle-dependent manner. Specifically, in *S. cerevisiae* the B subunit is phosphorylated early in S-phase and dephosphorylated at the exit from mitosis. The phosphorylation sites of the B subunit of human DNA pol α were mapped to Ser141, Ser147, Ser152 and Thr156 and shown to be targeted by different cyclin/cdk complexes. Interestingly, the functional consequences of these phosphorylation events were not equivalent. Phosphorylation by cylinA/cdk2 and cyclinA/cdk1 complexes resulted in a DNA pol α-primase holoenzyme incapable of promoting intiation of DNA replication in the SV40 system. On the opposite, phosphorylation by cyclinE/cdk2 resulted in higher competence in the initiation of DNA replication. Studies of the patterns of phosphorylation during the cell cycle revealed that the human B subunit is phosphorylated by cyclinE/cdk2 at the G1/S border, consistent with its requirement for the initiation of DNA replication, whereas it is phosphorylated by cyclinA/cdk2 in G2, when replication is completed. These phosphorylation events did not change the basic enzymatic properties of the DNA pol α-primase holoenzyme, nor did they promote its degradation. Thus, their functional significance is likely to modulate protein-protein interactions that take place at the replication fork. A multiprotein complex competent for DNA replication and containing DNA pol α-primase, DNA pol δ and DNA pol ε, together with other replication fork components, has been purified from HeLa cells at different stages of the cell cycle.[151] Intriguingly, its association with the chromatin (and hence its competence to function as the replication fork) was shown to be regulated by which particular cyclin/cdk complex was bound. When cyclinA/cdk2 was present, the complex was loaded onto chromatin, whereas when it was replaced by cyclinA/cdk1 or cyclinB/cdk2, the complex was found in the soluble nuclear fraction. Thus, cyclin/cdk phosphorylation events might help to regulate the chromatin association state of DNA pols.

DNA polymerase λ. DNA pol λ expression is regulated during embryonic development in mice, but the transcript levels do not appear to change during the cell cycle of non-embryonic proliferating cells.[200] However, DNA pol λ protein levels are regulated in a cell cycle-dependent manner through phosphorylation.[323] DNA pol λ was shown to physically interact with cdk2 *in vivo* and to be phosphorylated *in vitro* by cyclinE/cdk2, cyclinA/cdk2 and cyclinA/cdk1 complexes. The main phosphorylation sites for cdk were mapped to Ser167, Ser177 and Ser230 of the N-terminal proline-rich domain and Thr553 in the C-terminal catalytic domain. The phosphorylation status of DNA pol λ was shown to fluctuate during the cell cycle, with an hyperphosphorylated form present in the G2 phase. Experiments with phosphorylation-defective mutants[324] suggested that DNA pol λ is targeted to the proteasomal degradation pathway through ubiquitination unless the Thr553 residue is phosphorylated. In particular, DNA pol λ is stabilized during cell cycle progression in the late S and G2 phases. This most likely allows pol λ to correctly conduct repair of damaged DNA during and after S phase.

DNA polymerase η. As mentioned in Section 4.4.4, DNA pol η plays a major role in the TLS-dependent tolerance to UV-induced DNA damages. Upon UV irradiation, cellular mRNA or protein levels of DNA pol η do not change, but the protein was shown to be stabilized upon UV-treatment[325] and to relocalize to stalled DNA replication forks in irradiated cells.[264] DNA pol η translocation to stalled replication forks was shown also in response to non damaging treatments such as nucleotide deprivation of hydroxyurea treatment, suggesting that such a dynamical behaviour was part of a general response to the block of DNA replication. The intracellular translocation of DNA pol η is controlled through phosphorylation by protein kinase C (PKC) and the S-phase checkpoint kinase ATR.[326] DNA pol η phosphorylation was specifically increased by UV-treatment and inhibition of ATR or PKC intracellular activity reduced DNA pol η accumulation at stalled replication forks. In addition, mutations at two PKC phosphorylation sites (Ser587 and Thr617) in DNA pol η prevented the recruitment of the mutated protein to blocked DNA replication forks. XP-V cells are naturally sensitive to UV, since they are deprived of DNA pol η. Transient expression of the Ser587Ala and Thr617Ala mutants in such a DNA pol η null genetic background, failed to complement the UV-sensitivity, contrary to wild type DNA pol η. Thus, phosphorylation is an important mechanism regulating the subcellular localization of DNA pol η in response to DNA damage.

4.8　Chapter Summary

The diversification, throughout evolution, from single-cell to multicellular organisms, required the acquisition of more complex levels of regulation for the duplication of the genetic information of the cells, given the different proliferation rates during embryonic development as well as in the various tissues of an adult organism. In eukaryotic cells, DNA replication is both spatially and temporally regulated, being restricted to a specific phase of the cell cycle (the S phase). The exact duplication of the genetic information before cellular duplication is accomplished through a complex network of enzymes and proteins, which constitute the replisome. The core enzymes of the replisome are the DNA pols, the only enzymes capable of using a DNA strand to generate its complementary copy. The polymerization reaction is quite complex, involving different kinetic steps, whose ordered sequence is ensured by a combination of thermodynamic barriers along the reaction pathway. The different structural features among DNA pols can have important effects on these energetic barriers, thus resulting in different catalytic properties. Moreover, the existence of a series of distinct kinetic steps during the reaction is essential for the intrinsic fidelity of DNA pols, which are able to "sense" a distorted base-pair and to stop the polymerization process. Additional factors such as proofreading exonuclease activities and physical interactions with auxiliary proteins, increase the ability of replicative DNA pols to produce a faithful copy of the genetic information of the mother cell. Besides an essential role in DNA replication, DNA pols are also key players in the DNA repair mechanisms, where a certain amount of damaged information has to be replaced with an intact copy. Different DNA pols act in different types of DNA repair (see **Table 5.3**, Chapter 5). The extreme case is the one of translesion DNA synthesis, where different specialized DNA pols have evolved unconventional catalytic properties to selectively bypass particular DNA lesions. DNA pols are also tightly linked to the cell cycle regulatory pathways and to the cellular response mechanisms to DNA damage. This regulation operates both at the transcriptional and post-transcriptional level. In spite of their different properties and functions, all eukaryotic DNA pols share a common structural "core" module, with distinct subdomains (palm, fingers, thumb) which can be recognized also in prokaryotic and viral DNA polymerases, testifying their evolutionary diversification from a common ancestral enzyme.

References

1. Hubscher U, Maga G, Spadari S. 2002. *Annu Rev Biochem* 71: 133–63.
2. Koonin EV. 2006. *Biol Direct* 1: 39.
3. O'Keefe RT, Henderson SC, Spector DL. 1992. *J Cell Biol* 116: 1095–110.

4. Stillman B. 1996. *Science* 274: 1659–64.
5. Toone WM, Aerne BL, Morgan BA, Johnston LH. 1997. *Annual Review of Microbiology* 51: 125–49.
6. Lindahl T, Wood RD. 1999. *Science* 286: 1899–905.
7. Kelly TJ, Brown GW. 2000. *Annu Rev Biochem* 69: 829–80.
8. Lisby M, Barlow JH, Burgess RC, Rothstein R. 2004. *Cell* 118: 699–713.
9. Gabaldon T, Huynen MA. 2007. *PLoS Comput Biol* 3: e219.
10. Falkenberg M, Larsson NG, Gustafsson CM. 2007. *Annu Rev Biochem* 76: 679–99.
11. Scarpulla RC. 2008. *Physiol Rev* 88: 611–38.
12. Tsurimoto T, Melendy T, Stillman B. 1990. *Nature* 346: 534–9.
13. Nethanel T, Kaufmann G. 1990. *J Virol* 64: 5912–8.
14. Syvaoja J, Suomensaari S, Nishida C, Goldsmith JS, Chui GS, *et al.* 1990. *Proc Natl Acad Sci U S A* 87: 6664–8.
15. Bambara RA, Murante RS, Henricksen LA. 1997. *J Biol Chem* 272: 4647–50.
16. Hübscher U, Sogo J. 1997. *News Physiol Sci* 12: 125–31.
17. Nick McElhinny SA, Gordenin DA, Stith CM, Burgers PM, Kunkel TA. 2008. *Mol Cell* 30: 137–44.
18. Benedict CL, Gilfillan S, Thai TH, Kearney JF. 2000. *Immunol Rev* 175: 150–7.
19. Bertocci B, De Smet A, Weill JC, Reynaud CA. 2006. *Immunity* 25: 31–41.
20. Di Santo R, Maga G. 2006. *Curr Med Chem* 13: 2353–68.
21. Steitz TA. 1999. *J Biol Chem* 274: 17395–8.
22. Steitz TA. 1998. *Nature* 391: 231–2.
23. Swan MK, Johnson RE, Prakash L, Prakash S, Aggarwal AK. 2009. *Nat Struct Mol Biol* 16: 979–86.
24. Brautigam CA, Steitz TA. 1998. *Curr Opin Struct Biol* 8: 54–63.
25. Zhao Y, Jeruzalmi D, Moarefi I, Leighton L, Lasken R, Kuriyan J. 1999. *Struct Fold Des* 7: 1189–99.
26. Rodriguez AC, Park HW, Mao C, Beese LS. 2000. *J Mol Biol* 299: 447–62.
27. Franklin MC, Wang J, Steitz TA. 2001. *Cell* 105: 657–67.
28. Hashimoto H, Nishioka M, Fujiwara S, Takagi M, Imanaka T, *et al.* 2001. *J Mol Biol* 306: 469–77.
29. Asturias FJ, Cheung IK, Sabouri N, Chilkova O, Wepplo D, Johansson E. 2006. *Nat Struct Mol Biol* 13: 35–43.
30. Ramadan K, Shevelev I, Hubscher U. 2004. *Nat Rev Mol Cell Biol* 5: 1038–43.
31. Moon AF, Garcia-Diaz M, Batra VK, Beard WA, Bebenek K, *et al.* 2007. *DNA Repair (Amst)* 6: 1709-25.
32. Beard WA, Wilson SH. 2006. *Chem Rev* 106: 361–82.
33. Garcia-Diaz M, Bebenek K, Krahn JM, Blanco L, Kunkel TA, Pedersen LC. 2004. *Mol Cell* 13: 561–72.
34. Garcia-Diaz M, Bebenek K, Krahn JM, Kunkel TA, Pedersen LC. 2005. *Nat Struct Mol Biol* 12: 97–8.
35. Picher AJ, Garcia-Diaz M, Bebenek K, Pedersen LC, Kunkel TA, Blanco L. 2006. *Nucleic Acids Res* 34: 3259–66.
36. Garcia-Diaz M, Bebenek K, Krahn JM, Pedersen LC, Kunkel TA. 2006. *Cell* 124: 331–42.
37. Yang W. 2003. *Curr Opin Struct Biol* 13: 23–30.
38. Beard WA, Wilson SH. 2001. *Structure* 9: 759–64.
39. Friedberg EC, Fischhaber PL, Kisker C. 2001. *Cell* 107: 9–12.
40. Lipps G, Weinzierl AO, von Scheven G, Buchen C, Cramer P. *Nature Struct Mol Biol* 11: 157–62.
41. Iyer LM, Koonin EV, Leipe DD, Aravind L. 2005. *Nucleic Acids Res* 33: 3875–96.

42. Weiner BE, Huang H, Dattilo BM, Nilges MJ, Fanning E, Chazin WJ. 2007. *J Biol Chem* 282: 33444–51.
43. Fan L, Sanschagrin PC, Kaguni LS, Kuhn LA. 1999. *Proc Natl Acad Sci USA* 96: 9527–32.
44. Carrodeguas JA, Theis K, Bogenhagen DF, Kisker C. 2001. *Mol Cell* 7: 43–54.
45. Mäkiniemi M, Pospiech H, Kilpeläinen S, Jokela M, Vihinene M, Syväoja JE. 1999. *Trends Biochem Sci* 24: 14–6.
46. Nuutinen T, Tossavainen H, Fredriksson K, Pirila P, Permi P, *et al.* 2008. *Nucleic Acids Res* 36: 5102–10.
47. Loeb LA, Kunkel TA. 1982. *Annu Rev Biochem* 51: 429–57.
48. Johnson KA. 1993. *Annu Rev Biochem* 62: 685–713.
49. Kool ET. 2002. *Annu Rev Biochem* 71: 191–219.
50. Joyce CM, Benkovic SJ. 2004. *Biochemistry* 43: 14317–24.
51. Florian J, Goodman MF, Warshel A. 2003. *Biopolymers* 68: 286–99.
52. Shevelev IV, Hubscher U. 2002. *Nat Rev Mol Cell Biol* 3: 364–76.
53. Beard WA, Wilson SH. 2003. *Structure* 11: 489–96.
54. Freisinger E, Grollman AP, Miller H, Kisker C. 2004. *Embo J* 23: 1494–505.
55. Zhang Y, Yuan F, Wu X, Wang Z. 2000. *Mol Cell Biol* 20: 7099–108.
56. Washington MT, Johnson RE, Prakash L, Prakash S. 2004. *Mol Cell Biol* 24: 936–43.
57. Ye F, Carrodeguas JA, Bogenhagen DF. 1996. *Nucl Acids Res* 24: 1481–8.
58. Kaguni LS. 2004. *Annu Rev Biochem* 73: 293–320.
59. Takeuchi R, Kimura S, Saotome A, Sakaguchi K. 2007. *Plant Mol Biol* 64: 601–11.
60. Seow F, Sato S, Janssen CS, Riehle MO, Mukhopadhyay A, *et al.* 2005. *Mol Biochem Parasitol* 141: 145–53.
61. Carrillo N, Bogorad L. 1988. *Nucleic Acids Res* 16: 5603–20.
62. Sala F, Amileni AR, Parisi B, Spadari S. 1980. *Eur J Biochem* 112: 211–7.
63. Liu B, Liu Y, Motyka SA, Agbo EE, Englund PT. 2005. *Trends Parasitol* 21: 363–9.
64. Simpson L. 1987. *Annu Rev Microbiol* 41: 363–82.
65. Graziewicz MA, Longley MJ, Copeland WC. 2006. *Chem Rev* 106: 383–405.
66. Klingbeil MM, Motyka SA, Englund PT. 2002. *Mol Cell* 10: 175–86.
67. Wernette CM, Kaguni LS. 1986. *J Biol Chem* 261: 14764–70.
68. Wang Y, Kaguni LS. 1999. *J Biol Chem* 274: 28972–7.
69. Ropp PA, Copeland WC. 1996. *Genomics* 36: 449–58.
70. Yakubovskaya E, Lukin M, Chen Z, Berriman J, Wall JS, Kobayashi R, Kisker C, Bogenhagen DF. 2007. *Embo J* 26: 4283–91.
71. Kaguni LS, Olson MW. 1989. *Proc Natl Acad Sci U S A* 86: 6469–73.
72. Wernette CM, Conway MC, Kaguni LS. 1988. *Biochemistry* 27: 6046–54.
73. Graves SW, Johnson AA, Johnson KA. 1998. *Biochemistry* 37: 6050–8.
74. Longley MJ, Ropp PA, Lim SE, Copeland WC. 1998. *Biochemistry* 37: 10529–39.
75. Lim SE, Longley MJ, Copeland WC. 1999. *J Biol Chem* 274: 38197–203.
76. Pinz KG, Bodenhagen DF. 2000. *J Biol Chem* 275: 12509–14.
77. Pinz KG, Shibutani S, Bogenhagen DF. 1995. *J Biol Chem* 270: 9202–6.
78. Pinz KG, Bogenhagen DF. 1998. *Mol Cell Biol* 18: 1257–65.
79. Boyd J, Sakaguchi K, Harris P. 1990. *Genetics* 125: 813–9.
80. Seki M, Marini F, Wood RD. 2003. *Nucleic Acids Res* 31: 6117–26.
81. Maga G, Shevelev I, Ramadan K, Spadari S, Hubscher U. 2002. *J Mol Biol* 319: 359–69.
82. Arana ME, Seki M, Wood RD, Rogozin IB, Kunkel TA. 2008. *Nucleic Acids Res* 36: 3847–56.
83. Seki M, Masutani C, Yang LW, Schuffert A, Iwai S, *et al.* 2004. *Embo J* 23: 4484–94.
84. Seki M, Wood RD. 2008. *DNA Repair (Amst)* 7: 119–27.

85. Yoshimura M, Kohzaki M, Nakamura J, Asagoshi K, Sonoda E, *et al.* 2006. *Mol Cell* 24: 115–25.

86. Masuda K, Ouchida R, Hikida M, Kurosaki T, Yokoi M, *et al.* 2007. *J Biol Chem* 282: 17387–94.

87. Marini F, Kim N, Schuffert A, Wood RD. 2003. *J Biol Chem* 278: 32014–9.

88. Arana ME, Takata K, Garcia-Diaz M, Wood RD, Kunkel TA. 2007. *DNA Repair (Amst)* 6: 213–23.

89. Takata K, Shimizu T, Iwai S, Wood RD. 2006. *J Biol Chem* 281: 23445–55.

90. Mizuno T, Yamagishi K, Miyazawa H, Hanaoka F. 1999. *Mol Cell Biol* 19: 7886–96.

91. Arezi B, Kuchta RD. 2000. *Trends Biochem Sci* 25: 572–6.

92. Frick DN, Richardson CC. 2001. *Annu Rev Biochem* 70: 39–80.

93. Wong SW, Paborsky LR, Fisher PA, Wang TS, Korn D. 1986. *J Biol Chem* 261: 7958–68.

94. Wang TS, Wong SW, Korn D. 1989. *Faseb J* 3: 14–21.

95. Copeland WC, N.K. L, Wang TSF. 1993. *J Biol Chem* 268: 11041–9.

96. Copeland WC, Wang TSF. 1991. *J Biol Chem* 266: 22739–48.

97. Copeland WC, Wang TSF. 1993. *J Biol Chem* 268: 11028–40.

98. Copeland BC, Wang TSF. 1993. *J Biol Chem* 268: 26179–89.

99. Foiani M, Lucchini G, Plevani P. 1997. *Trends Biochem Sci* 22: 424–7.

100. Grossi S, Puglisi A, Dmitriev PV, Lopes M, Shore D. 2004. *Genes Dev* 18: 992–1006.

101. Cullmann G, Hindges R, Berchtold MW, Hubscher U. 1993. *Gene* 134: 191–200.

102. Moussy G, De Recondo AM, Baldacci G. 1995. *Eur J Biochem* 231: 45–9.

103. Zeng XR, Hao H, Jiang Y, Lee MY. 1994. *J Biol Chem* 269: 24027–33.

104. Park H, Francesconi S, Wang TS. 1993. *Mol Biol Cell* 4: 145–57.

105. Shikata K, Ohta S, Yamada K, Obuse C, Yoshikawa H, Tsurimoto T. 2001. *J Biochem (Tokyo)* 129: 699–708.

106. Hughes P, Tratner I, Ducoux M, Piard K, Baldacci G. 1999. *Nucleic Acids Res* 27: 2108–14.

107. Gerik KJ, Li X, Pautz A, Burgers PM. 1998. *J Biol Chem* 273: 19747–55.

108. Reynolds N, Watt A, Fantes PA, MacNeill SA. 1998. *Curr Genet* 34: 250–8.

109. Zuo S, Bermudez V, Zhang G, Kelman Z, Hurwitz J. 2000. *J. Biol. Chem.* 275: 5153–62.

110. Goulian M, Herrmann SM, Sackett JW, Grimm SL. 1990. *J Biol Chem* 265: 16402–11.

111. Xie B, Mazloum N, Liu L, Rahmeh A, Li H, Lee MY. 2002. *Biochemistry* 41: 13133–42.

112. Brown WC, Campbell JL. 1993. *J Biol Chem* 268: 21706–10.

113. Eissenberg JC, Ayyagari R, Gomes XV, Burgers PM. 1997. *Mol Cell Biol* 17: 6367–78.

114. Sun Y, Jiang Y, Zhang P, Zhang SJ, Zhou Y, *et al.* 1997. *J Biol Chem* 272: 13013–8.

115. Burgers PM, Gerik KJ. 1998. *J Biol Chem* 273: 19756–62.

116. Wu SM, Zhang P, Zeng XR, Zhang SJ, Mo J, *et al.* 1998. *J Biol Chem* 273: 9561–9.

117. Perderiset M, Maga G, Piard K, Francesconi S, Tratner I, *et al.* 1998. *Biochem J* 335: 581–8.

118. Mozzherin DJ, Tan C-K, Downey KM, Fisher PA. 1999. *J Biol Chem* 274: 19862–7.

119. Zhang SJ, Zeng XR, Zhang P, Toomey NL, Chuang RY, *et al.* 1995. *J Biol Chem* 270: 7988–92.

120. Tratner I, Piard K, Grenon M, Perderiset M, Baldacci G. 1997. *Biochem. Biophys. Research. Comm.* 231: 321–8 .

121. Zhou JQ, He H, Tan CK, Downey KM, So AG. 1997. *Nucleic Acids Res* 25: 1094–9.

122. Podust VN, Chang LS, Ott R, Dianov GL, Fanning E. 2002. *J Biol Chem* 277: 3894–901.

123. Li H, Xie B, Zhou Y, Rahmeh A, Trusa S, *et al.* 2006. *J Biol Chem* 281: 14748–55.

124. Meng X, Zhou Y, Zhang S, Lee EY, Frick DN, Lee MY. 2009. *Nucleic Acids Res* 37: 647–57.

125. Zhang S, Zhou Y, Trusa S, Meng X, Lee EY, Lee MY. 2007. *J Biol Chem* 282: 15330–40.

126. Podust V, Mikhailov V, Georgaki A, Hubscher U. 1992. *Chromosoma* 102: S133–41.

127. Podust LM, Podust VN, Floth C, Hubscher U. 1994. *Nucleic Acids Res* 22: 2970–5.

128. Podust VN, Podust LM, Muller F, Hubscher U. 1995. *Biochemistry* 34: 5003–10.

129. Zhou JQ, Tan CK, So AG, Downey KM. 1996. *J Biol Chem* 271: 29740–5.

130. Mozzherin DJ, McConnell M, Jasko MV, Krayevsky AA, Tan CK, *et al.* 1996. *J Biol Chem* 271: 31711–7.
131. Mozzherin DJ, Shibutani S, Tan CK, Downey KM, Fisher PA. 1997. *Proc Natl Acad Sci USA* 94: 6126–31.
132. Chen X, Zao S, Kelman Z, Hurwitz J, Goodman MF. 2000. *J Biol Chem* 275: 17677–85.
133. Einolf HJ, Guengerich FP. 2000. *J Biol Chem* 275: 16316–22.
134. Einolf HJ, Guengerich FP. 2001. *J Biol Chem* 276: 3764–71.
135. Hashimoto K, Shimizu K, Nakashima N, Sugino A. 2003. *Biochemistry* 42: 14207–13.
136. Tsurimoto T, Stillman B. 1991. *J Biol Chem* 266: 1961–8.
137. Blank A, Kim B, Loeb LA. 1994. *Proc Natl Acad Sci USA* 91: 9047–51.
138. Matsumoto Y, Kim K, Bogenhagen DF. 1994. *Mol Cell Biol* 14: 6187–97.
139. Zeng XR, Jiang Y, Zhang SJ, Hao H, Lee MY. 1994. *J Biol Chem* 269: 13748–51.
140. Giot L, Chanet R, Simon M, Facca C, Faye G. 1997. *Genetics* 146: 1239–51.
141. Longley MJ, Pierce AJ, Modrich P. 1997. *J Biol Chem* 272: 10917–21.
142. Torres-Ramos CA, Prakash S, Prakash L. 1997. *J Biol Chem* 272: 25445–8.
143. Kamath-Loeb AS, Johansson E, Burgers PM, Loeb LA. 1999. *Proc Natl Acad Sci USA* 97: 4603–8.
144. Kolodner RD, Marsischky GT. 1999. *Curr Opin Genet Dev* 9: 89–96.
145. Huang ME, de Calignon A, Nicolas A, Galibert F. 2000. *Curr Genet* 38: 178–87.
146. Kamath-Loeb AS, Loeb LA, Johansson E, Burgers PM, Fry M. 2001. *J Biol Chem* 276: 16439–46.
147. Matsumoto Y. 2001. *Prog Nucleic Acid Res Mol Biol* 68: 129–38.
148. Maloisel L, Bhargava J, Roeder GS. 2004. *Genetics* 167: 1133–42.
149. Francesconi S, Park H, Wang TS. 1993. *Nucleic Acids Res* 21: 3821–8.
150. Francesconi S, De Recondo AM, Baldacci G. 1995. *Mol Gen Genet* 246: 561–9.
151. Frouin I, Montecucco A, Biamonti G, Hubscher U, Spadari S, Maga G. 2002. *Embo J* 21: 2485–95.
152. Maga G, Hubscher U. 2003. *J Cell Sci* 116: 3051–60.
153. Podust VN, Hübscher U. 1993. *Nucleic Acids Res* 21: 841–6.
154. Bae SH, Seo YS. 2000. *J Biol Chem* 275: 38022–31.
155. Maga G, Villani G, Tillement V, Stucki M, Locatelli GA, *et al.* 2001. *Proc Natl Acad Sci U S A* 98: 14298–303.
156. Ayyagari R, Gomes XV, Gordenin DA, Burgers PM. 2003. *J Biol Chem* 278: 1618–25.
157. Jin YH, Ayyagari R, Resnick MA, Gordenin DA, Burgers PM. 2003. *J Biol Chem* 278: 1626–33.
158. Syvaoja JE. 1990. *Bioessays* 12: 533–6.
159. Araki H, Ropp PA, Johnson AL, Johnston LH, Morrison A, Sugino A. 1992. *Embo J* 11: 733–40.
160. Hübscher U, Thommes P. 1992. *Trends Biochem Sci* 17: 55–8.
161. Kesti T, Frantti H, Syvaoja JE. 1993. *J Biol Chem* 268: 10238–45.
162. Szpirer J, Pedeutour F, Kesti T, Riviere M, Syväoja JE, *et al.* 1994. *Genomics* 20: 223–6.
163. Chui G, Linn S. 1995. *J Biol Chem* 270: 7799–808.
164. D'Urso G, Nurse P. 1997. *Proc Natl Acad Sci USA* 94: 12491–6.
165. Huang D, Knuuti R, Palosaari H, Pospiech H, Syväoja JE. 1999. *Biochim Biophys Acta* 1445: 363–71.
166. Huang D, Pospiech H, Kesti T, Syväoja JE. 1999. *Biochem J* 339: 657–65.
167. Li Y, Asahara H, Patel VS, Zhou S, Linn S. 1997. *J Biol Chem* 272: 32337–44.
168. Jokela M, Makiniemi M, Lehtonen S, Szpirer C, Hellman U, Syvaoja JE. 1998. *Nucleic Acids Res* 26: 730–4.
169. Dua R, Edwards S, Levy DL, Campbell JL. 2000. *J Biol Chem* 275: 28816–25.
170. Li Y, Pursell ZF, Linn S. 2000. *J Biol Chem* 275: 31554–315.

171. Ohya T, Maki S, Kawasaki Y, Sugino A. 2000. *Nucleic Acids Res* 28: 3846–52.
172. Pursell ZF, Kunkel TA. 2008. *Prog Nucleic Acid Res Mol Biol* 82: 101–45.
173. Maga G, Jonsson ZO, Stucki M, Spadari S, Hubscher U. 1999. *J Mol Biol* 285: 259–67.
174. Vlatkovic N, Guerrera S, Li Y, Linn S, Haines DS, Boyd MT. 2000. *Nucl. Acids Res.* 28: 3581–6.
175. Makiniemi M, Hillukkala T, Tuusa J, Reini K, Vaara M, *et al.* 2001. *J Biol Chem* 6: 6.
176. Fuss J, Linn S. 2002. *J Biol Chem* 277: 8658–66.
177. Lee SH, Pan ZQ, Kwong AD, Burgers PM, Hurwitz J. 1991. *J Biol Chem* 266: 22707–17.
178. Budd ME, Campbell JL. 1993. *Mol Cell Biol* 13: 496–505.
179. Zlotkin T, Kaufmann G, Jiang YQ, Lee M, Uitto L, *et al.* 1996. *The EMBO Journal* 15: 2298–305.
180. Waga S, Masuda T, Takisawa H, Sugino A. 2001. *Proc Natl Acad Sci USA* 98: 4978–83.
181. Pursell ZF, Isoz I, Lundstrom EB, Johansson E, Kunkel TA. 2007. *Science* 317: 127–30.
182. Navas TA, Zhou Z, Elledge SJ. 1995. *Cell* 80: 29–39.
183. Navas TA, Sanchez Y, Elledge SJ. 1996. *Genes Dev* 10: 2632–43.
184. Dua R, Levy DL, Campbell JL. 1998. *J Biol Chem* 253: 30046–55.
185. Feng W, D'Urso G. 2001. *Mol Cell Biol* 21: 4495–504.
186. Kesti T, McDonald WH, Yates JR, 3rd, Wittenberg C. 2004. *J Biol Chem* 279: 14245–55.
187. Smith JS, Caputo E, Boeke JD. 1999. *Mol Cell Biol* 19: 3184–97.
188. Gan GN, Wittschieben JP, Wittschieben BO, Wood RD. 2008. *Cell Res* 18: 174–83.
189. Gibbs PE, McGregor WG, Maher VM, Nisson P, Lawrence CW. 1998. *Proc Natl Acad Sci U S A* 95: 6876–80.
190. Han KY, Chae SK, HAn DM. 1998. *FEMS Morcobiol Lett* 164: 13–9.
191. Xiao W, Lechler T, Chow BL, Fontaine T, Augustus M, *et al.* 1998. *Carcinogenesis* 19: 945–9.
192. Eeken JC, Romeijn RJ, de Jong AW, Pastink A, Lohman PH. 2001. *Mutat Res* 485: 237–53.
193. Murakumo Y, Roth T, Ishii H, Rasio D, Numata S, *et al.* 2000. *J Biol Chem* 275: 4391–7.
194. Murakumo Y, Ogura Y, Ishii H, Numata S, Ichihara M, *et al.* 2001. *J Biol Chem* 276: 35644–51.
195. Murakumo Y. 2002. *Mutat Res* 510: 37–44.
196. Johnson RE, Washington MT, Haracska L, Prakash S, Prakash L. 2000. *Nature* 406: 1015–9.
197. Haracska L, Unk I, Johnson RE, Johansson E, Burgers PM, *et al.* 2001. *Genes Dev* 15: 945–54.
198. Johnson RE, Haracska L, Prakash S, Prakash L. 2001. *Mol Cell Biol* 21: 3558–63.
199. Garg P, Stith CM, Majka J, Burgers PM. 2005. *J Biol Chem* 280: 23446–50.
200. Garcia-Diaz M, Dominguez O, Lopez-Fernandez LA, de Lera LT, Saniger ML, *et al.* 2000. *J Mol Biol* 301: 851–67.
201. Garcia-Diaz M, Bebenek K, Sabariegos R, Dominguez O, Rodriguez J, *et al.* 2002. *J Biol Chem* 277: 13184–91.
202. Garcia-Diaz M, Bebenek K, Gao G, Pedersen LC, London RE, Kunkel TA. 2005. *DNA Repair (Amst)* 4: 1358–67.
203. Garcia-Diaz M, Bebenek K, Kunkel TA, Blanco L. 2001. *J Biol Chem* 276: 34659–63.
204. Ramadan K, Shevelev IV, Maga G, Hubscher U. 2004. *J Mol Biol* 339: 395–404.
205. Maga G, Ramadan K, Locatelli GA, Shevelev I, Spadari S, Hubscher U. 2005. *J Biol Chem* 280: 1971–81.
206. Ramadan K, Shevelev IV, Maga G, Hubscher U. 2002. *J Biol Chem* 277: 18454–8.
207. Blanca G, Villani G, Shevelev I, Ramadan K, Spadari S, *et al.* 2004. *Biochemistry* 43: 11605–15.
208. Blanca G, Shevelev I, Ramadan K, Villani G, Spadari S, *et al.* 2003. *Biochemistry* 42: 7467–76.
209. Lee JW, Blanco L, Zhou T, Garcia-Diaz M, Bebenek K, *et al.* 2004. *J Biol Chem* 279: 805–11.
210. Nick McElhinny SA, Havener JM, Garcia-Diaz M, Juarez R, Bebenek K, *et al.* 2005. *Mol Cell* 19: 357–66.
211. Crespan E, Hubscher U, Maga G. 2007. *Nucleic Acids Res* 35: 5173–81.
212. Maga G, Villani G, Crespan E, Wimmer U, Ferrari E, *et al.* 2007. *Nature* 447: 606–8.

213. Maga G, Crespan E, Wimmer U, van Loon B, Amoroso A, *et al.* 2008. *Proc Natl Acad Sci U S A* 105: 20689–94.
214. Bertocci B, De Smet A, Flatter E, Dahan A, Bories JC, *et al.* 2002. *J Immunol* 168: 3702–6.
215. Braithwaite EK, Kedar PS, Lan L, Polosina YY, Asagoshi K, *et al.* 2005. *J Biol Chem* 280: 31641–7.
216. Braithwaite EK, Prasad R, Shock DD, Hou EW, Beard WA, Wilson SH. 2005. *J Biol Chem* 280: 18469–75.
217. Uchiyama Y, Kimura S, Yamamoto T, Ishibashi T, Sakaguchi K. 2004. *Eur J Biochem* 271: 2799–807.
218. Uchiyama Y, Takeuchi R, Kodera H, Sakaguchi K. 2009. *Biochimie* 91: 165–70.
219. Dominguez O, Ruiz JF, Lain de Lera T, Garcia-Diaz M, Gonzalez MA, *et al.* 2000. *Embo J* 19: 1731–42.
220. Covo S, Blanco L, Livneh Z. 2004. *J Biol Chem* 279: 859–65.
221. Ruiz JF, Dominguez O, Lain de Lera T, Garcia-Diaz M, Bernad A, Blanco L. 2001. *Philos Trans R Soc Lond B Biol Sci* 356: 99–109.
222. Zhang Y, Wu X, Yuan F, Xie Z, Wang Z. 2001. *Mol Cell Biol* 21: 7995–8006.
223. Bertocci B, De Smet A, Berek C, Weill JC, Reynaud CA. 2003. *Immunity* 19: 203–11.
224. Duvauchelle JB, Blanco L, Fuchs RP, Cordonnier AM. 2002. *Nucleic Acids Res* 30: 2061–7.
225. Zhang Y, Wu X, Guo D, Rechkoblit O, Taylor JS, *et al.* 2002. *J Biol Chem* 277: 44582–7.
226. Havener JM, McElhinny SA, Bassett E, Gauger M, Ramsden DA, Chaney SG. 2003. *Biochemistry* 42: 1777–88.
227. Mahajan KN, Nick McElhinny SA, Mitchell BS, Ramsden DA. 2002. *Mol Cell Biol* 22: 5194–202.
228. Beard WA, Prasad R, Wilson SH. 2006. *Methods Enzymol* 408: 91–107.
229. Ahn J, Kraynov VS, Zhong X, Werneburg BG, Tsai MD. 1998. *Biochem J* 331: 79–87.
230. Jezewska MJ, Rajendran S, Bujalowski W. 2001. *Biochemistry* 40: 3295–307.
231. Waser J, Hübscher U, Kuenzle CC, Spadari S 1979. *Eur J Biochem* 97: 361–8.
232. Hübscher U, Kuenzle CC, Spadari S. 1979. *Proc Natl Acad Sci U S A* 76: 2316–20.
233. Fortini P, Pascucci B, Parlanti E, Sobol RW, Wilson SH, Dogliotti E. 1998. *Biochemistry* 37: 3575–80.
234. Canitrot Y, Frechet M, Servant L, Cazaux C, Hoffmann JS. 1999. *FASEB J* 13: 1107–11.
235. Plug AW, Clairmont CA, Sapi E, Ashley T, Sweasy JB. 1997. *Proc Natl Acad Sci U S A* 94: 1327–31.
236. Hoffmann JS, Pillaire MJ, Maga G, Podust V, Hubscher U, Villani G. 1995. *Proc Natl Acad Sci U S A* 92: 5356–60.
237. Frechet M, Canitrot Y, Cazaux C, Hoffmann JS. 2001. *FEBS Lett* 505: 229–32.
238. Frechet M, Canitrot Y, Bieth A, Dogliotti E, Cazaux C, Hoffmann JS. 2002. *Oncogene* 21: 2320–7.
239. Sugo N, Aratani Y, Nagashima Y, Kubota Y, Koyama H. 2000. *EMBO J* 19: 1397–404.
240. Ernst P, Hahm K, Trinh L, Davis JN, Roussel MF, *et al.* 1996. *Mol Cell Biol* 16: 6121–31.
241. Ernst P, Hahm K, Cobb BS, Brown KE, Trinh LA, *et al.* 1999. *Cold Spring Harb Symp Quant Biol* 64: 87–97.
242. Thai TH, Kearney JF. 2005. *Adv Immunol* 86: 113–36.
243. Thai TH, Kearney JF. 2004. *J Immunol* 173: 4009–19.
244. Thai TH, Purugganan MM, Roth DB, Kearney JF. 2002. *Nat Immunol* 3: 457–62.
245. Delarue M, Boule JB, Lescar J, Expert-Bezancon N, Jourdan N, *et al.* 2002. *Embo J* 21: 427–39.
246. Mahajan KN, Gangi-Peterson L, Sorscher DH, Wang J, Gathy KN, *et al.* 1999. *Proc Natl Acad Sci U S A* 96: 13926–31.
247. Mickelsen S, Snyder C, Trujillo K, Bogue M, Roth DB, Meek K. 1999. *J Immunol* 163: 834–43.

248. McDonald JP, Rapic-Otrin V, A. EJ, Broughton BC, Wang X, *et al.* 1999. *Genomics* 60: 20–30.

249. Masutani C, Kusumoto R, Yamada A, Dohmae N, Yokoi M, *et al.* 1999. *Nature* 399: 700–4.

250. Kannouche P, Broughton BC, Volker M, Hanaoka F, Mullenders LH, Lehmann AR. 2001. *Genes Dev* 15: 158–72.

251. Masutani C, Kusumoto R, Iwai S, Hanaoka F. 2000. *EMBO J* 19: 3100–9.

252. Prakash S, Johnson RE, Washington MT, Haracska L, Kondratick CM, Prakash L. 2000. *Cold Spring Harb Symp Quant Biol* 65: 51–9.

253. Washington MT, Johnson RE, Prakash S, Prakash L. 2000. *Proc Natl Acad Sci USA* 97: 3094–9.

254. Haracska L, Yu S-L, Johnson RE, Prakash L, Prakash S. 2000. *Nature Genet* 25: 458–61.

255. Haracska L, Prakash S, Prakash L. 2000. *Mol Cell Biol* 20: 8001–7.

256. Yuan F, Zhang Y, Rajpal DK, Wu X, Guo D, *et al.* 2000. *J. Biol. Chem.* 275: 8233–9.

257. Zhang Y, Yuan F, Wu X, Rechkoblit O, Taylor JS, *et al.* 2000. *Nucleic Acids Res* 28: 4717–24.

258. Johnson RE, Haracska L, Prakash S, Prakash L. 2001. *Mol Cell Biol* 21: 3558–63.

259. Yu SL, Johnson RE, Prakash S, Prakash L. 2001. *Mol Cell Biol* 21: 185–8.

260. McCulloch SD, Kokoska RJ, Masutani C, Iwai S, Hanaoka F, Kunkel TA. 2004. *Nature* 428: 97–100.

261. Washington MT, Johnson RE, Prakash S, Prakash L. 1999. *J Biol Chem* 274: 36835–8.

262. Johnson RE, Washington MT, Prokash S, Prakash L. 2000. *J Biol Chem* 275: 7447–50.

263. Matsuda T, Bebenek K, Masutani C, Hanoaka F, Kunkel TA. 2000. *Nature* 404: 1011–3.

264. Kannouche P, Fernandez de Henestrosa AR, Coull B, Vidal AE, Gray C, *et al.* 2002. *Embo J* 21: 6246–56.

265. Haracska L, Kondratick CM, Unk I, Prakash S, Prakash L. 2001. *Mol Cell* 8: 407–15.

266. Haracska L, Johnson RE, Unk I, Phillips B, Hurwitz J, *et al.* 2001. *Mol Cell Biol* 21: 7199–206.

267. Kannouche PL, Wing J, Lehmann AR. 2004. *Mol Cell* 14: 491–500.

268. Acharya N, Yoon JH, Gali H, Unk I, Haracska L, *et al.* 2008. *Proc Natl Acad Sci U S A* 105: 17724–9.

269. Sabbioneda S, Green CM, Bienko M, Kannouche P, Dikic I, Lehmann AR. 2009. *Proc Natl Acad Sci U S A* 106: E20; author reply E1.

270. Kawamoto T, Araki K, Sonoda E, Yamashita YM, Harada K, *et al.* 2005. *Mol Cell* 20: 793–9.

271. McIlwraith MJ, Vaisman A, Liu Y, Fanning E, Woodgate R, West SC. 2005. *Mol Cell* 20: 783–92.

272. Rattray AJ, Strathern JN. 2005. *Mol Cell* 20: 658–9.

273. McDonald JP, Tissier A, Frank EG, Iwai S, Hanaoka F, Woodgate R. 2001. *Philos Trans R Soc Lond B Biol Sci* 356: 53–60.

274. Vaisman A, Tissier A, Frank EG, Goodman MF, Woodgate R. 2001. *J Biol Chem* 276: 30615–22.

275. Zhang Y, Yuan F, Wu X, Taylor J-S, Wang Z. 2001. *Nucleic Acids Res* 29: 928–35.

276. Haracska L, Johnson RE, Unk I, Phillips BB, Hurwitz J, *et al.* 2001. *Proc Natl Acad Sci U S A* 98: 14256–61.

277. Ohashi E, Bebenek K, Matsuda T, Feaver WJ, Gerlach VL, *et al.* 2000. *J Biol Chem* 275: 39678–84.

278. Levine RL, Miller H, Grollman A, Ohashi E, Ohmori H, *et al.* 2001. *J Biol Chem* 276: 18717–21.

279. Zhang Y, Yuan F, Wu X, Wang M, Rechkoblit O, *et al.* 2000. *Nucleic Acids Res* 28: 4138–46.

280. Suzuki N, Ohashi E, Kolbanovskiy A, Geacintov NE, Grollman AP, *et al.* 2002. *Biochemistry* 41: 6100–6.

281. Haracska L, Prakash L, Prakash S. 2002. *Proc Natl Acad Sci U S A* 99: 16000–5.

282. Nelson JR, Lawrence CW, Hinkle DC. 1996. *Nature* 382: 729–31.

283. Lin W, Xin H, Zhang Y, Yuan F, Wang Z. 1999. *Nucleic Acids Res* 27: 4468–75.

284. Lawrence CW. 2000. *Cold Spring Harb Symp Quant Biol* 65: 61–9.

285. Lawrence CW, Maher VM. 2001. *Biochem Soc Trans* 29: 187–91.

286. Lawrence CW. 2004. *Adv Protein Chem* 69: 167–203.

287. Zhu F, Zhang M. 2003. *World J Gastroenterol* 9: 1165–9.
288. Lawrence CW. 2002. *DNA Repair (Amst)* 1: 425–35.
289. Hübscher U, Maga G, Podust VN. 1996. In *DNA Replication in Eukaryotic Cells*, ed. ML De Pamphilis, pp. 525–43. Cold Spring Harbor: Cold Spring Harbor Laborytory Press.
290. Jonsson ZO, Hübscher U. 1997. *BioEssays* 19: 967–75.
291. Xu H, Zhang P, Liu L, Lee MYWT. 2001. *Biochemistry* 40: 4512–20.
292. Shamoo Y, Steitz TA. 1999. *Cell* 99: 155–66.
293. Maga G, Hubscher U. 1995. *Biochemistry* 34: 891–901.
294. Maga G, Villani G, Ramadan K, Shevelev I, Tanguy Le Gac N, *et al.* 2002. *J Biol Chem* 277: 48434–40.
295. Maga G, Blanca G, Shevelev I, Frouin I, Ramadan K, *et al.* 2004. *Faseb J* .
296. O'Day CL, Burgers PM, Taylor JS. 1992. *Nucleic Acids Res* 20: 5403–6.
297. Wold MS. 1997. *Annu Rev Biochem* 66: 61–92.
298. Maga G, Frouin I, Spadari S, Hubscher U. 2001. *J Biol Chem* 276: 18235–42.
299. Maga G, Shevelev I, Villani G, Spadari S, Hubscher U. 2006. *Nucleic Acids Res* 34: 1405–15.
300. Fortune JM, Stith CM, Kissling GE, Burgers PM, Kunkel TA. 2006. *Nucleic Acids Res* 34: 4335–41.
301. Iyengar B, Luo N, Farr CL, Kaguni LS, Campos AR. 2002. *Proc Natl Acad Sci U S A* 99: 4483–8.
302. Wahl AF, Geis AM, Spain BH, Wong SW, Korn D, Wang TS. 1988. *Mol Cell Biol* 8: 5016–25.
303. Wong SW, Wahl AF, Yuan PM, Arai N, Pearson BE, *et al.* 1988. *Embo J* 7: 37–47.
304. Nasheuer HP, Moore A, Wahl AF, Wang TS. 1991. *J Biol Chem* 266: 7893–903.
305. Pearson BE, Nasheuer HP, Wang TSF. 1991. *Mol Cell Biol* 11: 2081–95.
306. D'Urso G, Grallert B, Nurse P. 1995. *J Cell Sci* 108: 3109–18.
307. Foiani M, Liberi G, Lucchini G, Plevani P. 1995. *Mol Cell Biol* 15: 883–91.
308. Ferrari M, Lucchini G, Plevani P, Foiani M. 1996. *J Biol Chem* 271: 8661–6.
309. Mizuno T, Okamoto T, Yokoi M, Izumi M, Kobayashi A, *et al.* 1996. *J Cell Sci* 109: 2627–36.
310. Voitenleitner C, Fanning E, Nasheuer HP. 1997. *Oncogene* 14: 1611–5.
311. Wittmeyer J, Formosa T. 1997. *Mol Cell Biol* 17: 4178–90.
312. Desdouets C, Santocanale C, Drury LS, Perkins G, Foiani M, *et al.* 1998. *Embo J* 17: 4139–46.
313. Mimura S, Takisawa H. 1998. *Embo J* 17: 5699–707.
314. Martelli AM, Capitani S, Neri LM. 1998. *J Cell Biochem* 71: 11–20.
315. Kuroda K, Ueda R. 1999. *Biochem Biophys Res Commun* 254: 372–7.
316. Voitenleitner C, Rehfuess C, Hilmes M, O'Rear L, Liao PC, *et al.* 1999. *Mol Cell Biol* 19: 646–56.
317. Chen X, Li Q, Fischer JA. 2000. *Genetics* 156: 1787–95.
318. Matsumoto H, Sugino A, Araki H. 2000. *Mol Cell Biol* 20: 2809–17.
319. Michael WM, Ott R, Fanning E, Newport J. 2000. *Science* 289: 2133–7.
320. Tan S, Wang TS. 2000. *Mol Cell Biol* 20: 7853–66.
321. Dehde S, Rohaly G, Schub O, Nasheuer HP, Bohn W, *et al.* 2001. *Mol Cell Biol* 21: 2581–93.
322. Nakayama J, Allshire RC, Klar AJ, Grewal SI. 2001. *Embo J* 20: 2857–66.
323. Frouin I, Toueille M, Ferrari E, Shevelev I, Hubscher U. 2005. *Nucleic Acids Res* 33: 5354–61.
324. Wimmer U, Ferrari E, Hunziker P, Hubscher U. 2008. *EMBO Rep* 9: 1027–33.
325. Skoneczna A, McIntyre J, Skoneczny M, Policinska Z, Sledziewska-Gojska E. 2007. *J Mol Biol* 366: 1074–86.
326. Chen YW, Cleaver JE, Hatahet Z, Honkanen RE, Chang JY, *et al.* 2008. *Proc Natl Acad Sci U S A* 105: 16578–83.

CHAPTER 5

Global Functions of DNA Polymerases

5.1 Fifteen DNA Polymerases: Share of Workload and Redundancies

Complex cellular functions are carried out by networks of protein machineries. Therefore the DNA pols involved in a particular DNA transaction (see Box 5.1) have to act in a concerted way with many other proteins.[1,2] The dynamic action of DNA pols in DNA transaction machineries such as the replisome, the repairosomes, the recombination complexes and the mismatch repair is essential. Moreover, these machineries are not static since they will act upon request at the right time (e.g. during the cell cycle, after a genotoxic insult, during development), at the right place (e.g. nucleus, nucleolus, open chromatin and mitochondria), at the right concentration (e.g. up- or down-regulation in response to a DNA damage) and, finally, they should interact with the right partners. These interactions are often guaranteed by posttranslational modifications either within the DNA pols and/or their interacting partners. PCNA, the sliding clamp interacting with several DNA pols (see Chapter 2, **Table 2.7** for details) is a well-studied partner of several DNA pols.[3,4] Depending on the ubiquitination and SUMOylation status PCNA can help to favor error-prone translesion synthesis (monoubiquitination), error-free translesion synthesis (polyubiquitination), SUMOylation (inhibition of recombination if modified at lysine 164 or inhibition of cohesion establishment if modified at lysine 127, reviewed in Ref. 5). DNA pols themselves are the targets of many posttranslational modifications such as phosphorylation, acetylation, methylation, ubiquitination and SUMOylation. Even though our knowledge of these modifications and their consequences in the interactions of DNA pols with partners is still very limited, we will try to dissect a few aspects of these interactions in more details in this chapter.

In addition, we try to focus on the regulation of DNA pols (Section 5.5), their interaction at the replication forks (Section 5.2), and their engagements during

Box 5.1 DNA transactions requiring DNA polymerases

DNA replication:

This is the process in which the DNA is duplicated in the S-phase of the cell cycle before a cell goes into mitosis. One strand, called the leading strand, is synthesized in a continuous way and the other, called the lagging strand, in a discontinuous way. This is due to the unidirectional direction in which all DNA pols synthesize DNA, namely in the $5' \rightarrow 3'$ direction. The short stretches on the lagging strand are called Okazaki fragments and they have to be primed every 200 bases leading to over 10^7 initiation events.

Base excision repair (BER):

This is an essential repair pathway that deals with DNA alteration that frequently occur on one strand of the DNA. They include depurinations, depyrimidations, oxidations and deaminations. A group of different DNA glycosylases recognizes these alterations and remove the wrong base and renders the DNA a substrate for an efficient nuclease called apurinic endonuclease 1. In case of depurinations and depyrimidations leading to abasic sites, they are directly cut by apurinic endonuclease 1. Two types of DNA synthesis then occur: if only one nucleotide incorporation takes place the process is called short patch BER, while if strand displacement of 2–12 nucleotides with subsequent DNA synthesis by DNA pols occurs, the process is called long patch BER.

Nucleotide excision repair (NER):

When lesions cause serious helix distortions on one strand of the DNA the very complex NER has to act. These distortions are first recognized by specific proteins. Next the DNA is unwound followed by the excision of about 30 bases by specific endonucleases on each side. The resulting gap is filled by DNA pols and finally ligated.

Non-homologous end joining (NHEJ):

This is a repair pathway that can restore double-stranded breaks and this without requiring the information of a homologous DNA sequence. This pathway is therefore error-prone. The ends are recessed by specific exonucleases and the broken ends bound by specific proteins. Finally after DNA synthesis by DNA pols the ends are rejoined. This repair pathway is mainly active in the G1 phase of the cell cycle and is non-replication associated.

Box 5.1 (*Continued*)

Homologous recombination (HR):

In this repair pathway the sister chromatid can be employed for the repair of the broken strand. First a resection by an exonuclease occurs and the 3′-overhang can invade a homologous DNA stretch on the sister chromatid. DNA synthesis by DNA pols restores the missing information. Finally, the complex is resolved by cutting and ligation to restore the repaired DNA. This repair pathway is error free and mainly active in the late S- and the G2-phase of the cell cycle.

Mismatch repair (MMR):

This repair pathway can restore mismatches and gaps that can occur after DNA replication. The mismatches are recognized by specific proteins that guarantee the removal of the wrongly incorporated nucleotide by an exonuclease. The DNA pol δ or ε holoenzyme (DNA pol δ or ε, RF-C and PCNA) gets therefore a second chance to correctly incorporate the base.

Translesion DNA synthesis (TLS):

Lesion on the DNA that are in front of the replication DNA pols are resolved by specific translesion DNA pols that can first insert a nucleotide opposite the lesion. Even-though these translesion DNA pols are per se not accurate their relatively correct synthesis over a lesion might be good enough to minimize mutations. Next special DNA pols are required to extend from the lesion and to hand over the DNA to the replication machinery again.

DNA crosslink repair:

In situations where the two DNA strands are covalently linked a complex set of pathways have to act, in which the many Fanconi's anemia proteins come into action. They work together with other DNA repair and damage avoidance pathways such as HR, NER and TLS.

V(D)J recombination and somatic hypermutation:

V(D)J recombination is a physiological event during development of pre B cells to guarantee the enormous variability of the antibodies. This double-stranded break repair is as already observed in NHEJ inaccurate, but this is of advantage to create an enormous amount of different antibodies. In addition somatic hypermutation of

Box 5.1 (*Continued*)

the immunoglobulin genes requires the activity of DNA pols to finally introduce mutations at both A/T and C/G base pairs. In summary nature has borrowed the low fidelity of DNA pols to enlarge the variability of antibodies.

Chromosome end replication:

Due to the unidirectional DNA synthesis by all DNA pols a problem arises at the end of linear chromosomes. This end replication problem is solved by specific enzymes called telomerases.

translesion synthesis (Section 5.4). The different DNA repair events all need DNA synthesis and therefore DNA pols (Section 5.3). The physiological double-stranded break repair of V(D)J recombination requires DNA pols (Section 5.3). Since DNA replication very often encounters DNA damages during the synthesis event, next the interplay between DNA pols at the replication fork and translesion DNA pols will be summarized (Section 5.6). Finally the DNA pols themselves are targets for checkpoint control mechanisms (Section 5.7). **Table 5.1** overviews different DNA transactions and the DNA pols required in higher eukaryotes such as mammals. More details will be given in the paragraphs below.

Figure 5.1 shows a model that was proposed by us in 2003,[3] where we postulated replication foci that can store the DNA pols and the respective auxiliary proteins (see Chapters 2.6 and 2.7 for details) to be exchanged when a replication fork encounters a lesion on the DNA (see also Section 5.6 for more details

Table 5.1   Many of the known DNA transactions need more than one DNA polymerase[a]

DNA transaction[b]	DNA pols involved
DNA replication	DNA pols α, δ and ε
Base excision repair (short patch)	DNA pols β, λ and θ
Base excision repair (long patch)	DNA pols β, δ, ε and λ
Nucleotide excision repair	DNA pols δ, ε, η and κ
Non-homologous end joining	DNA pols λ and μ
Homologous recombination	DNA pol η
Mismatch repair	DNA pol δ
Translesion synthesis: Insertion[c] Extension[c]	DNA pols η, ι, κ, δ and rev1 DNA pols ζ and κ
Interstrand crosslink repair	DNA pols θ and υ
V(D)J recombination and somatic hypermutation	DNA pols μ, λ, η, θ and TdT
Chromosome end replication	telomerase, DNA pol α

a: For details see subsequent paragraphs and the references therein. b: The DNA transactions are explained in Box 5.1. c: Depends on the lesion type. For details see paragraphs below.

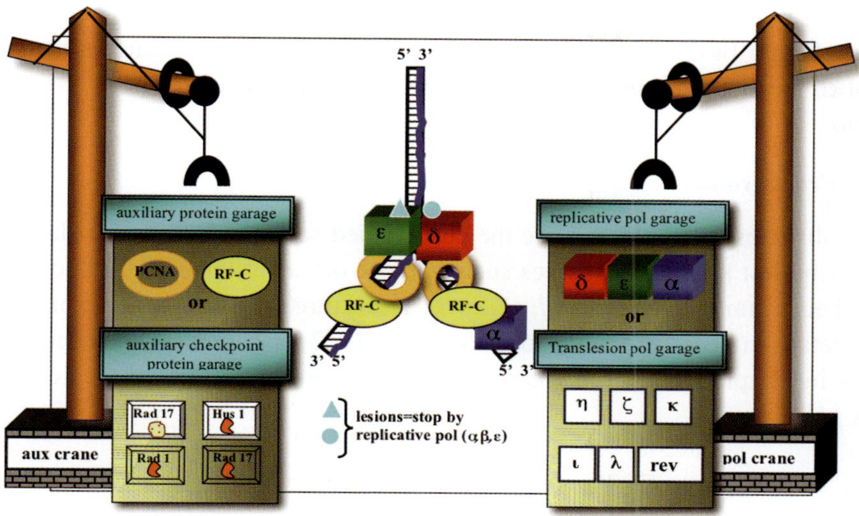

Figure 5.1   A model for the interplay between DNA polymerases involved in replication and translesion DNA synthesis. During DNA synthesis, the DNA replication machinery (schematically drawn as the DNA pol ε/δ holoenzymes (DNA pol ε/δ, RF-C, and PCNA)), eventually meets lesions on the DNA (represented with blue symbols). DNA pols ε and δ are unable to traverse DNA lesions and their arrest causes the block of the replication fork. For details of the DNA pol switch see text.

about the DNA pol switches in prokaryotes and eukaryotes). The current model predicts the existence of subnuclear compartments or foci (the "garages" in the drawing), where replicative or translesion (TLS) pols and their cognate auxiliary proteins are stored. When a replication fork stalls, checkpoints are activated leading to the recruitment of specific factors at the lesion through a yet unidentified machinery (drawn as an "auxiliary crane"). As a consequence, replicative pols are lifted off and replaced by specialized TLS pols by a "polymerase crane". After damage bypass, the normal replication machinery is reconstituted through an inverse mechanism. PCNA constitutes the common element to all these pathways, providing essential interactions with both replicative and TLS pols, as well as acting in the checkpoint process (see also posttranslational modification of PCNA in Section 5.6.2.1).

5.2   DNA Replication in Living Organisms Requires Three DNA Polymerase Molecules at the Replication Fork

Very recently a database of information and resources for the eukaryotic replication community has been established.[6] The database (http://www.dnareplication.net)

summarizes organism-sorted data on replication proteins in the categories of nomenclature, biochemical properties, motifs, interactions, modifications, structure, cell localizations and expression, as well as general comments.

5.2.1 *Prokaryotes*

The replisomes in prokariotes are the best studied so far.[7,8] Details are elaborated in the simplest known replisomes such as those of bacteriophages T7 and T4 and in the bacterium *Escherichia coli (E. coli)*. These three replisomes are explained in more details in Sections 5.2.1.1 for T7 and in 5.2.1.2 for *E. coli* (in addition to what shown in Chapter 3.3 and in Chapter 6.1). The priority will be on the actions of DNA pols.

5.2.1.1 *Bacteriophage T7: The simplest but best known replisome*

More than fifteen years ago the laboratory of Charles Richardson established for the first time the coordination of leading and lagging strand DNA synthesis at the replication fork of the *E. coli* bacteriophage T7.[9] This bacteriophage has minimal requirements for replication of double-stranded DNA (**Table 5.2**).

The T7 gene 5 codes for the DNA pol and the T7 gene 4 for the helicase and primase activities. The host organism *E. coli* provides a small protein called thioredoxin that acts in analogy to the β-subunit in *E. coli* and PCNA in eukaryotes as a processivity factor for the DNA pol. On a synthetic replication fork, primer synthesis inhibits helicase activity and the rate of lagging strand synthesis, by the combined action of helicase/primase and DNA pol/thioredoxin, dictates the rate of the leading strand advance. It was then proposed that leading and lagging

Table 5.2   Bacteriophage T7: Minimal requirements for the replication of a double-stranded DNA molecule

Gene product	Function[a]
T7 gene 5 protein	DNA pol[b]
E. coli thioredoxin	processivity factor for DNA pol[b]
T7 gene 4 protein	Helicase/Primase
	p63: helicase/primase
	p56: helicase only
T7 gene 2.5 protein	Single-stranded binding protein

a: For more functional details see Ref. 10. b: These two proteins constitute the DNA pol holoenzyme. This complex is commercially available under "Sequenase" for routine sequencing.

strand synthesis are coupled and that the replication proteins are recycled.[9] The T7 replisome can synthesizes DNA with a processivity greater that 17000 nucleotides added per cycle. This result was obtained by conventional measurements and by single molecule technology and indicated that a dynamic processivity is possible through two modes of interaction between the DNA pol and the helicase.[11] While during DNA synthesis the DNA pol and the helicase interact through a high affinity site, after 5000 bases the DNA pol dissociates from the DNA but stays bound to the helicase via electrostatic binding. The DNA pol finally transfers via electrostatic interaction around the hexameric helicase to trace another primer-template synthesized by the primase (**Figure 5.2**). A recent and very elegant work, again employing single molecule technology, revealed how the dynamics of the DNA replication loops dictate a temporal control of the lagging-strand synthesis.[12] Based on the initial "trombone model"[13] the formation and the release of the replication loops by individual T7 replisome molecules supported the coordinated DNA replication. In addition is was shown that the DNA primase acts as a molecular break in DNA replication.[14] Moreover, the initiation of primer synthesis and the completion of the Okazaki fragments can each serve as a trigger for the loop release.

This suggests that a safe two step mechanism guarantees the timely reset of the replisome after the synthesis of an Okazaki fragment. Finally, the single molecule technology was also applied to study DNA replication on a rolling

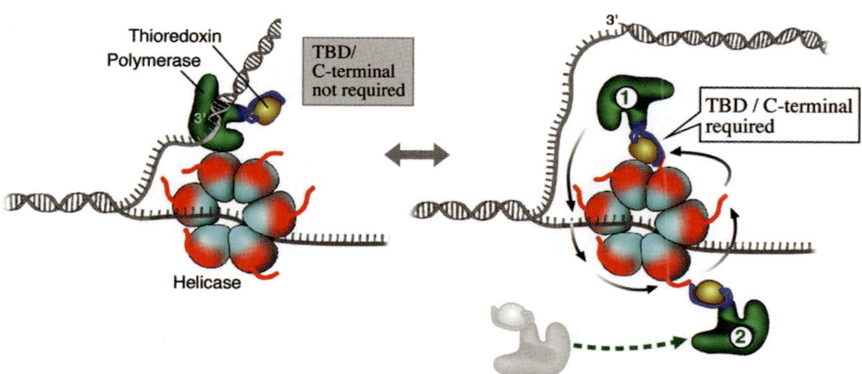

Figure 5.2   Two binding modes can mediate the DNA polymerase/thioredoxin and helicase interaction at the T7 replication fork. When synthesizing DNA, the DNA pol/thioredoxin interact with a tight polymerizing mode that is independent of the C-teminal tail of the helicase. After about 5000 bases have been synthesized, the DNA pol/thioredoxin dissociates from the DNA and binds in an electrostatic mode to the C-terminal tail of the helicase (red). The DNA pol/thioredoxin can more sequentially form on C-terminal tail to another on the helicase until it finds another primer-template. (Reproduced with permission from Ref. 11.)

circle model.[15] A recent review entitled "motors, switches and contacts in the replisome" gives more details on this simple but fascinating T7 replication apparatus.[10]

5.2.1.2 *The Escherichia coli replisome*

As indicated in Chapters 2.5 and 3.1, the *E. coli* replicase DNA pol III acts as a holenzyme. DNA pol III holoenzyme contains 10 different polypeptides (**Table 2.6.**).[2] These proteins are grouped into the three major functional units: (i) the DNA pol III core consisting of three subunits, (ii) the sliding clamp consisting as a homodimeric ring and (iii) the clamp loader/polymerase connector complex consisting of six different subunits (the τ subunit is a dimer). It was shown that the *E. coli* replisome contains two DNA pols and it was postulated, according to the "trombone model",[13] that they have asymmetric functions in replicating the two DNA strands.[16] As already mentioned for bacteriophage T7, the *E. coli* helicase, the hexameric Dna B protein which encircles the lagging strand, plays an important role for the asymmetry of the two DNA pol III molecules. Moreover the two DNA pol III assemblies act in a symmetrical way again supporting the "trombone model". The Dna B helicase contacts the τ-subunit of the DNA pol III holenzyme and thus, together with the β-clamp, holds the leading strand DNA pol III on the DNA,[17] allowing processive DNA synthesis. The Dna B helicase/τ-subunit interaction can speed up the helicase movement[18] and also generates the asymmetry in the dimeric DNA pol III acting on double-stranded DNA. In subsequent years the dynamics at the replication fork were refined (reviewed in Ref. 19) and are schematically drawn in **Figure 5.3**. While leading strand synthesis is continuous and needs only a few priming events, the lagging strand has to be initiated every 1000 bases and is consequently discontinuous. First, the primase synthesizes an RNA piece on the lagging strand and dissociates after synthesis (**Figure 5.3 a**). Second, the γ-complex, which is the clamp loader, uses the energy of ATP to open the ring-like β-clamp and loads it onto the primer placed at the lagging strand. The γ-complex leaves the DNA and lets the β-clamp to wait for the next DNA pol III (**Figure 5.3 b**). Third, the DNA pol III that has completed the previous Okazaki fragment leaves the DNA and cycles to the β-clamp that has been loaded by the γ-complex (**Figure 5.3 c**). Fourth, the lagging strand DNA pol III synthesizes the next Okazaki fragment (**Figure 5.3 d**). Finally, the next round of Okazaki fragment synthesis can start again.

Recent data, however, suggested that the DNA pol III holoenzyme appears to assemble into a replisome that contains three DNA pol III[8] (**Figure 5.4**). They all seem to have the capacity to act simultaneously.[20] In addition, the trimeric DNA

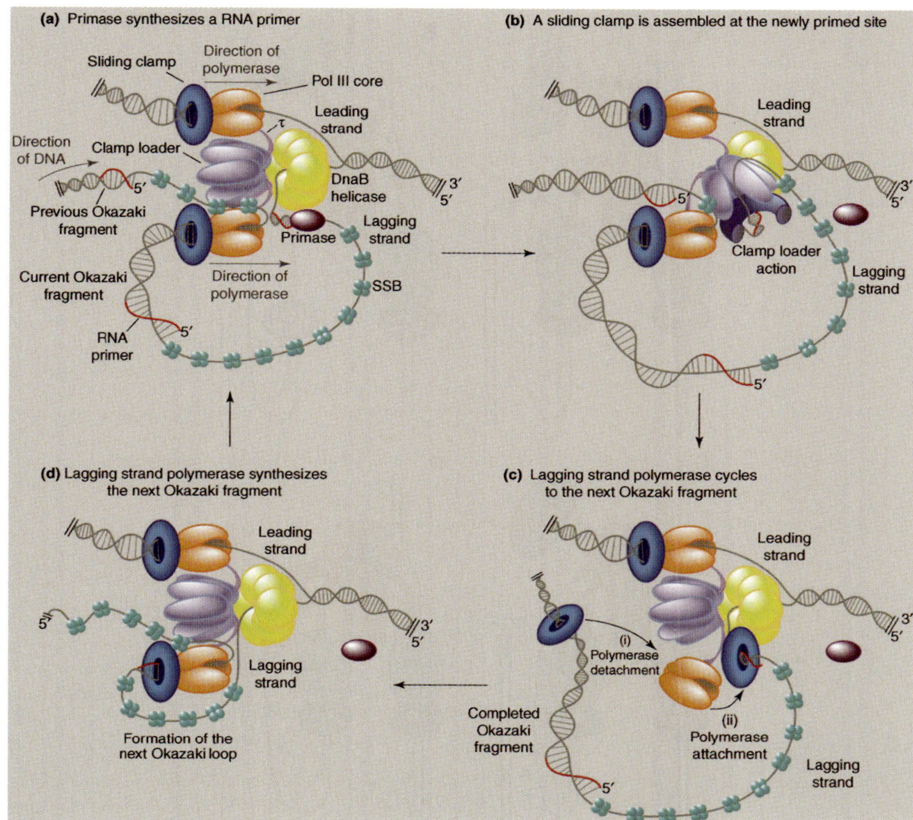

Figure 5.3 Protein dynamics at the *E. coli* replication fork. The leading strand synthesis requires a few priming events, while the lagging strand is initiated about every 1000 bases leading to discontinuous DNA pieces called Okazaki fragments. (a) Primase synthesizes a RNA primer and it dissociates after primer synthesis is completed. (b) The γ-complex clamp loader uses the energy of ATP to transiently open the β-clamp ring thus loading the β-clamp on the DNA. The γ-complex dissociates from the DNA thus leaving the β-clamp on the primer-template junction. (c) The lagging strand DNA pol III leaves, dissociates from the β-clamp as soon as an Okazaki fragment is completed and cycles to the next Okazaki fragment synthesized by the primase. (d) Finally the lagging strand DNA pol III synthesizes the next Okazaki fragment. (Reproduced with permission from Ref. 19.)

pol III functionally interacts with other replisome components such as the Dna B helicase, the Dna G primase and the β-subunit sliding clamp. The data suggested that two DNA pol III act as leading and lagging strand replicases, respectively. The third DNA pol III, even though not bound to DNA but to the replisome, may be a reserve enzyme for purposes such as when the lagging strand has problems to follow the speed of the leading strand or when the leading and/or lagging strand are stalled due to DNA damage (see also Section 5. 6).

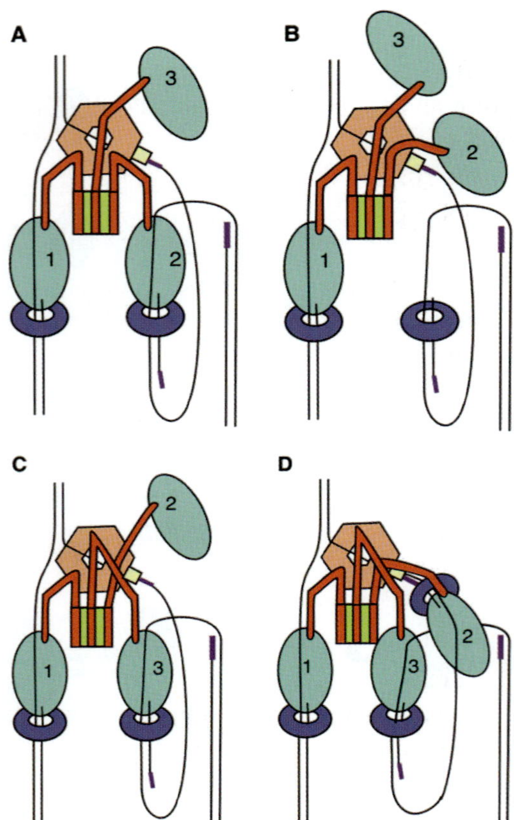

Figure 5.4   The three DNA polymerase III molecules and their dynamics at the *E. coli* replication fork. (A) The three DNA pol III molecules are held together and to the replicative Dna B helicase with three τ-subunits (orange) of the clamp loader complex (green). Primase (yellow) synthesizes RNA (purple). The DNA pol III engaged in DNA synthesis is complexed to the β-clamp (blue). (B) A transient dissociation of one DNA pol III leaves a β-clamp and 3′ end free to engage translesion DNA pols. (C) After translesion DNA synthesis the third DNA pol III binds again to the β-clamp thus (D) allowing three DNA pol III to act, namely one on the leading strand and two on the lagging strand of the replication fork. (Reproduced with permission from Ref. 8.)

Finally, single molecule studies of fork dynamics in *E. coli* DNA replication indicated[21] that when DNA pol III is coupled to the Dna B helicase it has a processivity of 10 500 bases. Upon addition of the Dna G primase the processivity is reduced three-fold. This reduction is caused by the protein-protein interaction of Dna B and Dna G and is not due to the primase activity of Dna G. In summary it was proposed that the modulation of the Dna B helicase activity by its interaction with Dna G primase prevents leading strand synthesis from outpacing the lagging strand due to the slower synthesis on the lagging strand.[21]

5.2.2 *Eukaryotes*

In eukaryotes three different DNA pols are engaged in DNA replication, being DNA pol α, DNA pol δ and DNA pol ε.[1] In a historical context (for details see Chapter 1) DNA pol α was the first DNA pol to have a role in eukaryotic DNA replication.[22] A few years later the primase activity was identified in one of the four subunit DNA pol α complex.[23] In the yeast *S. cerevisiae* DNA pol δ was found to be essential[24] and soon thereafter the essentiality of DNA pol ε was again documented in *S. cervisiae*.[25] These results suggested that three different DNA pols share the workload of DNA replication in the entire eukaryotic kingdom. While DNA pol α was assigned to be the initiating DNA pol, due to its associated primase, there was a debate over more than fifteen years about the *in vivo* functions of DNA pols δ and ε. Biochemical studies suggested a function of DNA pol δ at the lagging strand since it was able to idle at the lagging strand to provide a ligatable nick during lagging strand DNA replication.[26] Moreover, DNA pols δ and ε have both high processivity, but the effect of PCNA and their binding affinity to DNA was different, suggesting that DNA pol ε with its high affinity to DNA was more suited as the leading strand DNA pol, while DNA pol δ with a lower affinity to DNA might be a candidate for the frequent relocation at the lagging strand of the replication fork.[27]

The eukaryotic DNA replication mechanistic details were elaborated first in the simian virus 40 (SV 40) *in vitro* reconstituted system (reviewed in Ref. 28). This viral model suggested that DNA pol α is the initiating pol at both strands and that a DNA pol switch from DNA pol α to DNA pol δ allows the latter to replicate both strands. Moreover, no DNA pol ε was required for SV 40 replication *in vitro* (**Figure 5.5 A**). In *S. cerevisiae*, thanks to the power of genetics, convincing data were reported that DNA pol ε is the leading strand DNA pol[29] and that DNA pol δ the lagging strand DNA pol,[30] respectively. Special amino-acids were mutated separately in DNA pol ε and δ, that did not affect the DNA synthesis rate, but reduced their fidelity. By using these mutants it was possible to analyze the frequency of mutation in a reporter gene closed to the origin of replication and to qualify DNA pol ε as the leading strand and DNA pol δ as the lagging strand enzyme at the DNA replication fork in eukaryotic cells. Likely, these data can be transferred to higher eukaryotic cells including human (**Figure 5.5 B**).

Recent work addressed which DNA pols are recruited for chromosome end replication.[32] Data suggested that the leading and the lagging strand DNA pols arrived differently at the telomers. DNA pol ε, the leading strand enzyme, arrived earlier at the telomers than the lagging strand DNA pols α and δ. Furthermore the S-phase specific recruitment of Trt1, Pot1 and Sts1, components of the telomer replication machinery, matched the arrival of DNA pol α. These data suggested

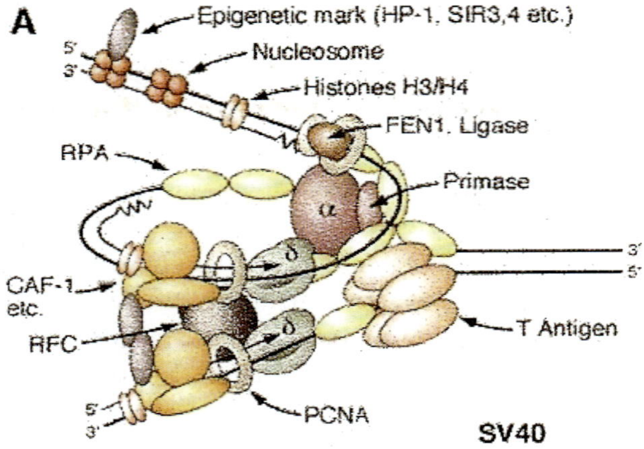

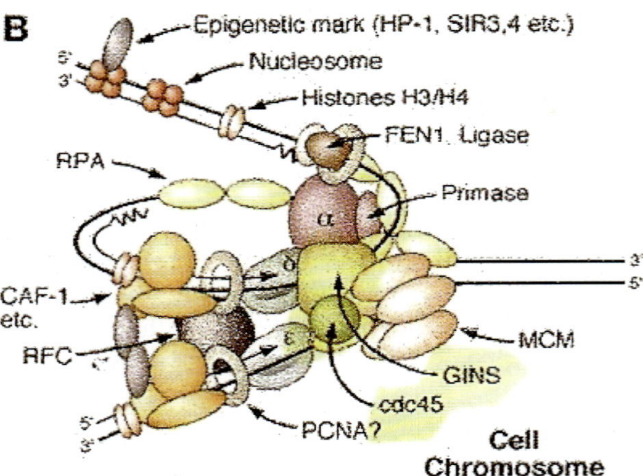

Figure 5.5   SV40 replication needs DNA polymerases α and δ, while the eukaryotic cellular replication fork relies on the three DNA polymerases α, δ and ε. Note that in the case of SV40 DNA pol δ replicates both strands (A), while in the cellular context DNA pol ε is the leading and DNA pol δ the lagging strand replicase (B), respectively. For details see text. (Reproduced with permission from Ref. 31.)

that the recruitment of the lagging strand replication machinery is delayed until the telomerase enzyme itself is recruited.

5.2.3 *Proofreader versus Non-proofreader DNA Polymerases*

DNA pols are not absolutely accurate *per se* (see also Chapter 2). An important activity that enables the replicative enzymes to reduce the misincorporation to a low level is the associated $3' \rightarrow 5'$ exonuclease, also called proofreading activity (reviewed in Ref. 33). The replication fidelity largely depends on this activity. DNA pol δ and ε each contains a proofreading activity but not the initiating DNA pol α.[34] Proofreading *in trans* is an attractive hypothesis in which a non-proofreading DNA pol will cooperate with a proofreading DNA pol.[35] A mutant of DNA pol α with normal activity and processivity, but reduced fidelity, has a strong mutator phenotype. This mutator phenotype was increased when the proofreading exonuclease of DNA pol δ, but not of DNA pol ε was inactivated. This led to the hypothesis that the $3' \rightarrow 5'$ exonuclease of DNA pol δ could proofread for the error-prone DNA pol α after initiation of an Okazaki fragment.[36] Such intramolecular proofreading might also be of relevance in DNA repair and translesion DNA synthesis.

5.3 Different DNA Repair Pathways Have Their Own DNA Polymerases, But Can also Borrow Them from the Replication Machinery

Thirty years ago, when only DNA pols α, β and γ were known, data were presented indicating that DNA pol β was a repair enzyme.[22,37] DNA pol β was isolated from rat adult neuronal nuclei and was shown to perform ultraviolet-induced repair DNA synthesis *in vitro*. Since DNA pol β was virtually the exclusive DNA pol in these postmitotic nuclei it was concluded that DNA pol β was responsible for the observed DNA repair. This was further substantiated by demonstrating a virtually complete suppression of DNA repair in irradiated nuclei by $2',3'$-dideoxyribosylthymine $5'$-triphosphate (d_2TTP), a potent DNA pol β inhibitor. It was therefore speculated that the ability to perform repair DNA synthesis is not unique to the neuronal DNA pol β but is a general function of DNA pol β in all tissues.[38] This statement is still true, but, as summarized in **Table 5.3**, has to be refined for different DNA repair pathways and in view of the discoveries of many new DNA pols since then.

The DNA repair pathways mentioned below are complex events that cannot be described in details here. We focus only on the re-synthesis step in all these pathways. For more detailed informations on the mechanisms of DNA repair, the reader is referred to the excellent book "DNA Repair and Mutagenesis" (Friedberg *et al.*, ASM Press Washington, 2006).

Base excision repair (BER): The mammalian genome suffers from many endogenous and exogenous insults that modify DNA leading to base loss or base alterations. It has been estimated that approximately 100 000 modifications occur daily in the DNA of a single cell.[39] In order to remove damage and to maintain the integrity of the genome, different DNA repair pathways have evolved. The major repair

Table 5.3   DNA polymerases involved in the different DNA repair pathways[a]

DNA repair pathway	DNA pols engaged
Base excision repair (short one nucleotide patch)	DNA pols β, λ and θ
Base excision repair (long 2–12 nucleotide patch)	DNA pols β, ε, δ and λ
Nucleotide excision repair	DNA pols ε, δ, η and κ
Mismatch repair	DNA pol δ
Non homologous end joining	DNA pols λ and μ
Recombination of immunoglobulin genes	DNA pols λ, μ, and TdT
Hypermutation of immunoglobulin genes	DNA pols θ and η
Homologous recombination	DNA pol η
Interstrand cross-link repair	DNA pols θ and υ

a: see also **Table 5.1**.

pathway protecting cells against a single-base DNA damage is base BER.[39] It is often initiated by a damage type specific DNA glycosylase that cleaves the *N*-glycosidic bond of the damaged base, leaving an apurinic/apyrimidinic site, also referred to as an abasic site or AP site. Subsequently, AP endonuclease 1 (APE 1) cleaves the sugar-phosphate backbone at the 5 -side of the AP site resulting in $3'$-hydroxyl and $5'$-deoxyribose phosphate (dRP) group flanking a one-nucleotide gap. DNA pol β inserts a first nucleotide into the gap, leaving nicked DNA with a $5'$-dRP flap. At this point, the repair can be accomplished via two different BER sub-pathways depending on the nature of AP site (reviewed in Refs. 39 and 40). In the case of regular AP sites, DNA pol β removes the $5'$-deoxyribose-5-phosphate ($5'$-dRP) group through its associated dRP-lyase activity, and the resulting nick is sealed by the DNA ligase III/XRCC1 complex. This sub-pathway is referred to as short-patch BER (SP-BER). However, if the sugar group is oxidized or reduced, DNA pol β cannot remove the $5'$dRP moiety and repair proceeds through the alternate long-patch BER (LP-BER) sub-pathway involving removal and replacement of 2–12 nucleotides. It has been found that, in the LP-BER pathway, the collaboration of DNA pols δ and ε, proliferating cell nuclear antigen (PCNA), replication factor C (RF-C) and flap endonuclease 1 (Fen 1) can displace the strand $3'$ to the nick and synthesize up to 12 nucleotides.[41,42] The resulting flap is cut by Fen 1 and the final nick is sealed by DNA ligase I. However, DNA pol β is also a key enzyme in LP-BER,[43] that not only initiates both sub-pathways but can also perform efficient strand displacement via the "hit and run" mechanism. In this model DNA pol β and Fen 1 act successively, followed by the action of DNA ligase I.[44]

Other DNA pols also appear to function in BER. First, dRP-lyase activities have been identified in DNA pol λ[45] and DNA pol θ,[46] suggesting a role in SP-BER.

DNA pol λ has also been described as a back-up BER activity in extracts from mouse embryonic fibroblasts.[47] The cooperation of DNA pols θ and β was suggested in BER of oxidative damage.[48] The adenine misincorporated by DNA pols δ and ε opposite 7,8-dihydro-8-oxoguanine (8-oxo-G), can be removed by a specific glycosylase MUTYH, leaving the lesion on the DNA. Subsequent incorporation of C opposite 8-oxo-G on the resulting one nucleotide gapped DNA is essential for the removal of the damaged base. By using model DNA templates, purified DNA pols β and λ and knockout cell extracts, it was demonstrated that the auxiliary proteins RP-A and PCNA act as molecular switches to activate the DNA pol λ-dependent highly efficient and faithful repair of A:8-oxo-G mismatches in human cells, while suppressing the action of the less faithful DNA pol β.[49] Recent data suggested that the switch from DNA pol β to DNA pol λ by RP-A and PCNA can be a switch from SP-BER (1 nucleotide by DNA pol β) to LP-BER (2 nucleotides by DNA pol λ) in faithful insertion of a C opposite 8-oxo-G.[50]

Nucleotide excision repair (NER). In NER the product left for DNA repair re-synthesis is a 30-base gap. Initially it was found that PCNA is required for NER suggesting that the two replicative DNA pols δ and ε are involved in NER re-synthesis.[51] Indeed the first complete NER *in vitro* system suggested the involvement for either DNA pol δ or ε.[52,53] That both DNA pols are involved makes sense since leading strand DNA pol ε could be borrowed for NER at the leading strand and lagging strand DNA pol δ at the lagging strand, respectively. It should be noted that the Xeroderma pigmentosum Variant gene product, identified as DNA pol η,[54] protects against UV-damage[55] (see also Section 5.4). DNA synthesis in eukaryotes requires generally two DNA pols: an inserter and an extender. DNA pol κ-deficient cells had a defect in UV damage repair that could be restored to normal sensitivity and NER, suggesting that DNA pol κ also plays a role in NER.[56]

Mismatch repair (MMR). Reconstitution of the MMR pathway *in vitro* clearly showed that DNA pol δ is the main enzyme involved in the resynthesis step after the proper recognition of the mismatch and the resection by the exonuclease 1 (EXO1).[57]

Non-homologous end-joining (NHEJ). NHEJ is the process that repairs double-stranded breaks as a consequence from DNA damage or in physiological way during V(D)J recombination (see below). DNA pols λ, DNA pol μ and TdT all have a BRCT-1 C terminal domain which was shown to interact with component of the NHEJ machineries such as Ku and XRRC4/DNA ligase IV.[58−62] Moreover, these three DNA pols were found after irradiation either in double-stranded break complexes or shown to co-localize with them. Furthermore it was found that when DNA pol λ or DNA pol μ were knocked out, recessed DNA ends decreased in NHEJ, suggesting again roles for these two DNA pols in NHEJ.[63,64] Finally, the same result was obtained in *S. cerevisiae* where DNA pol IV, the DNA pol λ-like counterpart and the only X-family DNA pol from yeast (see **Table 2.2** in Chapter 2), physically interacted with Dnl4-lif1 (the homologs of the human MRN and XRCC1/DNA ligase IV complexes, respectively) thus linking DNA synthesis to NHEJ.[65]

Recombination of immunoglobulin genes. V(D)J recombination is the process in which the variable (V), diversity (D) and joining (J) regions of the germline immunoglobulin genes are transiently broken and subsequently rejoined to allow a great variety of different immunoglobulin proteins being synthesized (reviewed in Ref. 66). First it was shown that DNA pol μ participates in kappa light chain gene rearrangement but not heavy chain gene rearrangement.[67] Next it was found that immunoglobulin heavy chain junctions from DNA pol λ-deficient animals have shorter length with normal N-additions, thus suggesting that DNA pol λ might be recruited during heavy chain rearrangement at a step that precedes the action of TdT.[68] In contrast to previous *in vitro* studies, analysis of animals with combined inactivation of these enzymes revealed no overlapping or compensatory activities for V(D)J recombination between these three DNA pols. This complex usage of DNA pols with distinct catalytic specificities might correspond to the specific function that the third hypervariable region assumes for each immunoglobulin chain, with DNA pol λ maintaining a large heavy chain junctional heterogeneity and DNA pol μ ensuring a restricted light chain junctional variability.[68] The role of these three X family DNA pols was confirmed by biochemical data in which a gradient of template dependencies was proposed to have distinct biological roles for DNA pol λ, DNA pol μ and TdT.[62]

Hypermutation of immunoglobulin genes. Besides the V(D)J recombination by NHEJ there is a second cycle of diversification that is called somatic hypermutation (SHM). During SHM base substitutions are introduced into a region that surrounds about 1000 bases of the antibody coding sequence.[69] SHM is induced by the enzyme activation-induced cytosine deaminase (called AID[70] and reviewed in Ref. 71). AID can deaminate cytosine to uracil in the DNA. The enzyme has a strong preference for 5′WRC sequences (where W = A or T and R = A or G) in single-stranded DNA and it can carry out multiple C deaminations on individual DNA strands, rather than jumping from one strand to another.[72] Soon after its discovery DNA pol μ was proposed as a candidate for SHM,[73] but transgenic knock out mice for DNA pol λ and DNA pol μ were viable, fertile and did not have an abnormal SHM phenotype.[74] Later it was demonstrated that DNA pol θ plays a dominant role in SHM, likely by introducing mismatches while bypassing abasic sites generated by the Uracil DNA glycosylase-mediated removal of dU residues produced by the deaminase activity of AID, and by extending past such abasic sites or by synthesizing DNA during dU-dG mismatch repair.[75] Later it was found that DNA pol η and DNA pol θ function in the same pathway to generate mutation at A/T during SHM[76] and DNA pol η appears to be the limiting factor for A/T mutations in Ig genes and might thus contribute to antibody affinity maturation.[77]

Homologous recombination (HR). Two studies suggested that DNA pol η can function to extend 3′-strands that have been exchanged during homologous recombination. The action of the strand transferase, the Rad51 protein, enables a D-loop formation that can act as a primer for DNA pols. DNA synthesis from such strand invasion intermediates of HR was exclusively possible with DNA pol η suggesting an *in vitro* role for this TLS DNA pol in HR.[78] While these studies were done with human proteins another study in chicken cells came to a similar conclusion. DT40 chicken cells can diversify their IgV genes though HR-mediated gene conversion. Cells lacking DNA pol η showed a decrease in Ig gene conversion and in double-stranded break induced HR. The defect could be rescued upon complementation with DNA pol η, suggesting that this enzyme has an important role in HR.[79]

Interstrand cross link repair (ICLR). Drosophila melanogaster harbours a gene called *Mus308*, which is implicated in resistance to interstrand-crosslinking agents. A homolog of this protein has been first cloned from human cells and was called DNA pol θ.[80] Then, with the aim to identify unconventional DNA pols from human cells, and using a special assay to fractionate HeLa extracts based on the ability to bypass DNA lesions, an aphidicolin-resistant DNA pol activity was obtained that was able to utilize either undamaged or abasic sites-containing DNA with the same efficiency. Biochemical characterization and immunoblot analysis allowed its identification as the human homolog of DNA pol θ.[81] Further studies suggested that two DNA pols could be implicated in ICLR, the first being DNA pol θ[82] and the second DNA pol υ.[83]

5.4 Translesion DNA Synthesis in Eukaryotes Generally Requires Two DNA Polymerases: An Inserter and an Extender

While we have introduced the translesion DNA synthesis (TLS) DNA pols in Chapter 1 (discoveries) and described in detail the properties of the prokaryotic TLS DNA pols in Chapter 3 and those of the eukaryotic counterparts in Chapter 4, their functions over different lesion will be described in this paragraph. Why do the TLS DNA pols have important functions in the DNA damage tolerance pathway? In contrast to the replication and the repair DNA pols, the TLS DNA pols have a much lower fidelity. Can it be of advantage when an unfaithful DNA pol acts on the DNA?

The problem: DNA is often chemically altered. Bases in the DNA can be chemically attacked by environmental agents such as UV-light, radioactivity, various toxic chemicals and reactive oxygen species. Moreover, many endogenous DNA damages are the consequence of hydrolysis (depurination, depyrimidation, spontaneous deamination (e.g. C leading to a U, 5-methyl-C leading to T), oxidation (8-oxo-G, thymine glycol, cytosine hydrates, lipid peroxidation products such as etheno-A and etheno-G) and nonenzymatic methylation (7-methyl-G, 3-methyl-A and O^6-methyl-G). These attacks are very frequent: it is estimated that 10 000-100 000 alterations in the 6×10^9 bases of human genome occur per cell per day! It is a logic consequence that nature has invented different strategies to deal with all these problems.

In view of their effect on DNA synthesis by DNA pols, the lesions can be classified into "coding", "blocking" and "miscoding". First, a lesion is still "coding" when the altered base in the DNA can base pair, like the undamaged base, with the correct nucleotide (G-C and A-T). Second, a lesion is "blocking" when the alteration in the DNA base dramatically changes its shape and size so that the access of the damaged base to the active site of the DNA pol is restricted. This then results in a stop of DNA synthesis. Third, and this might be the most dangerous case, "miscoding" lesions are those that base pair with an incoming nucleotide different from the one predicted by the Watson-Crick rules. **Table 5.4** shows three examples of base modifications and their consequence on eukaryotic DNA pols.

Table 5.4   Behavior of eukaryotic DNA polymerases on three different DNA lesions

DNA pol	Large modification (TT dimer)	Small modification (8-oxo-G)	Depurintion (AP site)	Error rate[a] ($\times 10^{-5}$)
DNA pol α	blocking	miscoding	blocking	10
DNA pol δ	blocking	miscoding	miscoding	1–0.1
DNA pol ε	blocking	miscoding	?	2–0.5
DNA pol γ	?	miscoding	blocking	1–0.1
DNA pol β	miscoding	miscoding	miscoding	10
DNA pol λ	?	miscoding[b]	miscoding	100
DNA pol μ	coding	miscoding	miscoding	10
DNA pol η	coding	miscoding[c]	miscoding	3500
DNA pol ι	blocking	miscoding	blocking	72 000[d]
DNA pol κ	blocking	miscoding	blocking	1000

a: Base substitution rate. b: DNA pol λ in the presence of PCNA and RP-A is coding.[49,84] c: DNA pol η in the presence of PCNA and RP-A is partially coding.[84] d: For the preferred T-G mismatch.

The TT dimer is representative of a large modification, 8-oxo-G of a small base modification and finally the AP site represents a complete lack of a coding base as a consequence of depurination. It is obvious that a given lesion (e.g TT dimer) can be blocking, miscoding, or coding for different DNA pols and a given DNA pol (e.g. DNA pol η) can be blocked, miscode or code in the presence of different lesions. **Table 5.4** therefore suggests the tremendous variability that we have with the many DNA pols having different properties.

The behavior of the DNA pols does not necessary correlate with the intrinsic fidelity of DNA pols on undamaged DNA (**Table 5.4**). The replicative family B DNA pols α, γ, δ and ε are the least tolerant against modified DNA. This makes sense in the view of their task to faithfully duplicate the genome. On the other hand the family Y DNA pols η, ι and κ together with the family X DNA pol λ are special enzymes that can act in the DNA damage tolerance pathway called TLS.[85] The TLS is an extremely important mechanism that supports DNA synthesis over lesions that could otherwise not be handled by the high fidelity replicative DNA pols α, γ, δ and ε. If a replication fork stalls as a lesion an event called S-phase checkpoint is triggered. If a fork is collapsing this might lead to DNA strand breaks, to aberrant recombination and even to programmed cell death (apoptosis). The latter might happen when the S-phase arrest is too long or if the accumulated DNA damages lead to aberrant mitosis. It is therefore important that the TLS pathways are immediately active by employing specialized DNA pols that are suited to efficiently cope with particular lesions, eventually allowing the replisome to continue DNA synthesis (**Figure 5.1**) Details about the DNA pol switch will be discussed in Section 5.6.

TLS occurs in two steps: first one or two bases, depending on the lesions, are inserted opposite the damage and, second, an extension from this lesion has to be carried out. We still only consider here these two fundamental TLS steps and will discuss the DNA pol switch from a replicative or a repair DNA pol in Section 5.6. Data from many laboratories suggested that many TLS pols are very proficient in incorporating nucleotides opposite certain lesions, but were unable to extend the resulting primer terminus.[85,86] On the other hand, DNA pols were discovered that were efficient in extending from a variety of mispaired of aberrant primer termini. For some lesions (eg. an AP, site or a T-glycol) even the faithful DNA pol δ is able to incorporate but not to elongate. The two categories were designated as TLS "inserter" DNA pols or as TLS "extender" DNA pols. While there are many "inserter" DNA pols (**Table 5.5**), the current knowledge is that in most cases DNA pol ζ and in rare cases DNA pol κ are "lesion extenders". Consequently we have an interplay, also called a DNA pol switch, between the "inserter" and the "extender" DNA pol. The outcome of the TLS can either be error-free or error-prone. The latter might lead to mutations in the next replication round.

Table 5.5   Insertion and extension over DNA lesions

DNA lesion	Inserter DNA pol	Extender DNA pol	Consequence
cys-syn TT	DNA pol ι	DNA pol ζ or κ	error-prone
6-4 TT	DNA pol η or ι	DNA pol ζ	error-prone
6-4 TC	DNA pol η	DNA pol ζ	error-free
	DNA pol ι	DNA pol ζ	error-prone
6-4 CC	DNA pol η	DNA pol ζ	error-free
	DNA pol ι	DNA pol ζ	error-prone
AP site	DNA pol ι	DNA pol ζ	error-prone
	DNA pol η	DNA pol ζ	error-prone
	DNA pol δ	DNA pol ζ	error-prone
T-glycol	DNA pol δ	DNA pol ζ	error-free
AAF adduct	DNA pol η	DNA pol ζ	error-free
N^2-dG acrolein adduct	DNA pol ι	DNA pol κ	error free
(+/−)BPDE-N^2dG	DNA pol κ	DNA pol κ	error-prone
8-oxo-G	DNA pol δ[a]	DNA pol δ	error-prone[a]
	DNA pol ε[a]	DNA pol ε	error-prone[a]
	DNA pol α[a]	DNA pol α	error-prone[a]
	DNA pol β[a]	DNA pol β	error-prone[a]
	DNA pol η	DNA pol η	error-prone
	DNA pol η, PCNA/RP-A	DNA pol η	partially error-free
	DNA pol λ	DNA pol λ	error-prone
	DNA pol λ/PCNA/RP-A	DNA pol λ	error-free

a: PCNA and RP-A do not render these DNA pols error free over 8-oxo-G.

From history to present days: about error prone versus error free TLS. UV irradiation of bacteriophages λ and ΦX174 resulted in killing but not mutagenesis of the phages after the infection of untreated host cells, whereas UV irradiation of the host cells caused mutagenesis of both untreated and irradiated phage.[87,88] Furthermore, UV irradiated single-stranded ΦX174 DNA extracted from untreated host cells was found to be replicated only to the first pyrimidine dimer.[89] Caillet–Fauquet *et al.*[90] found that UV irradiation of the host cells led to an enhancement of DNA synthesis on UV irradiated ΦX174 DNA *in vivo*, which led them to propose that pyrimidine dimers in the ΦX174 DNA can become mutagenic by causing misincorporation of deoxyribonucleotides opposite them.

These experiments laid down the foundation for the formulation by Miro Radman and Evelyn Witkin of the so called "SOS" (Save Our Soul) hypothesis that postulates that a complex cellular emergency response is triggered by unrepaired DNA lesions. Implicit in the SOS hypothesis was the fact that the damaged cell could save their life at the price of increased mutagenesis and, therefore, that whatever mechanism of replication of the DNA damage was involved, it was error prone. Indeed the subsequent discovery that the SOS response in *E. coli* controlled the induction of the error prone DNA pols IV and V, belonging to the recently discovered Y family of DNA pols (see Section 3.1.4 of Chapter 3), reinforced this idea. This family of DNA pols is characterized by the absence of an associated $3' \rightarrow 5'$ proofreading exonuclease activity and by a relaxed structure that allows to accommodate bulky DNA lesions, such as pyrimidine dimers, in their active site, and have been named Trans Lesion Synthesis or TLS DNA pols.

However, as usual, life is not so simple and the notion, supported by numerous data, that Y family DNA pols always act as low fidelity DNA pols during translesion synthesis was challenged by the fact that XP-V patients, who bear mutations in the Rad30/pol η alleles, are prone to skin and other cancers, suggesting that the function of the DNA pol η may be antimutagenic in humans (for a recent review see Ref. 91). Indeed DNA pol η has been shown to replicate cys-syn cyclobutane pyrimidine dimers or CPDs, one of the major lesions resulting from UV irradiation, with high accuracy and efficiency.[92] Furthermore, DNA pol η independent bypass of CPD, which is thought to involve other TLS DNA pols, appears to be significantly more mutagenic, presumably accounting for the increased frequency of cancers in XPV patients. Very recently the identity of some of the backup Y family DNA pols that carry out the mutagenic TLS of CPDs has been uncovered.[93] The situation is made more complicated, but even more interesting, by a recent finding in Livneh's group, in collaboration with others, that three main classes of TLS seem to exist in mammalian cells: two rapid and error free and a third slow and error prone.[94] These results indicate that a single DNA pol, namely DNA pol ζ, plays a pivotal role in all the classes of TLS and that discrete two-polymerases combinations between ζ and other TLS DNA pols dictate error-prone or error-free TLS across the same lesion.

Recently a novel role of DNA pol η was suggested, since it appears that DNA pol η is also required for common fragile site stability during unperturbed DNA replication.[95] When DNA pol η was depleted by short hairpin RNA–mediated depletion this affected unperturbed DNA replication. These data challenge the current idea that pol η is only responsible for TLS damage tolerance following exogeneous DNA damage.

Finally, auxiliary replicative factors such as PCNA and RP-A could modulate the fidelity of TLS by a given DNA pol. One example of such behavior is the finding that PCNA and RP-A allowed the correct incorporation of dCMP opposite the very mutagenic lesion 8-oxo-guanine 1200-fold more efficiently than the incorrect dAMP by the X family DNA pol λ.[84] This effect of PCNA and RP-A is not restricted to a particular type of DNA pol since it was also observed, although at a lower extent, for DNA pol η.

Clearly more genetic and biochemical studies are needed to begin to understand what makes a DNA pol capable, alone or in combination with others, to replicate in an error free or error prone mode a DNA lesion. Nevertheless a systematic cataloguing of how a specific DNA damage can be bypassed in a mutagenic or faithful way and by which DNA pol may have considerable consequences for the prevention or even cure of some human cancers.

5.5 Expression of DNA Polymerases

Regulation of the fifteen DNA pols is undoubtedly a very important issue. Too much of an unfaithful DNA pol as well as reduced levels of a faithful DNA pol can cause genomic instability. A few examples are:

(1) DNA pol β, a rather inaccurate DNA pol from the X family, has been shown to confer genetic instability when up-regulated in cells, thus resulting in muta-genesis, aneuploidy and tumorigenesis.[96] Moreover, overexpression of DNA pol β strengthens the mutagenicity of oxidative damages, concomitantly with a higher cellular sensitivity and increased apoptosis. Deregulated expression of DNA pol β was proposed as a predisposition factor for mutagenic effects of oxidative stress and thus have implication in the generation and/or evolution of cancer.[97]

(2) DNA pol κ, an error-prone DNA pol from the Y family is up-regulated in lung cancers, induces DNA breaks and stimulates DNA exchanges as well as ane-uploidy. Excess DNA pol κ appears to favor the proliferation of competent tumor cells as observed in immunodeficient mice, suggesting that altered regu-lation of DNA pol κ might be related to cancer-associated genetic changes and phenotype.[98]

(3) Downregulation of faithful DNA pols of the B family induces chomosome fragile site instability. This has been documented for DNA pol α and for DNA pol δ. Such findings suggested that a change in the amounts, either absolute or relative, of replicative DNA pols results in recombinogenic DNA lesions and double-stranded breaks. Even-though these results were obtained in yeast, they are probably also valid for mammalian cells.

A study testing 68 different tumours measured the expression levels of a variety of DNA pols.[99] When the two X family DNA pols β and λ, the two Y family DNA pols ι and κ and the three replicative DNA pol α, δ and ε were tested, the data indicated that in more than 45% of these 68 tumors at least one of the X and/or Y DNA pols was overexpressed. Of particular interest was that over 30% of all tumors tested had an overexpressed DNA pol β. DNA pols λ and ι were also overexpressed but less frequently than DNA pol β. Surprisingly DNA pol κ was underexpressed in many tumor samples. Finally, a few tumors also overexpressed the three replicative DNA pols α, δ and ε while in most cases they were underexpressed. All the 68 tumors were summarized in an overexpression scale that gave the following order: DNA pol β > DNA pol ι > DNA pol λ > DNA pol α > DNA pol δ > DNA pol κ > DNA pol ε. In summary these data clearly indicated that the expression patterns of DNA pols in normal and cancer tissues are very complex and the expression levels of the DNA pols are changed in tumours and can vary from on type of a tumor to another.

5.6   DNA Polymerases Switch between Different DNA Transactions

As already indicated in this and in previous Chapters, the interplay of different DNA pols in the various DNA transactions are complex but so far not yet completely understood at the mechanistic level. In this paragraph we try to address this issue first in prokaryotes (5.6.1) and second in eukaryotes (5.6.2). In the latter the regulation of DNA pols is first considered, followed by the consequences of posttranslational modifications of DNA pols and to some extent of the DNA pol clamp PCNA.

5.6.1   *Prokaryotes*

In *E. coli* the replication task is carried out by DNA pol III. It was thought for a long time that besides this replicative enzyme, DNA pol I is responsible to replace Okazaki fragments and to act in DNA repair. The three other DNA pols II, IV and V have important functions in TLS and are therefore designated as TLS DNA pols.[100] Their discoveries are described in Chapter 2. How and when are these TLS pols engaged in the bacterial cell?

The initial observation derived from studies in which DNA pol V was shown to interact on one side with the catalytic subunit of DNA pol III (the α-subunit) and with the β-clamp.[101] Moreover, it was found that mutagenesis by DNA pol IV and V depends on the interaction motif of these DNA pols with the β-clamp and a "toolbelt model" was proposed in which the β-clamp may attract several DNA pols thus allowing the DNA pol switch between the replicative DNA pol III and the TLS pols IV and V.[102] These findings were soon confirmed by structural analysis showing how a conserved motif at the carboxyterminal end of DNA pol IV can interact with the β-clamp.[103] Competition experiments suggested that DNA pol III has the highest affinity to the β-clamp and that a second DNA pol can bind to the β-clamp. Since the three DNA pols II, IV and V are up-regulated upon SOS response, it was proposed that these three TLS DNA pols can compete for the β-clamp after DNA damage.[104] The DNA pol switch is regulated so that the action of the TLS pol IV is limited, since only under conditions that cause DNA pol III to stall on the DNA the TLS DNA pol IV can take control on the primer-template. After the stall is relieved and DNA pol III is replicating again it becomes resistant to binding by DNA pol IV suggesting that under normal replication conditions the TLS DNA pols do not have access to the fork.[105] An elegant study addressed the question what happens when the leading as well as the lagging strand during replication contain a block. The replication fork is uncoupled since a block in the leading strand let lagging strand synthesis proceed suggesting that concurrent synthesis was uncoupled.[106] When TLS was reconstituted in the presence of DNA pol III and V and in the presence of auxiliary proteins, it was found that the replicative DNA pol III is disconnected from the lesion first, then DNA pol V binds to Rec A and the β-clamp, thus allowing TLS to proceed to generate a novel DNA primer long enough for DNA pol III to continue DNA replication thereafter.[107] These findings were nicely confirmed by showing that TLS by DNA pol V is regulated by Rec A nucleoprotein filaments binding to separate single-stranded DNA molecules in *trans*.[108]

DNA pol V is regulated to properly interact with different sets of proteins. After damage the homodimeric $UmuD_2$ is overexpressed and a slow autocleavage removes 24 amino-acids from the N-terminus. This fragment $UmuD'_2$ can interact with UmuC to result in the active trimeric DNA pol V.[109,110] Recent work suggested that $UmuD_2$ and $UmuD'_2$ have different interacting partners: $UmuD_2$ has a high affinity to bind DNA pol IV and the β-clamp while $UmuD'_2$ preferentially interacts with the catalytic subunit of DNA pol III (the α-subunit) and Umu C to form the active DNA pol V.[111]

Recently it was found that DNA pol II and IV can functionally interact with the DnaB helicase and can thus regulate the speed of unwinding by slowing it down several orders of magnitude. The two TLS DNA pols might do this by exchanging DNA pol III via the β-clamp to form "alternative" replisomes even before the DNA pol III stalls at a lesion. Since the levels of DNA pol II and IV are induced upon DNA damage the two TLS can "kick out" the replicative DNA pol III. Since they bind to the DnaB helicase they can "maintain" the architecture of the replisome, which moves now very slowly (**Figure 5.6**). Such a dynamic replisome action could give additional time for DNA repair even before the DNA pol III reaches the lesion on one hand and also allowing the TLS DNA pols themselves to carry out TLS.[112]

Over forty years ago it was proposed that UV-lesions in *E. coli* were skipped during replication resulting in discontinuities and gaps in the nascent DNA and this even in the continuous leading strand of the replication fork. The gaps were finally

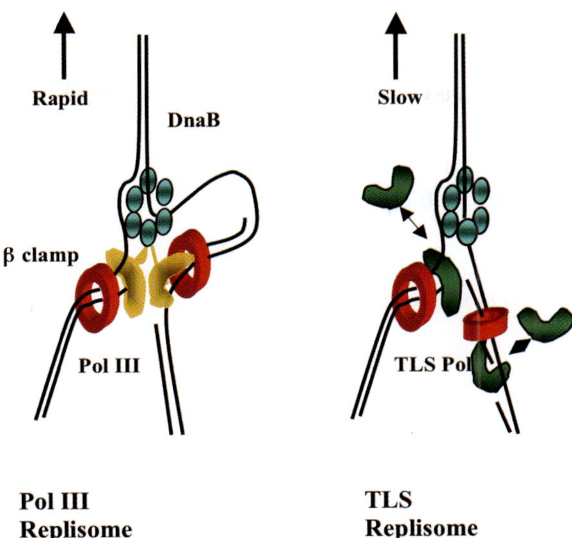

Figure 5.6   TLS DNA polymerases can form an alternative replisome. The left part shows the normal DNA pol III replisome. The τ-subunit of the clamp loader, which is not shown, dimerizes the DNA pol III for fast coupled leading and lagging strands DNA replication. The right part documents the alternative replisome in which the TLS DNA pol II and IV replace DNA pol III at the fork, interact with the DnaB helicase thus slowing down the speed of DNA synthesis. The model predicts that leading and lagging strands are uncoupled in the TLS replisome. (Adapted from Ref. 112.)

repaired.[113] Over three decades later replication fork reactivation downstream a blocked nascent leading strand was demonstrated *in vitro*. Biochemical reconstitution of the replication restart after replication block suggested that the block is bypassed and the downstream *de novo* priming event allows continuation of DNA replication but thus creates gaps in the nascent DNA.[114] These data clearly suggested that re-priming of DNA synthesis downstream of a blocked fork occurs not only on the lagging strand but also on the leading strand. So, contrary to what we generally see in the textbooks, leading strand DNA replication does not have to be continuous (reviewed in Refs. 115 and 116).

Finally, the flexibility and the dynamics of the replisome are also documented by the fact that, after a collision with the transcribing RNA polymerase, it can displace the RNA polymerase, continuing DNA synthesis using the transcript mRNA as a primer.[117]

5.6.2 *Eukaryotes*

In contrast to *E. coli* the eukaryotes have more sophisticated ways to regulate the DNA pol switches. First, they possess many more DNA pols: four from the

Y family, four from the X family, four from the B family and three from the A family. Besides this higher number of DNA pols, postranslational modifications (PTM) of DNA pols themselves and of their processivity factor PCNA, have a great influence in regulating the proper interactions of the different DNA pols in different DNA transactions. We therefore briefly discuss the PTM of PCNA (5.6.2.1) and the known DNA pol switches (5.6.2.2).

5.6.2.1 *Posttranslational modifications of PCNA*

In Chapter 2 (**Table 2.7**) we indicated that PCNA can interact with 10 different DNA pols from the families B, X and Y. By interacting with the leading strand replicase DNA pol ε and the lagging strand replicase DNA pol δ, PCNA provides processivity to these enzymes. The interaction with the BER enzymes DNA pols β and λ allows them to perform accurate repair synthesis. The translesion DNA pol ζ, η, κ, ι, and Rev1 all can interact with PCNA. For these PTM of both PCNA and ubiquitin binding motifs in DNA pols are necessary. Here we briefly describe the PTMs of PCNA.

Phosphorylation of PCNA. The first PTM identified for PCNA was phosphorylation. Association of PCNA with nuclear chromatin in human fibroblasts is related to its phosphorylation status, suggesting that binding of PCNA to DNA synthesis sites occurs after phosphorylation.[118]

Acetylation of PCNA. The transcriptional coactivator p300 interacts with PCNA *in vitro* and *in vivo* and forms a complex with PCNA that does not depend on the S phase of cell cycle suggesting a function in DNA repair synthesis.[119] Because p300 and histone deacetylase (HDAC1) were co-immunoprecipitated with PCNA, they are likely responsible for the acetylation and deacetylation of PCNA, respectively. Deacetylation reduced the ability of PCNA to bind to DNA pols β and δ.[120]

Ubiquitination and SUMOylation of PCNA. A key element in understanding how PCNA can interact with different partners including DNA pols came with the discovery that PCNA can be ubiquitinated[121,122] (reviewed in Ref. 4). Protein modification by ubiquitin is becoming important as a signal for many biological processes in eukaryotes, including regulated proteolysis, but also for non-degradative functions such as protein localization, maintenance of the genetic stability and regulation of chromatin structure. PCNA can be mono-ubiquitinated through RAD6 and RAD18 at lysine 164, modified by lysine-63-linked multi-ubiquitination, which in addition requires MMS2, UBC13 and RAD5, and it is conjugated to the small ubiquitin-related modifier (SUMO) by UBC9 at lysine 164. All three modifications affect the same lysine 164 residue of PCNA, suggesting that they label PCNA for alternative functions (**Figure 5.7**).

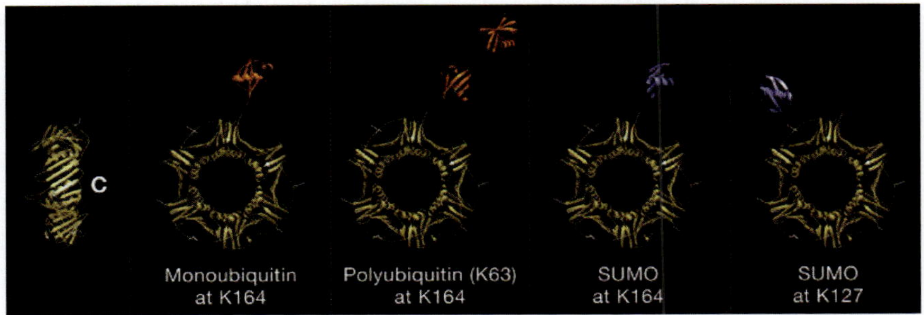

Figure 5.7　Ubiquitination and SUMOylation of PCNA. The structure of PCNA is shown in yellow. All three modifications are attached to lysine 164 of PCNA. The polyubiquitin chains are connected through lysine 63. PCNA from *S. cerevisiae* can be SUMOylated through lysine 127 which is in the interconnecting loop of PCNA where many proteins can bind.[3] (Reproduced with permission from Ref. 4.)

5.6.2.2　*DNA pol switches due to PCNA ubiquitination*

An extremely important issue now is the switch from the accurate replicative DNA pols ε and δ to the various TLS DNA pols. Cells can employ both transcriptional and postranslational regulatory mechanisms to allow the TLS DNA pols under tight control to act at the replication fork when required.[123] Depending on the lesions they can be "blocking", "miscoding" or "coding" for replicative and TLS DNA pols (see **Tables 5.4 and 5.5** for details). As we will see the DNA pol switch from the replicative to the TLS pols can either lead to error-prone or error free TLS. The error-free pathway could rely on the undamaged information of the sister duplex and the mechanism here would be template switching. Therefore two modes of switches have to be considered, first the DNA pol switches and second the DNA switch, which might be related to homologous recombination (HR).

As indicated in a review by Yang and Woodgate our understanding of TLS has been transformed in the last 10 years.[123] The biggest challenge that was mentioned in the outlook of this review is the ability to understand how the TLS DNA pols get access to a replication fork, a gap, or a D-loop, where they act in TLS, NER and HR. As will be discussed below it is the modification of PCNA that will likely play a key role for the DNA pol switches. How are DNA pols selected at the right time to the right place? Several findings support the hypothesis that the DNA pol best suited to perform TLS at a certain lesion will "win" the DNA pol switch (reviewed in Ref. 124 and references therein). These are:

(1)　DNA pols themselves can weakly interact.
(2)　There are weak interactions between DNA pols and PCNA that can be increased when PCNA is ubiquinated.

(3) The homotrimeric ring of PCNA could allow three DNA pols to bind simultaneously.

(4) The half life of DNA η and ι is very short.

In summary, it appears that dynamic interactions between different DNA pols bind to the PCNA clamp and that this binding is tighter and longer if PCNA is ubiquitinated. Below, some relevant examples of PCNA/DNA pols interactions are given.

An archaeon PCNA is a heterotrimer. While prokaryotes have a clamp with a homodimeric structure, the eukaryotes contain a homotrimer.[3] In the archaeon *Sulfolobus solfataricus*, belonging to the third kingdom of life a heterodimeric PCNA is present.[125] The three distinct PCNA subunits contact the DNA pol, DNA ligase and Fen1 and give a defined structure at the lagging strand of the replication fork suggesting a preformed scanning complex at the lagging strand of the replication fork.

TLS by DNA polymerase η. DNA pol η is the enzyme that is altered in the disease Xeroderma pigmentosum V. It is so far the best studied TLS pol. First it was found that the lesion bypass by DNA pol η is a multi-DNA pol process. In contrast to the accurate bypass of a TT cyclobutane dimers by DNA pol η, other lesions appear to be bypassed in an error-prone manner (e.g. 6-4TT and G-AAF lesions), suggesting the combined action of several TLS pols depending on the lesions.[126] When cells are treated with UV, PCNA becomes monubiquinated and this is dependent on the Rad18 protein. Specific interaction with DNA pol η is only possible after monubiquitination of PCNA suggesting that modification of PCNA induces the DNA pol switch.[127] Rad 18 furthermore guides DNA pol η to sites where the replication fork is stalled and it does this by physical interaction with PCNA.[128] Moreover, biochemical work suggested that ubiquitination of PCNA can activate the TLS DNA pol η and REV1. Thus ubiquitinated PCNA was proposed to promote mutagenic DNA replication.[129] The DNA pol switch between the replicative DNA pol δ and the TLS DNA pol η by monubiquitinated PCNA has to work back as well. This can only be achieved when PCNA is deubiquitinated.[130] DNA pol η has two modes to bind to PCNA; first to the interdomain connector loop of PCNA via the PCNA interacting protein (PIP) domain of DNA pol η and second, to the lysine 164 linked ubiquitin moiety of PCNA by the Ub-binding domain (UBD) of DNA pol η. It appears that the binding to the interdomain connector loop is more important for DNA pol η to act in TLS and that a direct binding of Ub modified lysine164 to the UBD of DNA pol η appears not to be required.[131] The UBD were generally suggested to regulate TLS for the Y family DNA pols[132] and at least for DNA pol η is responsible for correct localization.[133]

TLS by DNA polymerase ι. DNA pol ι can also bind to PCNA.[134] Moreover, monubiquitination of PCNA appears to be required as well as a UBD on DNA pol ι.[132] In general the cellular function of DNA pol ι remains unresolved. Recent studies on DNA pol ι have brought insights into its possible biological roles and suggested that DNA pol ι plays important functions in protecting humans from the deleterious consequences of exposure to both oxidative- and ultraviolet light-induced DNA damage (reviewed in Ref. 135).

TLS by DNA polymerase κ. DNA pol κ interacts with PCNA[136] and this interaction is stronger when PCNA is monoubiquitinated. The UBD is also present in DNA pol κ and is required for nuclear foci formation after UV-damage.[137]

TLS by DNA polymerase Rev1. The mouse Rev1 protein can interact with multiple TLS DNA pols. It interacts independently with Rev7 (a subunit of the B family TLS DNA pol ζ) and with the two Y family DNA pols η and ι, that bind to approximately the same 100 amino acid C-terminal region of Rev1. These data suggested that Rev1 plays a role in mediating protein-protein interactions among TLS DNA pols.[138] A UBD was also identified in Rev1 and TLS of Rev1 was enhanced upon ubiquitination of PCNA and it was suggested that this interaction was stabilized during DNA damage response.[139] In chicken DT40 cells the necessity for both the polymerase-interaction domain and the UBD was documented and the PCNA ubiquitination is essential for gap filling of postreplicative gaps. This is likely due to the fact that PCNA ubiquitination and REV1 play distinct roles in the coordination of DNA damage bypass and are temporally separated to the stalled replication fork.[140] Rev1 can promote complex formation of DNA pol ζ with the Pol 32 subunit of DNA pol δ. The Rev1 binding to Pol32 of DNA pol δ might thus have a significant function for DNA pol ζ in TLS and suggests a structural role for Rev1 in modulating the binding of DNA pol ζ with Pol32 in DNA pol δ stalled at a lesion site.[141] The TLS by Rev1 appears to be mediated by two separate domains in Rev1 which are involved in distinct modes of TLS that may depend on gap filling.[142] The early mode of TLS is mediated by the BRCT domain of Rev1. At later stages Rev1, but not its BRCT domain, is required for postreplicative gap filling.

How do DNA polymerases work together? A complete picture of how the different DNA pol switches are regulated is far from being complete. There is still controversy in the literature concerning mechanistic details. A work published in 2006 questions some issues raised above.[143] To be able to decipher the role of PCNA monoubiquitylation in the TLS process, PCNA modification was monubiquitinated *in vitro* from purified *S. cerevisiae* proteins. In addition to the requirement for Rad6-Rad18, the reaction depended on the loading of the PCNA homotrimeric ring onto the DNA by the clamp loader complex RF-C. All three PCNA monomers were efficiently ubiquitylated. Next the effects of this PCNA modification on DNA synthesis by DNA pols δ, η, ζ and Rev1 were tested. Contrary to published data that postulate a role for PCNA ubiquitylation in the disruption of DNA pol δ binding to PCNA or in the enhancement of the binding affinity of the TLS pols for PCNA, it was found that PCNA ubiquitylation did not affect any of these processes. These observations suggested a role for PCNA monoubiquitylation in disrupting the PCNA binding of a protein(s) that otherwise is (are) inhibitory to the binding of PCNA by TLS pols.

Livneh's group has recently shown that DNA pol ζ cooperates with DNA pols κ and ι in error-prone TLS across pyrimidine dimers in cells from XPV patients.[93] DNA pols ζ and κ but not DNA pol ι can protect XPV cells against the cytotoxic effect of UV and this is independent of NER. These results suggested that a benefit-risk balance might be inherent in TLS DNA pols, with the outcome that cells are protected against UV, but with the cost of increased mutations.[93]

Figure 5.8 summarizes the DNA pol switch intricate dynamics. DNA pol δ and PCNA are stalled at a thymidine dimer. PCNA is monoubiquitinated after the recruitment of Rad18-Rad16 and cleavage of USP1. Monoubiquitination attracts TLS pols (here exemplified by DNA pols η and ι), which perform TLS and then dissociate and replication can restart. Such scenarios are possible with the many other TLS pols known.

5.6.2.3 *Other ways to regulate DNA polymerases*

DNA polymerase β. An elegant study suggested that DNA pol β can be polyubiquitinated by the E3 ubiquitin ligase CHIP (carboxy terminus of HSP70 interacting protein) when not bound to the DNA repair complexes and all non-employed DNA pol β molecules are subsequently rapidly degraded.[144] Moreover, the steady state level of DNA pol β might be regulated by another E3 ligase called Mule (also known as ARF-binding protein or HectH9) and its partner ARF[145]. Monoubiquitination at Lys 41, 61 and 81 renders DNA pol β a substrate for polyubiquitination by CHIP with subsequent degradation. If bound to AFR, Mule is inactive and thus DNA pol β cannot be ubiquitinated and is ready to perform repair in the nucleus. This is important in the view that in over 30% of the tumours tested DNA pol β is over-expressed.[99] Two further modifications regulate the functions of DNA pol β. First, arginine methylation regulates the DNA pol β itself[146] and its binding to PCNA[147] and second acetylation of DNA pol β regulates its end-trimming activity (dRPlyase).[148]

DNA polymerases η, ι and Rev 1. Phosphorylation can regulate the activity and the translocation of DNA pol η. After UV damage DNA pol η gets phosphorylated and treatment with caffeine, siRNA against the ATP kinase reduced its accumulation at stalled replication forks.[149] In *Caenorhabditis elegans* during the DNA damage response DNA pol η is degraded by the Cul4-Ddb1-Cdt2 SUMO pathway.[150] It was furthermore found that DNA pols η and ι can associate with

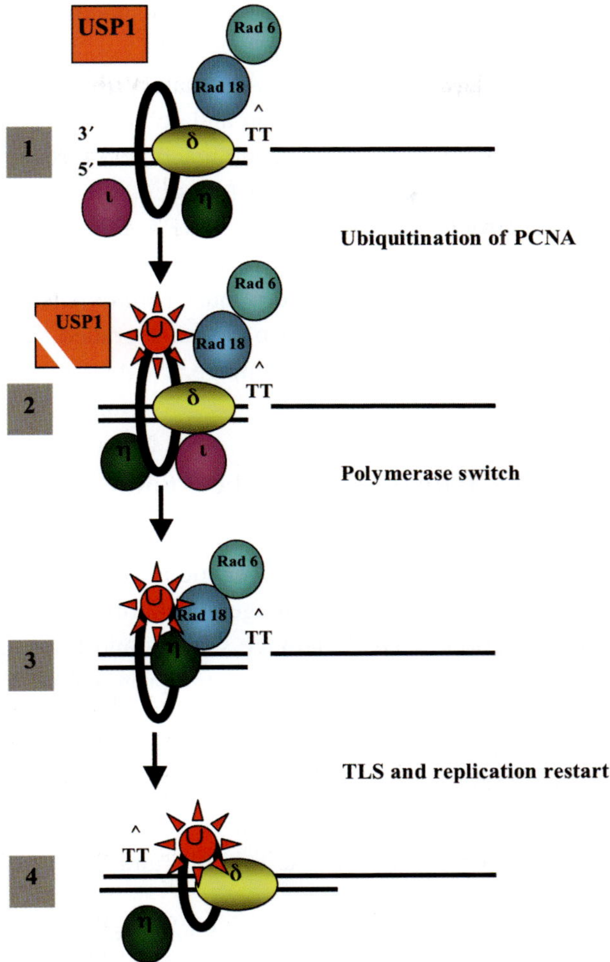

Figure 5.8   Translesion DNA synthesis and DNA polymerase switch. For details see text.

the replication machinery in the nuclei at stalled replication fork following DNA damage. Very interestingly the two DNA pols η and ι can physically interact during this process suggesting a coordinated function of both these TLS DNA pols.[151] Finally the interaction of DNA pol η is required for the nuclear accumulation of the Rev1 DNA pol. This interaction can suppress spontaneous mutations in human cells.[152]

DNA polymerase λ. Human DNA pol λ is degraded *in vivo* by the proteasome-ubiquitination pathway; however, it can be stabilized upon phosphorylation at T553. This stabilization seems to occur in particular in late S and G2 phase of the cell cycle, which probably enables the enzyme to fulfill its specific functions in DNA damage repair during these stages.[153]

Replicative and repair DNA polymerases. A human BER complex was found to be physically associated with DNA replication and cell cycle regulatory proteins, suggesting that a constitutively active DNA repair machinery is ready to be recruited when base damages occur at DNA replications forks.[154]

Alternative subunit structure of DNA polymerase δ. DNA pol δ contains four subunits called p125, p68, p50 and p12 (for details see Chapters 2 and 3). Upon genotoxic stress treatment of human cells the four-subunits DNA pol δ was converted to a three-subunits enzyme, since the p12 was degraded,[155] suggesting that this might be an adaptive response to DNA damage.[156]

5.7 Functions of DNA Polymerases in Checkpoint Control

Cell cycle events have to be coordinated in time and order and these steps are critical for a high fidelity transmission of the genetic information and the successful cell duplication. Activation of DNA damage checkpoint and DNA replication checkpoint require the action of DNA damage sensors and transducers.[157] Some of these factors, such as ataxia telangectasia mutated protein (ATM), ATM related protein (ATR), ATR interacting protein (ATRIP), Rad17, Rad9, Rad1 and Hus1, are thought to be involved in triggering DNA repair processes. Data appearing in this field and connecting DNA pols to checkpoint control appeared in the last 15 years. As already discussed in Section 5.2.2 the first DNA pols engaged in DNA replication are DNA pol α with its primase subunits and DNA pol ε the leading strand DNA pol.

Functional uncoupling of the replicative DNA helicase from replicative DNA pols activates the ATR-dependent checkpoint. It was found that uncoupling of the MCM helicase and Cdc45 from the replication fork resulted in checkpoint activation.[158] Successful checkpoint activation also required DNA synthesis by DNA pol α. This uncoupling renders DNA single-stranded and thus RP-A can bind. RP-A itself can activate the checkpoint[159] and does this by recruiting Rad17 a component of the alternate clamp loader[160] (see Chapter 2 for details).

DNA polymerase α *and checkpoint control.* The first data came from genetic experiments performed in the fission yeast *Schizosaccharomyces pombe* and these experiments suggested a role for DNA pol α in checkpoint control.[161] Later it was found that the two primase subunits of DNA pol α could affect the ChK1 activation (due to the p49 homolog of mammalian DNA pol α) and Cds1 kinase activation (due to the p58 homolog of mammalian DNA pol α).[162,163] At about the same time it was found that activation of the DNA replication checkpoint was initiated by

RNA synthesis by the primase of DNA pol α.[164] This was followed by studies that elaborated the role of single-stranded DNA and DNA pol α in establishing the ATR, Hus1 DNA replication checkpoint.[165]

DNA polymerase ε and checkpoint control. Already in 1995 it was realized, thanks to genetical studies in *S. cerevisiae*, that DNA pol ε links the DNA replication machinery to the S phase checkpoint[166] and it was proposed that DNA pol ε acts as a sensor of DNA replication that coordinates the transcriptional and the cell cycle responses to replication block. Subsequently Dbp11 was found to interact with DNA pol ε and could be part of the leading strand enzyme to act in checkpoint control during the S phase of the cell cycle to sense stalled replication forks.[167] Again in *S. cerevisiae* it was found that RAD9 and DNA pol ε form parallel sensory branches for transducing the DNA damage checkpoint signal.[168] The checkpoint has been identified in the C-teminal part of DNA pol ε[166] and this domain was further characterized by identifying putative zinc finger domain in the S/M phase checkpoint.[169] Next it was found that the catalytic domain of DNA pol ε was dispensable for DNA replication, DNA repair and cell viability, but the C-teminal part of DNA pol ε is the responsible part for viability.[170,171]

These data were questioned by a later study indicating that *Schizosaccharomyces pombe* DNA pol ε is required for chromosomal replication but not for the S phase checkpoint. It was proposed that the checkpoint signal operating in S phase depends on the assembly of the replication initiation complex and that this signal is generated prior to the elongation stage of DNA synthesis, e.g. through the primase action of DNA pol α.[172] Indeed the data discussed above for DNA pol α support this prediction.

DNA polymerase ζ, Rev1and checkpoint control. It has recently been shown in *S. cerevisiae* that TLS can occur in the absence of checkpoint proteins in nucleotide excision repair (NER) proficient cells. In the absence of NER, checkpoint protein-mediated Rev1 phosphorylation by Mec1 contributed to increasing the proficiency of DNA pol ζ dependent TLS.[173]

5.8 Chapter Summary

In Chapter 5 we undertook the task to look at the global functions of the several DNA pols acting in living cells. It is clear that most DNA transactions require more than one DNA pol. This backup strategy could guarantee the insurance for keeping changes in the genome at a very low level, either during DNA replication or for the various DNA repair pathways. DNA replication in living organisms requires three DNA molecules at the replication fork. The replisome of the *E. coli* bacteriophage T7 is the simplest and so far best-known replisome. Details of its mechanism are available mainly due to its simple component composition and single molecule technology. The *E. coli* replisome relies on the DNA pol III, but it appears that three molecules of DNA pol III might possibly act in concert at the fork. In eukaryotes from yeast to human the three different DNA pols α, ε and δ share the workload at the DNA replication fork in initiating, leading strand and lagging strand synthesis, respectively. Proofreading is very important to minimize mistakes during DNA synthesis. We know proofreader DNA pols, such as the replicating DNA pols δ and ε, but there are also strategies that proofreader $3' \rightarrow 5'$ exonuclease activities of DNA pols δ and ε act for non-proofreader DNA pols, such as DNA pol α.

The different DNA repair pathways have their own DNA pols, but can also borrow DNA pols from the replication machinery. Pathways to restore the integrity of DNA include BER, NER, MNR, HR, NHEJ and ICLR and they rely together on at least 10 different DNA pols. In most cases more than one DNA pol is employed for each pathway. In translesion DNA synthesis in eukaryotes generally two different DNA pols are required. One is responsible for the insertion and the other for the extension from the lesion. On an even more complex level, they all have to act in concert with the replicative DNA pols. Considering their actions in the cellular context the DNA pols have to switch between different DNA transactions.

Postranslational modifications play an essential role in regulating the actions of many DNA pols. They include phosphorylation, acetylation, methylation and ubiquitination reactions (monoubiquitination, polyubiquitination and SUMOylation) of DNA pols as well as similar postranslational modifications of the DNA pol clamp PCNA. Moreover, expression levels of DNA pols at the right time and the right place are essential to have DNA pols at the site of action when they are needed. Finally, regulation through checkpoint control events is dependent on "early" DNA pols in initiation of DNA replication, where DNA pols α and ε are likely involved in important control events in cell metabolism.

References

1. Hübscher U, Maga G, Spadari S. 2002. *Annu Rev Biochem* 71: 133–63.
2. Johnson A, O'Donnell M. 2005. *Annu Rev Biochem* 74: 283–315.
3. Maga G, Hubscher U. 2003. *J Cell Sci* 116: 3051–60.
4. Moldovan GL, Pfander B, Jentsch S. 2007. *Cell* 129: 665–79.
5. Bergink S, Jentsch S. 2009. *Nature* 458: 461–7.
6. Cotterill S, Kearsey SE. 2009. *Nucleic Acids Res* 37: D837-9.
7. Bell SD. 2006. *Nature* 439: 542–3.
8. Lovett ST. 2007. *Mol Cell* 27: 523–6.
9. Debyser Z, Tabor S, Richardson CC. 1994. *Cell* 77: 157–66.
10. Hamdan SM, Richardson CC. 2009. *Annu Rev Biochem* 78: 205–43.

11. Hamdan SM, Johnson DE, Tanner NA, Lee JB, Qimron U, *et al.* 2007. *Mol Cell* 27: 539–49.
12. Hamdan SM, Loparo JJ, Takahashi M, Richardson CC, van Oijen AM. 2009. *Nature* 457: 336–9.
13. Sinha NK, Morris CF, Alberts BM. 1980. *J Biol Chem* 255: 4290–3.
14. Lee JB, Hite RK, Hamdan SM, Xie XS, Richardson CC, van Oijen AM. 2006. *Nature* 439: 621–4.
15. Tanner NA, Loparo JJ, Hamdan SM, Jergic S, Dixon NE, van Oijen AM. 2009. *Nucleic Acids Res* 37: e27.
16. Yuzhakov A, Turner J, O'Donnell M. 1996. *Cell* 86: 877–86.
17. Kim S, Dallmann HG, McHenry CS, Marians KJ. 1996. *J Biol Chem* 271: 21406–12.
18. Kim S, Dallmann HG, McHenry CS, Marians KJ. 1996. *Cell* 84: 643–50.
19. Pomerantz RT, O'Donnell M. 2007. *Trends Microbiol* 15: 156–64.
20. McInerney P, Johnson A, Katz F, O'Donnell M. 2007. *Mol Cell* 27: 527–38.
21. Tanner NA, Hamdan SM, Jergic S, Schaeffer PM, Dixon NE, van Oijen AM. 2008. *Nat Struct Mol Biol* 15: 170–6.
22. Hubscher U, Kuenzle CC, Spadari S. 1977. *Nucleic Acids Res* 4: 2917–29.
23. Conaway RC, Lehman IR. 1982. *Proc Natl Acad Sci U S A* 79: 2523–7.
24. Boulet A, Simon M, Faye G, Bauer GA, Burgers PM. 1989. *EMBO J* 8: 1849–54.
25. Morrison A, Araki H, Clark AB, Hamatake RK, Sugino A. 1990. *Cell* 62: 1143–51.
26. Garg P, Stith CM, Sabouri N, Johansson E, Burgers PM. 2004. *Genes Dev* 18: 2764–73.
27. Chilkova O, Stenlund P, Isoz I, Stith CM, Grabowski P, *et al.* 2007. *Nucleic Acids Res* 35: 6588–97.
28. Waga S, Stillman B. 1998. *Annu Rev Biochem* 67: 721–51.
29. Pursell ZF, Isoz I, Lundstrom EB, Johansson E, Kunkel TA. 2007. *Science* 317: 127–30.
30. Nick McElhinny SA, Gordenin DA, Stith CM, Burgers PM, Kunkel TA. 2008. *Mol Cell* 30: 137–44.
31. Stillman B. 2008. *Mol Cell* 30: 259–60.
32. Moser BA, Subramanian L, Chang YT, Noguchi C, Noguchi E, Nakamura TM. 2009. *EMBO J* 28: 810–20.
33. Shevelev IV, Hubscher U. 2002. *Nat Rev Mol Cell Biol* 3: 364–76.
34. Kunkel TA. 2004. *J Biol Chem* 279: 16895–8.
35. Albertson TM, Preston BD. 2006. *Curr Biol* 16: R209–11.
36. Pavlov YI, Frahm C, Nick McElhinny SA, Niimi A, Suzuki M, Kunkel TA. 2006. *Curr Biol* 16: 202–7.
37. Hubscher U, Kuenzle CC, Spadari S. 1979. *Proc Natl Acad Sci U S A* 76: 2316–20.
38. Waser J, Hubscher U, Kuenzle CC, Spadari S. 1979. *Eur J Biochem* 97: 361–8.
39. Lindahl T, Wood RD. 1999. *Science* 286: 1897–905.
40. Fortini P, Dogliotti E. 2007. *DNA Repair (Amst)* 6: 398-409.
41. Pascucci B, Stucki M, Jonsson ZO, Dogliotti E, Hübscher U. 1999. *J Biol Chem* 274: 33696–702.
42. Stucki M, Pascucci B, Parlanti E, Fortini P, Wilson SH, *et al.* 1998. *Oncogene* 17: 835–43.
43. Dianov GL, Prasad R, Wilson SH, Bohr VA. 1999. *J Biol Chem* 274: 13741–3.
44. Liu Y, Beard WA, Shock DD, Prasad R, Hou EW, Wilson SH. 2005. *J Biol Chem* 280: 3665–74.
45. Garcia-Diaz M, Bebenek K, Kunkel TA, Blanco L. 2001. *J Biol Chem* 276: 34659–63.
46. Prasad R, Longley MJ, Sharief FS, Hou EW, Copeland WC, Wilson SH. 2009. *Nucleic Acids Res* 37: 1868–77.
47. Braithwaite EK, Prasad R, Shock DD, Hou EW, Beard WA, Wilson SH. 2005. *J Biol Chem* 280: 18469–75.
48. Yoshimura M, Kohzaki M, Nakamura J, Asagoshi K, Sonoda E, *et al.* 2006. *Mol Cell* 24: 115–25.

49. Maga G, Crespan E, Wimmer U, van Loon B, Amoroso A, *et al.* 2008. *Proc Natl Acad Sci U S A* 105: 20689–94.
50. van Loon B, Hubscher U. 2009. *Proc Natl Acad Sci U S A* 106: 18201–6.
51. Shivji KK, Kenny MK, Wood RD. 1992. *Cell* 69: 367–74.
52. Aboussekhra A, Biggerstaff M, Shivji MK, Vilpo JA, Moncollin V, *et al.* 1995. *Cell* 80: 859–68.
53. Shivji MK, Podust VN, Hubscher U, Wood RD. 1995. *Biochemistry* 34: 5011–7.
54. Masutani C, Kusumoto R, Yamada A, Yuasa M, Araki M, *et al.* 2000. *Cold Spring Harb Symp Quant Biol* 65: 71–80.
55. Stary A, Kannouche P, Lehmann AR, Sarasin A. 2003. *J Biol Chem* 278: 18767–75.
56. Ogi T, Lehmann AR. 2006. *Nat Cell Biol* 8: 640–2.
57. Longley MJ, Pierce AJ, Modrich P. 1997. *J Biol Chem* 272: 10917–21.
58. Fan W, Wu X. 2004. *Biochem Biophys Res Commun* 323: 1328–33.
59. Mahajan KN, Gangi-Peterson L, Sorscher DH, Wang J, Gathy KN, *et al.* 1999. *Proc Natl Acad Sci U S A* 96: 13926–31.
60. Mahajan KN, Nick McElhinny SA, Mitchell BS, Ramsden DA. 2002. *Mol Cell Biol* 22: 5194–202.
61. Lee JW, Blanco L, Zhou T, Garcia-Diaz M, Bebenek K, *et al.* 2004. *J Biol Chem* 279: 805–11.
62. Nick McElhinny SA, Havener JM, Garcia-Diaz M, Juarez R, Bebenek K, *et al.* 2005. *Mol Cell* 19: 357–66.
63. Capp JP, Boudsocq F, Bertrand P, Laroche-Clary A, Pourquier P, *et al.* 2006. *Nucleic Acids Res* 34: 2998–3007.
64. Capp JP, Boudsocq F, Besnard AG, Lopez BS, Cazaux C, *et al.* 2007. *Nucleic Acids Res* 35: 3551–60.
65. Tseng HM, Tomkinson AE. 2002. *J Biol Chem* 277: 45630–7.
66. Schatz DG. 2004. *Immunol Rev* 200: 5–11.
67. Bertocci B, De Smet A, Berek C, Weill JC, Reynaud CA. 2003. *Immunity* 19: 203–11.
68. Bertocci B, De Smet A, Weill JC, Reynaud CA. 2006. *Immunity* 25: 31–41.
69. Lebecque SG, Gearhart PJ. 1990. *J Exp Med* 172: 1717–27.
70. Honjo T, Nagaoka H, Shinkura R, Muramatsu M. 2005. *Nat Immunol* 6: 655–61.
71. Neuberger MS, Di Noia JM, Beale RC, Williams GT, Yang Z, Rada C. 2005. *Nat Rev Immunol* 5: 171–8.
72. Pham P, Bransteitter R, Petruska J, Goodman MF. 2003. *Nature* 424: 103–7.
73. Dominguez O, Ruiz JF, Lain de Lera T, Garcia-Diaz M, Gonzalez MA, *et al.* 2000. *Embo J* 19: 1731–42.
74. Bertocci B, De Smet A, Flatter E, Dahan A, Bories JC, *et al.* 2002. *J Immunol* 168: 3702–6.
75. Zan H, Shima N, Xu Z, Al-Qahtani A, Evinger Iii AJ, *et al.* 2005. *EMBO J* 24: 3757–69.
76. Masuda K, Ouchida R, Hikida M, Kurosaki T, Yokoi M, *et al.* 2007. *J Biol Chem* 282: 17387–94.
77. Masuda K, Ouchida R, Yokoi M, Hanaoka F, Azuma T, Wang JY. 2008. *Eur J Immunol* 38: 2796–805.
78. McIlwraith MJ, Vaisman A, Liu Y, Fanning E, Woodgate R, West SC. 2005. *Mol Cell* 20: 783–92.
79. Kawamoto T, Araki K, Sonoda E, Yamashita YM, Harada K, *et al.* 2005. *Mol Cell* 20: 793–9.
80. Sharief FS, Vojta PJ, Ropp PA, Copeland WC. 1999. *Genomics* 59: 90–6.
81. Maga G, Shevelev I, Ramadan K, Spadari S, Hubscher U. 2002. *J Mol Biol* 319: 359–69.
82. Seki M, Marini F, Wood RD. 2003. *Nucleic Acids Res* 31: 6117–26.
83. Marini F, Kim N, Schuffert A, Wood RD. 2003. *J Biol Chem* 278: 32014–9.
84. Maga G, Villani G, Crespan E, Wimmer U, Ferrari E, *et al.* 2007. *Nature* 447: 606–8.
85. Prakash S, Johnson RE, Prakash L. 2005. *Annu Rev Biochem* 74: 317–53.
86. Prakash S, Prakash L. 2002. *Genes Dev* 16: 1872–83.
87. Weigle JJ. 1953. *Proc Natl Acad Sci U S A* 39: 628–36.

88. Defais M, Caillet-Fauquet P, Fox M , M R. 1974. *Mol Gen Genet* 135: 19–27.
89. Benbow RM, Zuccarelli AJ, Sinsheimer RL. 1974. *J Mol Biol* 88: 629–51.
90. Caillet-Fauquet P, Defais M, Radman M. 1977. *J Mol Biol* 117: 95–110.
91. Waters LS, Minesinger BK, Wiltrout ME, D'Souza S, Woodruff RV, Walker GC. 2009. *Microbiol Mol Biol Rev* 73: 134–54.
92. Johnson RE, Prakash S, Prakash L. 1999. *Science* 283: 1001–4.
93. Ziv O, Geacintov N, Nakajima S, Yasui A, Livneh Z. 2009. *Proc Natl Acad Sci U S A* 106: 11552–7.
94. Shachar S, Ziv O, Avkin S, Adar S, Wittschieben J, *et al.* 2009. *EMBO J* 28: 383–93.
95. Rey L, Sidorova JM, Puget N, Boudsocq F, Biard DS, *et al.* 2009. *Mol Cell Biol* 29: 3344–54.
96. Bergoglio V, Canitrot Y, Hogarth L, Minto L, Howell SB, *et al.* 2001. *Oncogene* 20: 6181–7.
97. Frechet M, Canitrot Y, Cazaux C, Hoffmann JS. 2001. *FEBS Lett* 505: 229–32.
98. Bavoux C, Leopoldino AM, Bergoglio V, J OW, Ogi T, *et al.* 2005. *Cancer Res* 65: 325–30.
99. Albertella MR, Lau A, O'Connor M J. 2005. *DNA Repair (Amst)* 4: 583–93.
100. Wagner J, Etienne H, Janel-Bintz R, Fuchs RP. 2002. *DNA Repair (Amst)* 1: 159–67.
101. Sutton MD, Opperman T, Walker GC. 1999. *Proc Natl Acad Sci U S A* 96: 12373–8.
102. Becherel OJ, Fuchs RP, Wagner J. 2002. *DNA Repair (Amst)* 1: 703–8.
103. Bunting KA, Roe SM, Pearl LH. 2003. *EMBO J* 22: 5883–92.
104. Lopez de Saro FJ, Georgescu RE, Goodman MF, O'Donnell M. 2003. *EMBO J* 22: 6408–18.
105. Indiani C, McInerney P, Georgescu R, Goodman MF, O'Donnell M. 2005. *Mol Cell* 19: 805–15.
106. Pages V, Fuchs RP. 2003. *Science* 300: 1300–3.
107. Fujii S, Fuchs RP. 2004. *EMBO J* 23: 4342–52.
108. Schlacher K, Cox MM, Woodgate R, Goodman MF. 2006. *Nature* 442: 883–7.
109. Reuven NB, Arad G, Maor-Shoshani A, Livneh Z. 1999. *J Biol Chem* 274: 31763–6.
110. Tang M, Shen X, Frank EG, O'Donnell M, Woodgate R, Goodman MF. 1999. *Proc Natl Acad Sci U S A* 96: 8919–24.
111. Simon SM, Sousa FJ, Mohana-Borges R, Walker GC. 2008. *Proc Natl Acad Sci U S A* 105: 1152–7.
112. Indiani C, Langston LD, Yurieva O, Goodman MF, O'Donnell M. 2009. *Proc Natl Acad Sci U S A* 106: 6031–8.
113. Rupp WD, Howard-Flanders P. 1968. *J Mol Biol* 31: 291–304.
114. Heller RC, Marians KJ. 2006. *Nat Rev Mol Cell Biol* 7: 932–43.
115. Langston LD, O'Donnell M. 2006. *Mol Cell* 23: 155–60.
116. Lehmann AR, Fuchs RP. 2006. *DNA Repair (Amst)* 5: 1495–8.
117. Pomerantz RT, O'Donnell M. 2008. *Nature* 456: 762–6.
118. Prosperi E, Scovassi AI, Stivala LA, Bianchi L. 1994. *Exp Cell Res* 215: 257–62.
119. Hasan S, Hassa PO, Imhof R, Hottiger MO. 2001. *Nature* 410: 387–91.
120. Naryzhny SN, Lee H. 2004. *J Biol Chem* 279: 20194–9.
121. Hoege C, Pfander B, Moldovan GL, Pyrowolakis G, Jentsch S. 2002. *Nature* 419: 135–41.
122. Stelter P, Ulrich HD. 2003. *Nature* 425: 188–91.
123. Yang W, Woodgate R. 2007. *Proc Natl Acad Sci U S A* 104: 15591–8.
124. Lehmann AR, Niimi A, Ogi T, Brown S, Sabbioneda S, *et al.* 2007. *DNA Repair (Amst)* 6: 891–9.
125. Dionne I, Nookala RK, Jackson SP, Doherty AJ, Bell SD. 2003. *Mol Cell* 11: 275–82.
126. Bresson A, Fuchs RP. 2002. *EMBO J* 21: 3881–7.
127. Kannouche PL, Wing J, Lehmann AR. 2004. *Mol Cell* 14: 491–500.
128. Watanabe K, Tateishi S, Kawasuji M, Tsurimoto T, Inoue H, Yamaizumi M. 2004. *EMBO J* 23: 3886–96. Epub 2004 Sep 09.
129. Garg P, Stith CM, Majka J, Burgers PM. 2005. *J Biol Chem* 280: 23446–50.

130. Zhuang Z, Johnson RE, Haracska L, Prakash L, Prakash S, Benkovic SJ. 2008. *Proc Natl Acad Sci U S A* 105: 5361–6.
131. Acharya N, Brahma A, Haracska L, Prakash L, Prakash S. 2007. *Mol Cell Biol* 27: 7266–72.
132. Bienko M, Green CM, Crosetto N, Rudolf F, Zapart G, *et al.* 2005. *Science* 310: 1821–4.
133. Sabbioneda S, Green CM, Bienko M, Kannouche P, Dikic I, Lehmann AR. 2009. *Proc Natl Acad Sci U S A* 106: E20; author reply E1.
134. Haracska L, Johnson RE, Unk I, Phillips BB, Hurwitz J, *et al.* 2001. *Proc Natl Acad Sci U S A* 98: 14256–61.
135. Vidal AE, Woodgate R. 2009. *DNA Repair (Amst)* 8: 420–3.
136. Haracska L, Unk I, Johnson RE, Phillips BB, Hurwitz J, *et al.* 2002. *Mol Cell Biol* 22: 784–91.
137. Guo C, Tang TS, Bienko M, Dikic I, Friedberg EC. 2008. *J Biol Chem* 283: 4658–64.
138. Guo C, Fischhaber PL, Luk-Paszyc MJ, Masuda Y, Zhou J, *et al.* 2003. *Embo J* 22: 6621–30.
139. Wood A, Garg P, Burgers PM. 2007. *J Biol Chem* 282: 20256–63.
140. Edmunds CE, Simpson LJ, Sale JE. 2008. *Mol Cell* 30: 519–29.
141. Acharya N, Johnson RE, Prakash S, Prakash L. 2006. *Mol Cell Biol* 26: 9555–63.
142. Jansen JG, Tsaalbi-Shtylik A, Hendriks G, Gali H, Hendel A, *et al.* 2009. *Mol Cell Biol* 29: 3113–23.
143. Haracska L, Unk I, Prakash L, Prakash S. 2006. *Proc Natl Acad Sci U S A* 103: 6477–82.
144. Parsons JL, Tait PS, Finch D, Dianova, II, Allinson SL, Dianov GL. 2008. *Mol Cell* 29: 477–87.
145. Parsons JL, Tait PS, Finch D, Dianova, II, Edelmann MJ, *et al.* 2009. *EMBO J.*
146. El-Andaloussi N, Valovka T, Toueille M, Steinacher R, Focke F, *et al.* 2006. *Mol Cell* 22: 51–62.
147. El-Andaloussi N, Valovka T, Toueille M, Hassa PO, Gehrig P, *et al.* 2007. *Faseb J* 21: 26–34.
148. Hasan S, El-Andaloussi N, Hardeland U, Hassa PO, Burki C, *et al.* 2002. *Mol Cell* 10: 1213–22.
149. Chen YW, Cleaver JE, Hatahet Z, Honkanen RE, Chang JY, *et al.* 2008. *Proc Natl Acad Sci U S A* 105: 16578–83.
150. Kim SH, Michael WM. 2008. *Mol Cell* 32: 757–66.
151. Kannouche P, Fernandez de Henestrosa AR, Coull B, Vidal AE, Gray C, *et al.* 2002. *EMBO J* 21: 6246–56.
152. Akagi JI, Masutani C, Kataoka Y, Kan T, Ohashi E, *et al.* 2009. *DNA Repair (Amst)* 8: 585–99.
153. Wimmer U, Ferrari E, Hunziker P, Hubscher U. 2008. *EMBO Rep* 9: 1027–33.
154. Parlanti E, Locatelli G, Maga G, Dogliotti E. 2007. *Nucleic Acids Res* 35: 1569–77.
155. Zhang S, Zhou Y, Trusa S, Meng X, Lee EY, Lee MY. 2007. *J Biol Chem* 282: 15330–40.
156. Meng X, Zhou Y, Zhang S, Lee EY, Frick DN, Lee MY. 2009. *Nucleic Acids Res* 37: 647–57.
157. Melo J, Toczyski D. 2002. *Curr Opin Cell Biol* 14: 237–45.
158. Byun TS, Pacek M, Yee MC, Walter JC, Cimprich KA. 2005. *Genes Dev* 19: 1040–52.
159. Zou L, Elledge SJ. 2003. *Science* 300: 1542–8.
160. Zou L, Liu D, Elledge SJ. 2003. *Proc Natl Acad Sci U S A* 100: 13827–32.
161. Bhaumik D, Wang TS. 1998. *Mol Biol Cell* 9: 2107–23.
162. Tan S, Wang TS. 2000. *Mol Cell Biol* 20: 7853–66.
163. Griffiths DJ, Liu VF, Nurse P, Wang TS. 2001. *Mol Biol Cell* 12: 115–28.
164. Michael WM, Ott R, Fanning E, Newport J. 2000. *Science* 289: 2133–7.
165. You Z, Kong L, Newport J. 2002. *J Biol Chem* 277: 27088–93.
166. Navas TA, Zhou Z, Elledge SJ. 1995. *Cell* 80: 29–39.
167. Araki H, Leem SH, Phongdara A, Sugino A. 1995. *Proc Natl Acad Sci U S A* 92: 11791–5.
168. Navas TA, Sanchez Y, Elledge SJ. 1996. *Genes Dev* 10: 2632–43.
169. Dua R, Levy DL, Campbell JL. 1998. *J Biol Chem* 273: 30046–55.
170. Kesti T, Flick K, Keranen S, Syvaoja JE, Wittenberg C. 1999. *Mol Cell* 3: 679–85.
171. Dua R, Levy DL, Campbell JL. 1999. *J Biol Chem* 274: 22283–8.
172. D'Urso G, Nurse P. 1997. *Proc Natl Acad Sci U S A* 94: 12491–6.
173. Pages V, Santa Maria SR, Prakash L, Prakash S. 2009. *Genes Dev* 23: 1438–49.

CHAPTER 6

Viral DNA Polymerases

6.1 Bacteriophage T4 DNA Polymerase

The DNA replication system of the *E. coli* infecting bacteriophage T4 has served as model for the type of reactions and protein–protein interactions needed to carry out and coordinate the synthesis of leading and lagging DNA strands. The approach of reconstituting a "replisome" from individual proteins assembled *in vitro* was pioneered by Bruce Alberts' laboratory and led to the concept of the so-called "trombone" model of DNA replication in which the lagging and leading strand DNA polymerase (DNA pol) complexes are connected physically, requiring a looping of the lagging strand template.[1–3] This arrangement allows coordination of the two replication complexes making them advance in the same direction. Eight proteins are required to reconstitute a T4 basic replisome that can form at and propagate a replication fork (see **Table 6.1**). Leading and lagging strand templates are copied by two holoenzyme complexes each containing the gp43 DNA pol and the processivity clamp gp45. The trimeric gp45 clamp protein is loaded on DNA by the clamp loader complex gp44/62 in an ATP-dependent mode.

The discontinuous replication of the lagging strand is initiated by a "primosome", a subassembly of replisome composed of the hexameric gp41 helicase and of the gp61primase. The primase gp61 synthesizes the ribopentamer necessary for initiation of Okazaki fragments and gp41 unwinds the double-stranded DNA ahead of the replication fork in an ATP or GTP dependent fashion. Loading of the helicase gp41 necessitates the helicase accessory factor gp59.

The single-stranded DNA produced by the helicase unwinding of the double-stranded DNA is coated by the binding protein gp32, which is also involved in coupling leading and lagging strand synthesis. A model of the replication fork of T4 is shown in **Figure 6.1**.

Electron microscopy experiments carried out in Jack Griffith laboratory have uncovered the DNA structures generated during coupled leading and lagging strand

199

Table 6.1   Bacteriophage T4 and T7 DNA replication proteins

Protein	T4 gene	T7 gene	Functions
DNA pol	gp43	gp5	Leading and lagging strand DNA synthesis DNA fidelity DNA recombination and repair
Single-stranded DNA binding protein	gp32	gp2.5	Cooperative binding to single-stranded DNA Destabilizes hairpins Stimulates lagging strand synthesis, Coordinates leading and lagging strand synthesis DNA recombination and repair
Processivity factor	gp45	trxA[a]	Confers processivity to DNA polymerase
Clamp loader	gp44/62	—	Loads processivity factor
Primase	gp61	gp4	RNA primer synthesis
DNA helicase	gp41	gp4	5′ to 3′ replicative helicase
Gene 59 protein	gp59	—	Stimulates primase and helicase activities

a: The processivity factor of gp5 is thioredoxin, the product of the trxA gene of *E. coli*

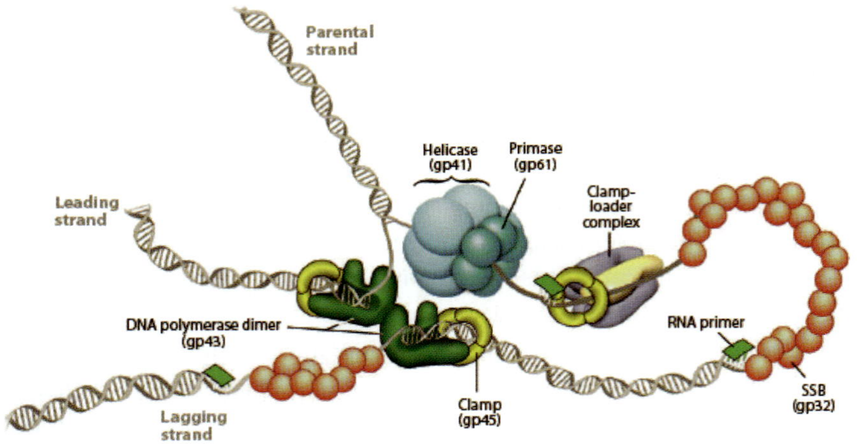

Figure 6.1   Model for the replication fork of bacteriophage T4. The dimeric DNA polymerase complex allows coordinated leading/lagging strand synthesis. The helicase-primase complex couples strand displacement to RNA primer synthesis on the lagging strand. The clamp loader complex is represented on the lagging strand laying down a new gp45 clamp to provide an attachment site to the DNA pol for the next Okazaki fragment synthesis. (Reproduced with permission from Ref. 38.)

synthesis by the T4 replisome.[4,5] These studies have confirmed the existence of a lagging strand loop, validating the original trombone model suggested by Alberts. Interestingly, biochemical analysis with highly purified human replicative accessory proteins RF-C and PCNA have demonstrated functions that are completely analogous to the functions of bacteriophage T4 DNA pol accessory proteins.[6] These results demonstrate a striking conservation of the DNA replication apparatus in human cells and bacteriophage T4.

Several excellent reviews have been published on the biochemistry of T4 replicative multi-proteins complex[7–12]; therefore in the rest of Chapter 6 we will mainly focus on the biochemical characteristic and functions of the T4 DNA pol and briefly refer to the available crystal structures of some of its interacting proteins.

The T4 encoded DNA pol was first purified many years ago[13] and identified as the product of the well-analyzed genetic locus gene 43.[14] The DNA pol is a single polypeptide with a calculated MW of 103 572 Da[15] and has more than 60% identity of amino acid sequence with the DNA pol from phage Rb69, both enzymes belonging to the B family of DNA pols. It was found that short single- and double-stranded DNA molecules bind a single T4 DNA pol in a nonspecific mode with moderate affinity ($K_d \sim 150$ nM) and a binding size of ~ 10 nucleotides for single-stranded DNA and ~ 13 bp for double-stranded DNA. In contrast the polymerase binds in a site-specific mode and significantly more tightly ($K_d \sim 5$ nM) to DNA constructs carrying a primer-template junction, with the enzyme covering ~ 5 nucleotides downstream and 6–7 bp upstream the $3'$-primer terminus.[16] The T4 DNA pol functions both as leading and lagging strand replicase and, in absence of accessory proteins, is only moderately processive[17,18]; addition of the sliding clamp gp45 and clamp loader gp44/62 proteins greatly stimulates its processivity from 800 to 3000 nucleotide per initiation event.[19] A subsequent study has shown that the carboxy (C-) terminus of the DNA pol is inserted into the subunit interface of the sliding clamp, thereby conferring processivity to the DNA pol.[20] Recently, an elegant study by Benkovic and co-workers using a catalytically inactive polymerase as trap, has shown that, within the T4 replisome, individual DNA pols molecules are exchanged rapidly in a reaction mediated by the processivity clamp.[21] These results have been interpreted by the authors as indicating a "dynamic processivity" of the DNA pols during replication where their recycling can take place without slowing down DNA synthesis. This process mimics the switching recently suggested during the translesion DNA synthesis in other organisms, implies the multiple functions of the clamp in replication, and may play a potential role in overcoming replication barriers by the T4 replisome. In an interesting development of this concept, it has been later suggested that such DNA pol exchange may also have mechanistic implication on the polymerase-to-exonuclease active site switching for proofreading by the polymerase.[22]

In addition to its DNA pol activity, the T4 DNA pol possesses an extraordinarily active $3' \rightarrow 5'$ proofreading exonuclease activity. Early experiments from Bessman and its colleagues with T4 DNA pol forms mutated in the exonuclease activity firmly established, for the first time, a strict correlation between the amount of $3' \rightarrow 5'$ exonuclease activity associated to the enzymes and the mutator or antimutator phenotypes of the bacteriophages, thus proving the crucial role of the proofreading activity in the *in vivo* mutagenesis.[23,24] A subsequent study indicated that the exonucleolytic proofreading improved nearly 10^3 fold the fidelity of the T4 DNA pol, further stressing the power of this activity in determining T4 phage fidelity.[25]

The proofreading activity associated to the T4 DNA pol seems also to play a role in the capacity of the enzyme to replicate past an abasic site in DNA, one of the most frequent endogenous lesions. In fact *in vitro* experiments have shown that, under given experimental conditions, the progression of the wild type form of the DNA pol is arrested at the base preceding the abasic site, while the exonuclease deficient mutant form can incorporate and extend past such a lesion, with dAMP being the deoxyribonucleoside monophosphate preferentially incorporated in front of it.[26–28] These data are in agreement with the idea that enzymatic idling occurs when an exonuclease proofreading activity efficiently removes the same base that is preferentially incorporated by the DNA pol in front of a lesion. Thus, the DNA pol associated $3' \rightarrow 5'$ exonuclease activity can also act as a kinetic barrier to translesion synthesis by preventing stable incorporation of bases opposite DNA damage.[29,30] Interestingly, a recent work reported a role for T4 DNA pol accessory factors sliding clamp and clamp loader in the TLS of an abasic site by an exonuclease deficient form of T4 DNA pol. When replication is performed with a large excess of DNA template over the DNA pol in the absence of accessory factors, the exonuclease deficient DNA pol inserts one nucleotide opposite the AP site but does not extend past the lesion. Addition of the clamp processivity factor and the clamp loader complex restores primer extension across the AP lesion.[31]

T4 DNA pol also participates in the rescue of a stalled replication fork via a series of steps that include fork regression, template switching and fork restoration. It has been shown that, together with the phage single-stranded recombinase UVsX and the T4-encoded helicase Dda, the T4 DNA pol can redirect *in vitro* DNA synthesis via two sequential template-switching reactions that allow T4 DNA replication to tolerate a non-coding DNA lesion.[32]

The crystal structure of the T4 DNA pol is not yet available, but other proteins of the replisome, such as the processivity clamp gp45, the helicase assembly protein gp59 and the single-stranded binding protein gp32 have been crystallized. The gp45 protein forms a trimeric ring with overall dimensions similar to those of bacterial and eukaryotic processivity factors. Each monomer of gp45 contains two domains that are very similar in chain fold to those of *E. coli* β-clamp and PCNA. Despite the overall negative charge, the inner surface of the ring is in a region of positive electrostatic potential, consistent with a mechanism in which DNA is threaded through the ring.[33] The crystal structure of the full length 217-residues monomeric gp59 helicase assembly protein reveals a novel α-helical bundle fold with two domains of similar size. Surface residues are predominantly basic but exposed hydrophobic residues indicate sites for potential contact with DNA and other proteins molecules.[34] The same study suggests that gp59 loads T4 gp41 helicase specifically at replication forks. The crystal structure of the gp32 binding domain complexed with single-stranded DNA uncovers that the DNA binding cleft comprises regions from three structural subdomains and includes a positively charged surface. Although only weak electro density is seen for the single-stranded DNA, this indicates that the phosphate backbone contacts an electropositive cleft of the protein, placing the bases in contact with the hydrophobic pockets. The DNA mobility implies that the weak electron density may reflect the role of gp32 as a sequence–independent single-stranded DNA chaperone, allowing the largely unstructured single-stranded DNA to slide freely through the cleft.[35]

6.2 Bacteriophage T7 DNA Polymerase

As for the bacteriophage T4, the DNA replication system of the *E. coli* infecting bacteriophage T7 has served as model for the type of reactions and protein-protein interactions needed to carry out coordinate replication of DNA leading and lagging strands. The approach that led to the reconstitution of a replisome from individual proteins was pioneered and developed in the laboratory of Charles Richardson. The replication machinery of T7 appears to be somewhat simpler than the one of T4 since only five proteins are required to reconstitute the basic replisome that can form at and propagate a replication fork (see **Table 6.1** and **Figure 6.2** and also **Table 5.2** in Chapter 5). On the leading strand the gp5 DNA pol interacts with its processivity factor thioredoxin (Trx), the product of *trx* gene of *E. coli* that binds tightly to the thumb subdomain of gp5 in a one-to-one complex. Both proteins then bind to a primer-template in a process that, unlike the situation in T4 or *E. coli*, does not need a clamp-loader for assembly. Trx is the only host protein that is a part of the T7 replisome. The unwinding of the duplex DNA to create a single-stranded DNA template for gp5/trx is accomplished by the DNA helicase activity located in the C-terminal half of the T7 gene 4 protein gp4. Gp4 assembles as a hexamer on the lagging strand and translocates unidirectionally $5' \rightarrow 3'$ on the single-stranded DNA in a reaction fuelled by hydrolysis of dTTP. The helicase has a high affinity for the gp5 DNA pol/trx complex and this interaction provides the high processivity necessary for replication of T7 genome. During replication of T7 DNA the tetraribonucleotides primers are synthesized by the primase located in the N-terminal half of gp4. The lagging strand folds back on itself so that the lagging

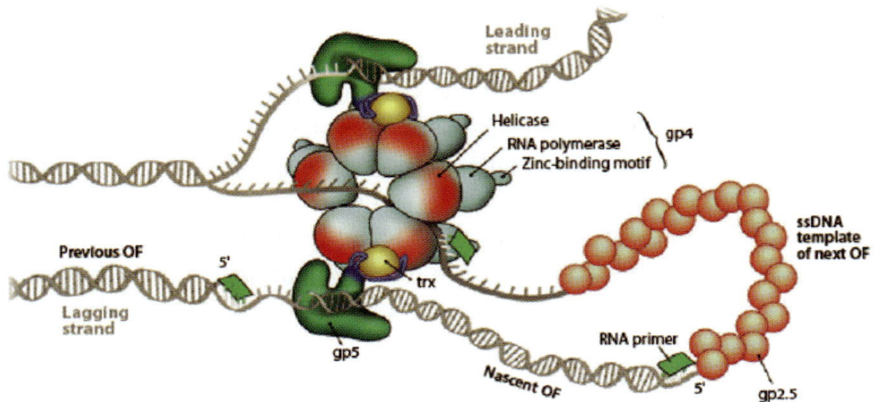

Figure 6.2   Model for the replication fork of bacteriophage T7. The gp4 helicase-primase couples DNA unwinding to RNA primer deposition. The gp5 DNA polymerase performs both lagging and leading strand DNA synthesis. The host processivity factor Trx associates to gp5 without the contribution of additional loading factors. (Reproduced with permission from Ref. 38.)

strand DNA pol can interact with the helicase. A model of the replication fork of T7 is shown in **Figure 6.2**.

The single-stranded DNA exposed during these operations is covered by the gp2.5 single-stranded DNA–binding protein. It should be noted that the C-terminus of gp2.5 also interacts with both gp5/trx and with the DNA helicase gp4 and that this interaction is essential for coordinating leading and lagging strand synthesis. Similarly to what observed for the T4 replisome, the T7 DNA pol can exchange with another DNA pol in solution without interfering with the overall processivity of DNA synthesis[36,37] (see also Chapter 5).

The above presented brief survey on the protein machinery that replicates T7 DNA is based on the content of an excellent and comprehensive review very recently published on the subject of Motors, Switches and Contacts in the Replisome[38] and we invite the readers to refer to it for more details. In the rest of this chapter we will mainly focus on the biochemical characteristic and functions of the T7 DNA pol and briefly refer to the crystal structures of the polymerase and of some of its interacting proteins.

The T7 encoded DNA pol was partially purified many years ago and identified as the product of the genetic locus gene 5.[14,39] Phage mutants defective in gene 5 are unable to replicate T7 DNA *in vivo*.[40] The protein was later purified to near homogeneity[41–43] and shown to be composed of two subunits, the 84 000-Da gene 5 protein and the 12 000-Da thioredoxin (Trx) a host protein that co-purified with a stoichiometry of 0.88 mol of Trx protein/mol of gp5 protein.[41,42,44] The two proteins form a complex with a dissociation constant (K_D) of 5 nM.[45] The gp5/Trx complex, in addition to polymerase activity, possesses a $3' \rightarrow 5'$ exonuclease activity that is active on single- and double-stranded DNA.[43,46,47] The gene 5 isolated subunit has been obtained either by purification from an *E. coli trx* deficient strain or by dissociation of the gp/Trx by guanidine[43] and its properties investigated. Both studies indicated that the isolated subunit retained no more that 1–2% of its polymerase activity and lost its exonuclease activity toward double-stranded DNA while retaining the $3' \rightarrow 5'$ exonuclease activity specific for single-stranded DNA. Addition of Trx led to recovery of both polymerase and double-stranded DNA exonuclease activity and increased the processivity of gp45 from 1–15 nucleotides per binding event to several hundreds.[48] The exact mechanism by which thioredoxin confers processivity on gp45 is not completely understood but kinetic studies revealed that Trx increases the processivity of gp5 by stabilizing its interaction with the primer-template.[49] The T7 DNA pol belongs to the A family of DNA pols and the crystal structure of gp45 and Trx bound to a DNA primer-template with an incoming dNTP in the active site has been solved.[50] To prevent the reaction from occurring in the crystal, the authors used a primer containing a dideoxynucleotide lacking the 3′OH. It was found that the right hand structure of the polymerase included the classical palm, thumb and fingers sub-domains and supported a "two metal ion" mechanism of nucleotide addition which was proposed by analogy to the nearly identical mechanism of the $3' \rightarrow 5'$ exonuclease of DNA pol I of *E. coli*.[50] A loop formed at the tip of the thumb, not found in other members of the A family DNA pols, provides the binding site for Trx and has been called Thioredoxin-Binding–Domain or TBD. T7 DNA pol is quite inefficient in lesion bypass of DNA distorting lesions, such as cis-syn thymine dimers (cis-syn TT), or not templating lesions, such as abasic sites.[51] An explanation of this behavior, at least as far as cis-syn TT lesion is concerned, came from structural studies with the T7 DNA pol. These studies showed that, when the 5′ T of a cis-syn TT is the templating base, the dimer lies within the polymerase active site and it can pair with the incoming dATP; however when 3′ T is the templating base, it is rotated out of the binding pocket of the polymerase, precluding normal base pairing with an incoming dATP.[52] It should be noted that, as already seen for the T4 DNA pol, the $3' \rightarrow 5'$ exonuclease associated to the T7 DNA pol plays a critical role in further diminishing the low level of bypass observed.[51] The situation appears to be different in the case of non distorting DNA lesions, such as the common oxidative lesion 7,8-dihydro-8-oxo-guanine (8-oxo-G). In fact, crystal structures have been solved demonstrating the existence of active complexes between T7 DNA pol and DNA containing 8-oxo-G, either during nucleotide insertion or in post-insertional complexes, showing 8-oxo-G paired with dC or dA at the 3′ end of the primer.[53] The presence of the $3' \rightarrow 5'$ exonuclease did not affect the bypass and mutational spectrum of 8-oxo-G by the T7 DNA pol, and the crystal structure explains how the 8-oxo-G·dA base pair defeats the normal error detection mechanisms that would otherwise suppress mutations during DNA synthesis. Along this line, further crystallographic studies have illustrated how the major, DNA distorting adduct formed by the carcinogen 2-acetylaminofluorene (dG-AAF) blocks replication by the T7 DNA pol while another less distorting DNA lesion (dG-AF) produced by the compound is less inhibitory.[54] These studies also provide a possible structural explanation for frame-shift mutagenesis observed during replication bypass of a dG-AAF adduct.[55]

The T7 DNA pol and helicase move separately at different rates and processivity but their interaction enormously increase their processivity during replication of the leading strand synthesis, indicating that the interaction of DNA pol with the helicase has dramatic effect on the stability of the proteins and on the overall processivity.[56,57] The T7 encoded gp4 DNA helicase is unique in that it contains both helicase and primase activities within the same polypeptide. The gene 4 encodes two colinear 63-kDa and 56-kDa proteins, expressed in 1:1 molar ratio.[58] The 56-kDa protein lacks the 63 amino acids present in the N terminus of the 63-kDa form that contain the Cys_4 zinc-binding motif required for primase activity. Both proteins bind and translocate on single-stranded DNA in a reaction coupled to dTTP hydrolysis.[59,60] Biochemical and structural data showed that gp4 forms both hexamers and heptamers, but only the hexameric form binds and translocates on single-stranded DNA.[61] Recent work has probed kinetics of individual ring-shaped T7 DNA helicase molecules as they unwound double-stranded DNA or translocated on single-stranded DNA. The results obtained fit with an active molecular motor mechanism of DNA unwinding by gp4. The sequences encoding the T7 helicase and primase have been cloned and active helicase and primase purified separately.[62,63] The crystal structure of the primase has been solved[64] and reveals two domains and the presence of two Mg^{2+} ions bound to the active site. NMR and biochemical data showed that the two domains remain separated until the primase binds to DNA and nucleotide. The zinc binding domain alone can stimulate primer extension by T7 DNA pol. These findings suggest that the zinc binding domain couples primer synthesis with primer utilization by securing the DNA template in the primase active site and then delivering the primed DNA template to the DNA pol. The gp2.5 protein interacts with both gp5[65–67] and gp4, stimulating the activity of each and probably coordinating leading and lagging strand DNA synthesis. In addition, gp2.5 participates in homologous DNA pairing during recombination[68–70] and in repair of double strand breaks.[71] The crystal structure of gp2.5 has been determined[72] and showed that it has a conserved OB-fold (oligosaccharide/oligonucleotide binding fold) that is well adapted for interactions with single-stranded DNA. Superposition of the OB-folds of gp2.5 and other single-stranded DNA binding proteins revealed a conserved patch of aromatic residues that stack against the bases of single-stranded DNA in the other crystal structures, suggesting that gp2.5 binds to single-stranded DNA in a similar manner. An acid C-terminal extension of the gp2.5 protein, which is required for dimer formation and for interactions with the T7 DNA pol and the primase-helicase, appears to be flexible and may act as a switch that modulates the DNA binding affinity of gp2.5.

6.3 HSV-1 DNA Polymerase

Herpesviruses are a large family of viruses morphologically very similar, with a large double-stranded linear DNA genome (130–230 kbp) and a virion consisting of an icosahedral nucleocapsid surrounded by a lipid bilayer envelope. Herpes Simplex Virus (HSV) is the prototype member of the nine (considering that HHV6 has two subtypes, A and B) known human herpesviruses. For this reason, among the Herpesviridae, replication of HSV-1 is the most extensively studied. The complete sequence of HSV-1 genome (152 kbp), with 67% GC content, is known and contains approximately 75 open-reading frames (ORFs) and 100 transcripts, indicating a low level of splicing. The genome consists of 6 important regions: a 108-kb unique long (U_L) and a 13-kb unique short (U_S) region, each of them bounded by two short inverted repeat (IR) regions of 9 and 6.6 kbp respectively. During replication, inversions of the U_S and U_L sequences occur by breakage and reunion giving four isomeric infectious rearrangements of the unique regions, but the biological significance of this is unknown.

Genes for the DNA replication enzymes/proteins are encoded in the 108 000 bp long U_L region. There are three origins (ORI) of replication: ORI_L in the middle of U_L region, and ORI_S, present in two copies, in each of the 6600 bp short repeats (R_S) flanking the U_S region. Either ORI_L or one copy of ORI_S appears dispensable. The viral genes necessary and sufficient for viral DNA replication are: the HSV DNA pol (U_L30), the DNA binding proteins (U_L42 and UL29 or ICP8), the ORI binding protein (U_L9), and the helicase/primase complex (U_L5, 8, and 52). But cellular enzymes might replace some of them and, among others, host RNA polymerase, DNA topoisomerases and perhaps DNA ligase IV are certainly needed for the molecular events leading to the reactivation of the virus from latency and to its replication.

HSV-1 gene expression leading to HSV DNA replication and growth in the host cell nucleus requires, for instance, cellular RNA polymerase II for transcription of immediate early (α) genes followed by early (β) genes which include thymidine kinase, ribonucleotide reductase, and all the enzymes/proteins necessary for viral DNA replication and gene expression regulation, and finally the late (γ) genes.

Host macromolecules biosynthesis is on the contrary shut off leading to death of the infected cell.

The detailed mechanism of HSV DNA replication is not known. Genome isomerizations, mentioned above, and homologous recombination events that accompany DNA replication certainly complicate the interpretation of the precise pattern of replication. However the action of exonucleases upon redundant terminal regions of the genome most probably leads to cohesive ends and, likely upon intervention of the host cell DNA ligase IV,[73] to conversion of linear to circular forms before theta type DNA replication is initiated at one or more of the three HSV-1 origins. Then HSV replicates its DNA via a rolling circle mechanism as suggested by the detection of multigenome-sized intermediates that are then cleaved and packaged into infectious viral particles. Origin sequences are functionally equivalent and contain a central AT-rich hairpin sequence (**Figure 6.3a**) where the double helix initially opens allowing the binding, on either side, of U_L9, the origin binding protein (**Figure 6.3b**). U_L9 acts as a dimer and cooperatively binds to two 8-bp boxes adjacent to the palindrome. Very much like SV40 T antigen and *E. coli* DnaA, U_L9 hydrolyzes ATP and with the $3' \rightarrow 5'$ helicase activity causes a localized melting of DNA duplex, recruiting ICP8 (**Figure 6.3c**) and also directs the entry of the tree-subunit *helicase-primase* complex of $U_L5/U_L8/U_L52$ (*primosome*) that uses ATP (or GTP) hydrolysis to further unwind the DNA helix, to translocate in the $3' \rightarrow 5'$ direction, and

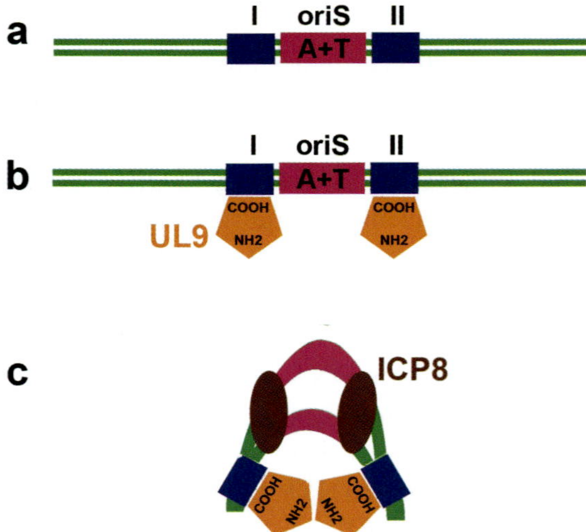

Figure 6.3   HSV-1 DNA replication origin binding and melting. Blue boxes (I and II) represent the UL9 binding sites. Upon binding (b), UL9 dimerizes, inducing a torsional stress which unwinds the A+T rich region of OriS, which can be bound by ICP8 (c).

to synthesize RNA primers for discontinuous strand DNA synthesis. U_L8 has no enzymatic activity whereas U_L5 and U_L52 are the helicase and the primase respectively.

The main actor in HSV DNA replication is of course the HSV DNA pol which is a heterodimer of U_L30, the catalytic subunit, and U_L42, the processivity subunit whose crystal structures have been solved.[74,75] The crystal structure of U_L42 bound to the C-terminal 36 residues of U_L30 shows that, despite a lack of sequence homology with PCNA, U_L42 has a remarkably similar structural fold. However, contrary to PCNA, U_L42 binds DNA as a monomer, and does not require ATP or accessory proteins for binding. The structure of U_L42 also revealed a surface of four conserved arginine residues on the side opposite the U_L30 binding site and substitution of any arginine residues with alanines decreased the binding affinity for DNA and processivity of DNA pol.[76]

The interaction of U_L30 C-terminus with U_L42 increases the polymerase processivity. Following the denaturation of the DNA molecule at the replication origin by U_L9, the helicase/primase (UL5/UL8/UL52) and the single-stranded DNA binding protein (U_L29) associate to the DNA pol/UL42 to begin and allow the predominant rolling circle mode of DNA synthesis.

Based on sequence similarity, HSV-1 DNA pol belongs to the B-family of DNA pols (see Chapters 2 and 4) and its crystal structure has been obtained.[75] The overall architecture of UL30 is very similar to that of other B-family DNA pols (**Figure 6.4**), in spite of low sequence homology (16–50%). There are six domains: a pre-N-terminal, N-terminal and $3' \rightarrow 5'$ exonuclease domains are located in the N-terminal half, whereas the C-terminal part adopts the usual DNA pol folding in palm, fingers and thumb domains. These domains are arranged to form a disk with a central hole from which three channels or grooves originate. A first groove delineates the interface between the N-terminal and exo domains and is lined with positively charged residues. It is predicted to interact with the phosphate backbone of the single-stranded part of the DNA template. The second groove is located between the exo domain and the tip of the thumb and provides access to the exonuclease active site. This is predicted to be the binding site for the melted primer strand in the editing mode. The third groove, located in the palm and thumb subdomains provides the binding site for the double stranded part of the DNA template. The pre-N-terminal domain consists of a 250-aa extension, unique to the herpesvirus DNA pol family. Its function is unknown, but the presence of a highly conserved box (the -FYNPYL- motif) directly upstream of the binding groove for the single-stranded DNA template suggested a role of this domain in the interaction with the viral helicase-primase complex. The N-terminal domain contains a structural motif of the type $\beta\alpha\beta\beta\alpha\beta$, commonly found in RNA binding protein, forming an RNA

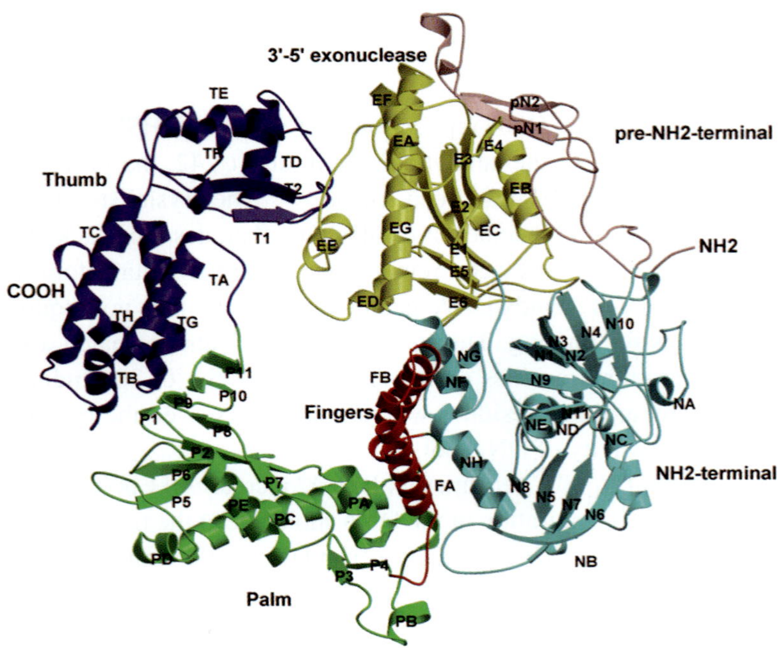

Figure 6.4   Structure of the HSV-1 DNA pol (UL 30). The different domains are represented in different colors. α-helices are indicated with letters and β-sheets with numbers. (Reproduced with permission from Ref. 75.)

binding cleft. Analysis of the overall structure of this domain led to the hypothesis that this could be the catalytic site for the RNase H activity of U_L30, even though experimental confirmation is still awaited.

The HSV-1 DNA pol also possesses a $3' \rightarrow 5'$ proofreading exonuclease activity that can ensure a high fidelity of DNA replication as well as an RNase H activity that can remove RNA primers in Okazaki fragments. However, despite the associated $3' \rightarrow 5'$ exonuclease activity, HSV-1 DNA pol is less faithful than other replicative DNA pols and was found to poorly discriminate between correct and incorrect nucleotide, with selectivity values in the order of 10^{-2}–10^{-3} (as a comparison, other exo-proficient replicative DNA pols show selectivity values in the order of 10^{-6}–10^{-9}).[77] Kinetic analysis showed that the major determinant of selectivity was at the level of ground-state binding affinities (K_d values) for dNTPs, which were in the 0.1 μM–0.5 μM range for correct base pairs and in the 175 μM–500 μM for mismatched nucleotides. Mismatch extension rates constituted a less effective kinetic barrier. In particular, during processive DNA synthesis under forced misincorporation conditions (i.e. in the presence of only three dNTPs), U_L30 was able to efficiently generate and elongate mismatches. Intriguingly, the wild type protein was more efficient in mismatch extension than the corresponding exo-mutant. The processivity factor U_L42 was found to increase the fidelity of U_L30 by decreasing both the misincorporation and mismatch extension rates.[78]

6.4 Protein Primed DNA Replication: Adenoviruses and Bacteriophage φ29

As outlined in Chapter 1 and 2, a common feature of all DNA pols is their requirement for a 3′-hydroxyl group to act as a primer for DNA synthesis. The majority of organisms use a short RNA or DNA oligonucleotide complementary to a sequence in the template strand for the initiation reaction. A remarkable exception to this rule is constituted by DNA pols which use the 3′-hydroxyl group of an amino acid (usually a serine) to prime the nucleotide incorporation reaction. The two best known examples of such enzymes are the adenovirus and the bacteriophage φ29 DNA pols.

6.4.1 *Adenovirus DNA Polymerase*

Adenoviruses infect mammals and birds and, occasionally, reptiles and frogs. They are DNA viruses whose genome consists of a 30–36 kb double-stranded DNA molecule. They are named after the human adenoids, a lymphoid gland in the nasopharynx, which represent the most common site of persistent infection and the first tissue from which adenoviruses have been isolated. Even though they do not cause serious illnesses, because of their widespread diffusion and their ability to replicate in cultured human cells, many adenoviruses have been the subject of intense studies. Recently, adenoviruses have been exploited as vectors for human gene therapy and recombinant vaccines.

The AdDNA replication was one of the first replicative systems reconstituted *in vitro*.[79] Replication of AdDNA requires three virally encoded proteins: the precursor Terminal protein (TP), the AdDNA pol and the DNA binding protein (DBP). In addition, cellular proteins, such as the transcription factors NFI, Oct-1, have been shown to participate in the initiation reaction.[80] AdDNA replication is initiated by a protein priming mechanism and proceeds through a "jumping-back" step followed by strand displacement-dependent DNA synthesis. The AdDNA replication cycle is schematically drawn in **Figure 6.5**.

The AdDNA has a TP molecule covalently attached to both 5′-ends (**Figure 6.5a**). Replication starts in the 102-bp long inverted terminal repeats (ITR) located at ends of the genome. These ITR contain at their extremities an essential triplet repeat (GTAGTA), which is followed by a conserved region that binds the TP-Ad DNA pol complex and an auxiliary origin recognized by the cellular

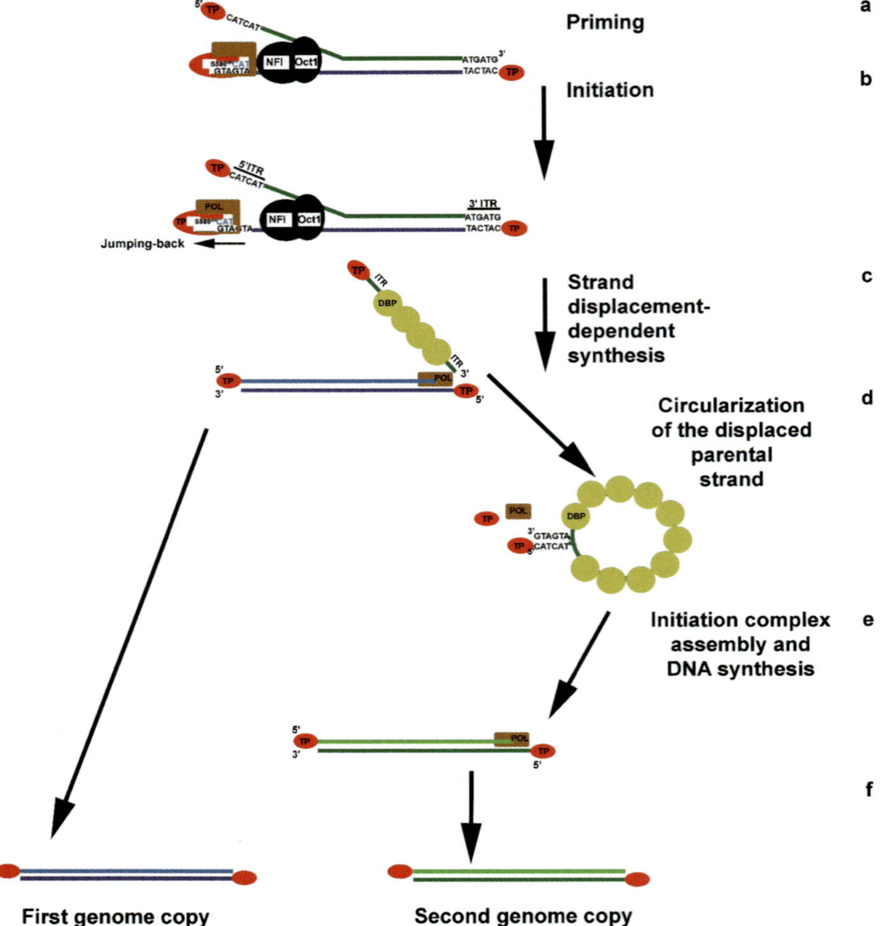

Figure 6.5   Schematic representation of the Adenoviral replication cycle. An initiation complex is formed by association of the Ad DNA pol–TP complex at the 3′-end of the genome, which is facilitated by binding of the transcription factors NFI and Oct 1. Ad DNA pol uses TP Ser580 to prime DNA synthesis (a). After the first CAT triplet is synthesized, a jumping back step reanneals it to the GTA at positions 1–3 of the template strand (b). Strand-displacement synthesis starts and the displaced strand is covered by the DBP protein (c). Annealing of the ITRs forms a panhandle structure (d) and synthesis can be initiated by the bound TP protein (e), leading to complete duplication of the genome (f).

transcription factors NFI and Oct-1. An initiation complex is formed by association of the Ad DNA pol-TP complex at the 3′-end of the genome (**Figure 6.5a**). This step is facilitated by the binding of the transcriptional factors NFI and Oct-1 to the adjacent sequences and their physical association to the Ad DNA pol-TP complex. Next, Ad DNA pol starts replication with the covalent coupling of the first nucleotide, a dCMP residue, to the hydroxyl group of S580 of the TP, followed by incorporation of two additional nucleotides, A and T, forming a TP-CAT intermediate (**Figure 6.5a**). For this reaction, the second GTA sequence (positions 4–6) of the triplet is used as template. The resulting trinucleotide product then realigns with the template strand at nucleotides 1–3 of the triple, through the so-called "jumping-back" mechanism, and functions as the primer for the subsequent elongation step (**Figure 6.5b**). Synthesis proceeds through an ATP-independent strand displacement mechanism, where the single-stranded parental DNA strand which is displaced by the advancing Ad DNA pol is covered by the viral DBP (**Figure 6.5c**). The ITRs present at both extremities of the DBP-coated singe-stranded parental DNA can re-anneal forming a typical panhandle structure (**Figure 6.5d**). The terminal double-stranded DNA region bearing a TP bound to its $5'$-end can then function as an origin to prime a new round of replication (**Figure 6.5e**). In such a way, semiconservative DNA replication is ensured, even though the two DNA strands are not replicated within the same fork (**Figure 6.5f**). The protein-priming mechanism and strand displacement-coupled DNA synthesis eliminate the need for a DNA primase and a replicative helicase.

The 140-kDa AdDNA pol belongs to the family B DNA pol (see Chapter 4). Site-directed mutagenesis and structural homology studies have revealed the existence of the highly conserved α-like DNA pol domains (see Chapter 4.2) in Ad DNA pol. In particular, regions I, II and IV have been shown to be essential for DNA synthesis.[81–84]. A conserved motif (I/Y)xG(G/A) present in region II was also shown to be important for the protein priming mechanism. Early biochemical characterization[85] showed that Ad DNA pol utilizes *in vitro* a variety of homopolymeric template-primers including poly(dC)/oligo(dG), poly(dA)/oligo(dT), poly(dT)/oligo(dA), and poly(dT)/oligo(rA). DNA synthesis by the Ad DNA pol was also stimulated as much as 100-fold by DBP, indicating that the Ad DNA pol can be a highly processive enzyme. Ad DNA pol is highly processive and possesses a $3' \rightarrow 5'$ proofreading exonuclease activity, which shows preference for mismatched $3'$-terminating nucleotides. Interestingly, it has been shown that at mismatched primers, exonucleolytic cleavage and subsequent incorporation occur without enzyme dissociation, suggesting that an intramolecular mechanism operates between the C-terminal DNA pol domain and the N-terminal exonuclease active site of Ad DNA pol.[86] Biochemical studies have shown that the Ad DNA pol covers approximately 14–15 nt and that the distance between the entrance of the template binding cleft and the polymerase active site is of 5 nt, whereas the exonuclease domain seems to be located approximately 6 nt away from the primer.[87] These very similar distances, thus, account for the observed rapid intramolecular switch between the polymerase and editing domains. In addition to Ser580, other amino acids of the viral TP protein play a role in stimulating the initiation of DNA replication. In particular, Asp578 and Asp582 are important in modulating the apparent affinity (K_m) and the catalytic efficiency (k_{cat}/K_m) of the first dCTP residue incorporation.[88] Affinity for dCTP in the priming reaction was in the low micromolar range ($K_m = 2\,\mu M$) and the catalytic efficiency was $2.2\,M^{-1}\,s^{-1}$.[88] The close proximity of these two Asp residues to the 3′-hydroxyl group of Ser580 suggests that they can play a role in coordinating the Mg^{2+} activating ions with the 3′-primer residue and the phosphates of the incoming nucleotides. Thus, the TP protein contributes to the architecture of the active site used for the initiation reaction by Ad DNA pol.

6.4.2 *Bacteriophage φ29 DNA Polymerase*

The complete *in vitro* φ29 DNA replication system has been reconstituted[89–91] allowing its detailed biochemical characterization. The overall mechanism is analogous to the one described for Adenovirus. The φ29 bacteriophage possesses a 19.3-kb linear double stranded DNA genome with an origin of replication at each end. The φ29 DNA pol and the viral terminal protein (TP) form a heterodimer, even in the absence of DNA.[92] DNA replication initiates by the addition of a dAMP residue to the 3′-hydroxyl group of the TP Ser 232,[93,94] using as the template the second base from the 3′-end of the origin of replication.[95] Next, this first dAMP is paired to the first nucleotide at the 3′-end of the genome, through a "sliding back" mechanism[95] and replication proceeds through strand-displacement DNA synthesis. The TP dissociates from φ29 DNA pol after 6–10 nucleotides have been incorporated.

The bacteriophage φ29 DNA pol is a 66-kDa monomeric enzyme encoded by the viral gene 2. Crystallization of a TP-φ29 DNA pol complex, allowed a detailed analysis of the protein-protein interactions essential for the protein-primed mechanism.[96] In the structure, the conformation of the φ29 DNA pol, complexed with TP is very similar to the unbound enzyme. However, the TP protein folds into a three recognizable domains, an N-terminal domain, an intermediate domain interacting with the TPR1 domain of the DNA pol and a so-called priming domain. The latter mimics a DNA double strand both in his electrostatic profile and binding to the φ29 DNA pol. In fact, comparison of the palm domain of φ29 DNA pol, with that of other B-family DNA pols, showed that the location of the priming domain overlaps that of the primer-template DNA bound to the DNA pol. Based on this structure, a mechanistic model has been proposed for the priming reaction (**Figure 6.6**).

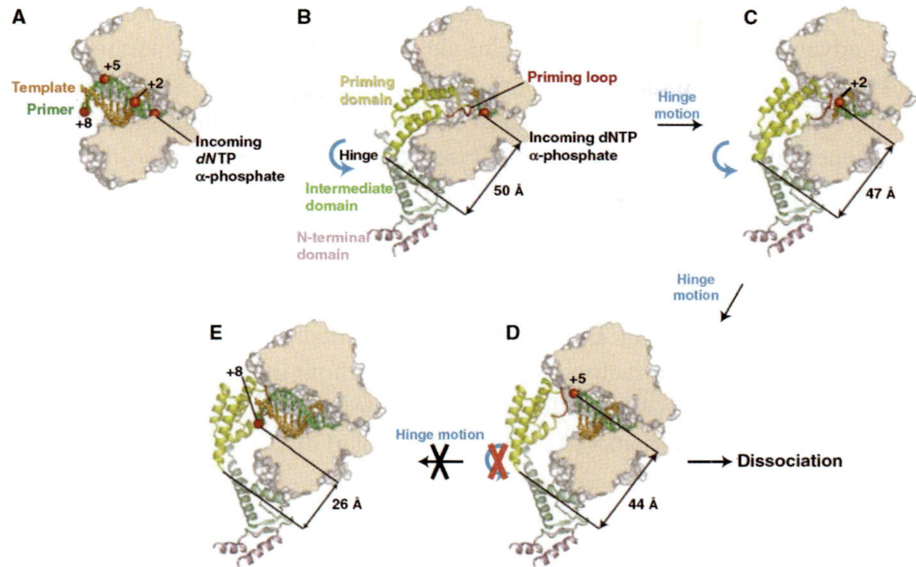

Figure 6.6   Structural model for the φ29 DNA pol protein-priming mechanism. For details see text. (Reproduced with permission from Ref. 96.)

The first nucleotide (dAMP) is attached to Ser232 of TP in complex with φ29 DNA pol (**Figure 6.6A**). During subsequent nucleotide additions (**Figure 6.6B–C**), this Ser residue traces a spiral path, due to the structural constraints, so that after incorporation of five nucleotides, Ser232 has rotated by 180°. Moreover, due to the advancement of the DNA pol, the Ser residue has translocated by 17 Å, placing the attached dAMP in close proximity with the thumb domain of φ29 DNA pol. The rotation of the priming domain occurs around a hinge connecting it to the intermediate domain of TP. This hinge motion allows only about six nucleotides to be added. When the nascent DNA chain becomes longer, the translocation of the TP-φ29 DNA pol complex will cause a steric clash between the α-phosphate of the dAMP residue linked to the Ser232, with the hinge domain (**Figure 6.6D–E**). This explains why TP dissociated from the complex after about six incorporation events. The dissociation of TP marks the transition from the initiation to the elongation step of the reaction.

The φ29 DNA pol has two sequences: TPR1 and TPR2, which are found only in protein-primed DNA pols. While TPR1 is a major site of interaction with the TP protein, the TPR2 domain is important for the strand displacement DNA synthesis during the elongation phase. Homology modeling of the enzyme-substrate complex suggested that the TPR2 domain forms a narrow channel where the template strand has to insert before reaching the active site. Since this channel is not large enough to accommodate a double-stranded DNA, the exclusion of the complementary strand by the advancing DNA pol provides a model for efficient strand separation during DNA synthesis.

The ϕ29 DNA pol is also characterized by an extremely high processivity in the absence of any auxiliary factors. Based on structural and homology model, it has been proposed that the ϕ29 DNA pol encircles the double-stranded DNA product with its polymerase and exonuclease domains, in a manner reminiscent of sliding clamp proteins such as PCNA, thus explaining the high processivity of DNA synthesis. This processivity of ϕ29 DNA pol has led to its use in the isothermal amplification of DNA. The system, called TempliPhi or GenomiPhi, can lead to amplification factors as high as 10^7 without the need of thermocyclers and thermoresistant DNA pols. Its major advantage is the possibility to directly amplify plasmid DNA, thanks to the excellent strand displacement activity of ϕ29 DNA pol.

The ϕ29 DNA pol is a multifunctional enzyme, endowed with at least three catalytic activities: (1) deoxynucleotidylation, which catalyzes the linkage of a dAMP residue to the TP Ser232 3′-hydroxyl group; (2) DNA polymerization and (3) 3′ → 5′ exonuclease proofreading activity, which degrades processively DNA substrates longer than six nucleotides with a catalytic rate as high as $500\,\text{s}^{-1}$ and a preference for mismatched vs. matched primer termini. The exonuclease activity resides in the N-terminal domain, whereas, as outlined above, the active site for the deoxynucleotidylation activity comprises the priming domain of the TP protein and the palm and thumb domains of ϕ29 DNA pol. The polymerase domain responsible for the intrinsic DNA synthetic activity of ϕ29 DNA pol resides in the C-terminal part of the protein. This domain is formed by the three highly conserved boxes: motif A ($\text{D}x_2\text{SLYP}$), where x is any amino acid, motif B ($\text{K}x_3\text{NS}x\text{YG}$) and motif C ($\text{Y}x\text{DTDS}$), plus two additional eukaryotic-type polymerase specific motifs: $\text{T}x_2\text{GR}$ and $\text{K}x\text{Y}$. The functions of each of these domains have been determined in ϕ29 DNA pol by site-directed mutagenesis. A summary of the results obtained, with the identified functions of specific residues within each domain is presented in **Table 6.2**.

As in the case of Ad DNA pol, also ϕ29 DNA pol must also possess a mechanism for the coordination between the DNA pol and the exonuclease active sites, establishing the transition between the "editing" and the "polymerizing" modes. Site directed mutagenesis studies, indicated that the $\text{Y}x\text{G(G/A)}$ motif located between the exonuclease and polymerization domain, plays a role in the stabilization of a correctly paired primer terminus and is involved in "sensing" a mismatched nucleotide through a conformational change which promotes switching from the polymerizing to the editing mode.[97]

Table 6.2   Functions of residues in the conserved domains of the φ29 DNA polymerase

Domain	Residues	Function
Dx_2SLYP	D249	Catalysis
	S252	DNA binding
	Y254	dNTP binding
Kx_3NSxYG	N387, G391, F393	DNA binding
	Y390	dNTP binding
YxDTDS	D456, D458	Catalysis
	Y454	dNTP binding
Tx_2GR	T434, R438	DNA binding
KxY	K498, Y500	DNA binding
YxG(G/A)	Y226, F230	DNA binding pol-exo switching

6.5   African Swine Virus DNA Polymerase

African Swine Fever Virus (ASFV) is an enveloped DNA virus which causes fatal disease in domestic pigs. It encodes for a B-family replicative DNA pol, but, contrary to the majority of other viruses, it also encodes for an additional DNA pol, DNA pol X. The ASFV DNA pol X has several unique features: it is the smallest DNA polymerase known (174 aa, ≈20 kDa) and it is the only known viral member of the DNA pol family X (See Chapter 4). In addition, it is the least accurate viral DNA pol described so far. All these facts prompted accurate investigations of its functional and structural properties. Early biochemical characterization[98] identified it as a *bona fide* template dependent DNA pol, characterized by high distributivity, gap-filling activity, lack of proofreading exonuclease and lack of terminal transferase activity (see Chapter 4). Further studies[99] showed that ASFV DNA pol X has different affinities for dNTP incorporation, which were dependent on the particular base-pair involved. In general, purines were preferred over pyrimidines as incoming nucleotides. Detailed kinetic studies[100] showed that ASFV DNA pol X, contrary to all the other known DNA pols, follows a different order of substrate binding, with Mg-dNTP as the first substrate, followed by the template/primer. Binding of DNA to the free enzyme results in a non productive binary enzyme-DNA complex which is unable to bind the nucleotide. Conversely, association of Mg-dNTP first facilitates subsequent formation of the catalytically competent ternary complex. ASFV DNA pol X exhibits low affinity for dNTPs (in the 80–900 µM range) and peculiar nucleotide insertion specificity, so that some misincorporation events are much more favored than others.[101] In particular, the G:G and G:T mismatches are generated with only 2-fold and 16-fold lower efficiency than the corresponding G:C correct base pair. Pre-steady state kinetic analysis showed that the catalytic rate

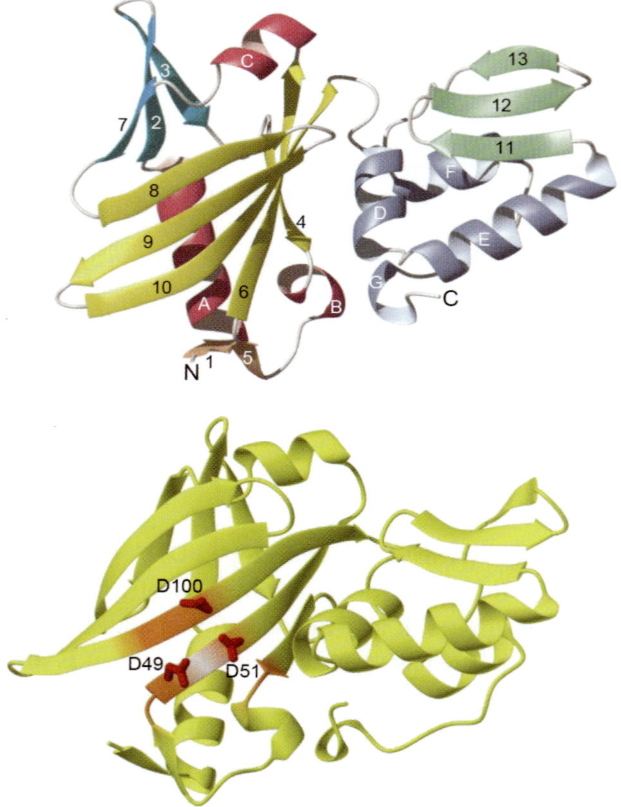

Figure 6.7 **NMR structure of the ASFV DNA polymerase X.** Top: shown are the palm (yellow, cyan, orange, purple) and the C-terminal (green, blue) subdomains. Bottom: the catalytic triad (D49, D51, D100) is shown in red. (Reproduced from Ref. 102 with permission.)

(kpol) for correct dCTP incorporation opposite a template G was $0.11\,s^{-1}$, while the kpol for dGTP incorporation opposite G was $0.009\,s^{-1}$. In the presence of 140 mM KCl, misincorporation was strongly favored over correct dCTP incorporation (kpol values of $0.05\,s^{-1}$ and $0.0007\,s^{-1}$ for G:C and G:G basepairs, respectively). The structure of ASFV DNA pol X has been determined by NMR studies,[102,103] and reveals a completely novel fold (**Figure 6.7**).

Unlike any other DNA pol, the ASFV DNA pol X is composed only of a palm and a C-terminal domain. The palm of ASFV DNA pol X also adopts a different fold with respect to other family X enzymes, with a positively charged helix which facilitates binding to DNA. The nucleotide binding site is located between the palm and C-terminal domain. Upon complex formation with the Mg-dNTP substrate, ASFV DNA pol X undergoes a substantial structural rearrangement, in order to properly coordinate the incoming nucleotide with the catalytic residues. Molecular dynamics simulations reveal that the formation of a G:G mispair occurs with the incoming dGTP in the *syn* configuration through Hoogsteen hydrogen bonding, and causes large-scale conformational changes similar to that observed for the correct base pair (G:C), resulting in an overall geometry of the G:G mispair that is well poised for catalysis.[104]

The ASFV DNA pol X has been shown to possess an abasic site (AP) lyase activity,[99] being able to endonucleolytically cut the phosphodiester bond at the 5′-side of an AP site. This activity is common to the Base Excision Repair (BER) enzymes DNA pol β. These results, coupled with the high efficiency of ASFV DNA pol X in filling one nucleotide gapped DNA substrates, led to the proposal that this enzyme might be involved in some virus-specific BER pathway. Interestingly, ASFV encodes its own ATP dependent DNA ligase. *In vitro* reconstitution experiments of the ASFV-specific BER reaction with purified proteins, indicated that ASFV DNA ligase is able to ligate the products of the gap filling catalyzed by ASFV DNA pol X, even though they contain mismatched nucleotides. Thus, it appears that this peculiar viral BER pathway has been positively selected during ASFV evolution because it increased the intrinsic mutation rate of the virus, accelerating the generation of viral variants able to evade the host immune system.[105] This hypothesis, while suggestive, is still controversial and any significance for this highly error-prone BER for the ASFV life cycle has yet to be proven.

6.6 RNA-Dependent DNA Synthesis: Reverse Transcriptases

The so called "central dogma" of Biology, was first enunciated in 1958 by Francis Crick,[106] Nobel Prize winner and co-discoverer together with Rosalind Franklin and James Watson, of the double-helix structure of the DNA.[107–109] In its essence, the dogma predicted that the flow of genetic information followed three obligatory steps from DNA to RNA to proteins. None of these steps was predicted to be reversible, nor was any direct transfer of information from DNA to proteins believed to take place. While the latter statement still holds true, a reversion of the first step, namely the flow of information from RNA to DNA, was demonstrated to occur by Howard Temin and David Baltimore,[110,111] in 1970, with their discovery of a novel enzyme called reverse transcriptase (RT), which allowed them to win the Nobel Prize in 1975, together with Renato Dulbecco. To date, RT has been shown to be essential in the life cycle of several animal RNA viruses, called Retroviruses, converting their RNA genomes into double-strand DNA molecules to be subsequently integrated within the host cell genome. Additionally, RT also participates in the duplication of transposable genetic elements called retrotransposons.

6.6.1 *HIV-1 Reverse Transcriptase*

After the discovery in 1983 by Luc Montagnier[112] and Robert Gallo,[113,114] that the causative agent of AIDS was the retrovirus HIV-1, its RT has become the major target for antiretroviral chemotherapy. After 25 years, the current anti-HIV chemotherapy is still largely based on small molecules targeting HIV-1 RT. The interest in HIV-1 RT as a chemotherapeutic target fostered biochemical, structural and enzymatic studies which made HIV-1 RT one of the most thoroughly investigated polymerases to date.

6.6.1.1 *The retro-transcription reaction*

HIV-1 RT has two enzymatic activities, DNA pol and RNase H, that cooperate to convert the linear, single-stranded RNA into a linear double-stranded DNA molecule (called provirus) in the cytoplasm of the infected cell. Like most DNA pols, RT incorporates nucleotides in a strictly template-dependent manner. However, a distinguishing feature of this enzyme is its ability to utilize both RNA and DNA as templates. The RNA-dependent DNA pol activity (RDDP) comes into play in the first phase of the retro-transcription reaction, when the positive sense (+) RNA is copied into a negative sense (−) DNA strand, generating an RNA/DNA hybrid. The DNA-dependent DNA pol activity (DDDP) is involved in the second step of the reaction, in which the (−) DNA strand is copied into a (+) DNA strand, to generate a double-stranded DNA molecule. The overall reaction is fairly complex and involves distinct coordinated steps for its completion, as outlined in **Figure 6.8**.

Initiation and minus-strand DNA synthesis. The initiation step of the retro-transcription reaction is highly specific and requires a cellular tRNA primer. HIV-1 RT uses exclusively the cellular tRNA^{Lys} to prime minus-strand DNA synthesis. The tRNA^{Lys} is selectively packaged into viral particles through a complex network of interactions involving the Gag-Pol polyprotein and the cellular lysyl-tRNA synthetase.[115−118] It was shown that the thumb and connection subdomains of the viral RT are required both for tRNA^{lys} incorporation into the virions and for its precise pairing at the PBS.[118] The specificity for this tRNA is given by the presence in the HIV-1 genome (positions +182–199) of a sequence called Primer Binding Site (PBS), complementary to the 3′-terminal 18 nucleotides of the tRNA^{lys}. Upstream to the PBS, in the U5 region of the HIV-1 genome (position +123–130) there is another sequence (Primer Activation Sequence, PAS) which is complementary to the TψC arm of the tRNA and has been shown to participate in the initiation reaction.[119] Primer binding likely proceeds through a biphasic mechanism, whereby the tRNA first anneals to the PBS, facilitating the opening of the TψC arm and then promoting a second set of base pairing interactions of the tRNA with the PAS

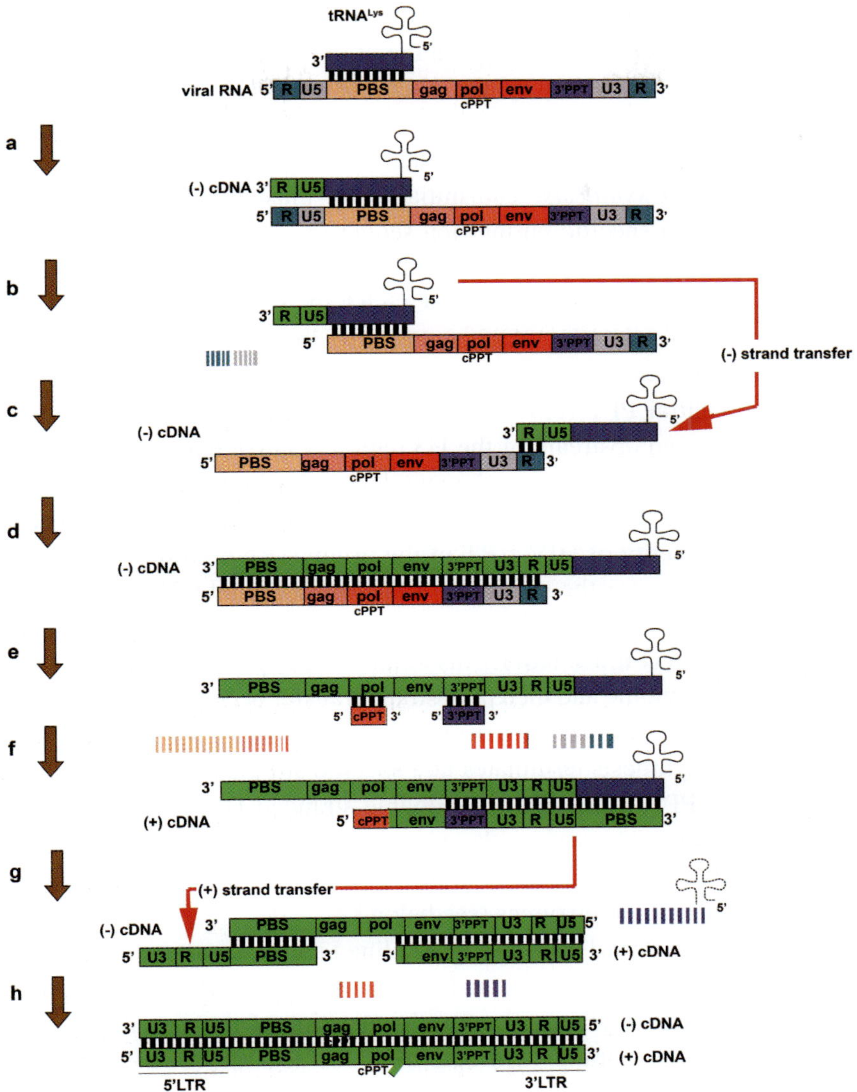

Figure 6.8   Scheme of the HIV-1 retrotranscription reaction. For details see text.

motif.[120] The conformational change necessary for PAS binding is mediated by the viral nucleocapsid protein, which acts as an RNA chaperone.[121] Upon formation of a stable initiation complex between RT, the viral RNA and the tRNA primer, synthesis of the minus-strand starts. The first product corresponds to the 5′-end of the viral genome (U5 and R regions, **Figure 6.8a**). Then synthesis stops (minus-strand strong stop), the RNAse H activity of RT degrades the RNA strand which has been copied (U5 and R, **Figure 6.8b**) and the newly synthesized DNA fragment relocates to the 3′-end of the RNA genome, annealing at the R element downstream the U3 region (**Figure 6.8c**). In this way, the 3′-end of the DNA fragment is used as a primer for the generation of a full-length minus DNA strand (**Figure 6.8d**), which will be used as a template for the plus-strand DNA synthesis.

Plus-strand DNA synthesis. The initiation of plus-strand DNA synthesis by HIV-1 RT requires a specific purine-rich sequence, designated as the polypurine tract (PPT). This region is highly conserved in human and simian lentiviruses and consists of the sequence 5′-AAAGAAAGGGGGGA-3′, flanked at the 5′-side by the uridine-rich region 5′-UUU(U/A)U-3′. Other retroviruses and retroelements contain similar purine-rich sequences. Plus-strand DNA synthesis initiates at two specific PPT sites: the central PPT (cPPT) located in the integrase part of the *pol* gene and the 3′PPT located upstream of the U3 region (**Figure 6.8e**). Precise initiation at the 3′PPT is essential, since this element defines the boundaries of the 5′ end of the viral genome and the U3 region. The 3′PPT primer is generated upon cleavage by the RNase H activity of HIV-1 RT of the original RNA strand (the one which served as the template for the synthesis of the minus-strand DNA) between the G and A nucleotides at the 3′ end of the element itself and about 15–19 nt upstream of the PPT-U3 junction (**Figure 6.8f**). Plus-strand DNA synthesis proceeds up to the 3′-end of the viral genome and then plus-strand transfer occurs (**Figure 6.8g**). Plus-strand DNA synthesis also occurs from the cPPT. After plus-strand transfer, the plus-strand DNA synthesis terminates at a specific site (CTS) located about 90 nt downstream the cPPT, thus within the double-stranded DNA region synthesized in the previous round. This causes the generation of a 90 nt "flap" structure at the center of the HIV-1 genome (**Figure 6.8h**), which is subsequently recognized and cut by specialized cellular enzymes (see below).

Minus- and Plus-strand transfers. As outlined above, synthesis of a full length proviral DNA requires two strand transfer events (**Figure 6.8b,g**). These are essential steps of the reaction, since correct re-localization of the partially synthesized minus- and plus- DNA strands will determine the accuracy of the entire reverse transcription reaction.[122] These steps are also a potential source of recombination (see below). Since two copies of the viral genome are present in each viral particle, the strand transfer reaction can be either intramolecular (i.e. involving re-localization of the cDNA on the same template RNA strand) or intermolecular (i.e. involving annealing of the cDNA to the complementary region of the other RNA strand). *In vivo*, intra- and inter-molecular reactions seem to occur at similar frequencies during the minus-strand transfer, whereas the plus-strand transfer is mainly intramolecular.[123] During DNA synthesis, several environmental factors can increase the probability for a strand transfer event to occur, such as pausing sites, RT synthesis rate and RNase H activity. A model called "dynamic copy choice" has been proposed to take into account these different contributions.[124] In this model, it is proposed that the relative ratio between the polymerase and RNase H activities of RT determines the efficiency of strand transfer. A slower polymerization rate (for example at sequence-specific pausing sites along the RNA template) causes an increase in the RNase H activity and accelerates the chance of strand transfer. Conversely, a reduced RNase H cleavage allows more efficient polymerization decreasing the rate of transfer events. RNase H cleavage of the template RNA from the nascent DNA strand makes it available for annealing to the homologous region of the acceptor strand. Indeed, RNase H-deficient RT mutants were unable to catalyze minus-strand transfer events.[125] The plus-strand transfer occurs when RT, after initiating plus-strand DNA synthesis from the 3′PPT reaches a specific termination site located within the tRNA primer which is still present at the 5′ end of the minus template strand (**Figure 6.8g**). This termination event is called plus-strand strong stop and is determined by the presence at position 58 of the tRNA^{Lys} of methylated adenine residue, which blocks RT polymerization. Some run-off products can be generated which stop either at the pseudouridine 55 or at another modified adenine at position 37.[126,127] Thus, the tRNA^{Lys} sequence is also important in specifying the termination site during the first step of plus-strand synthesis.

Completion. The end product of the reverse transcription reaction is the proviral DNA, which is the substrate for the integration reaction catalyzed by the viral enzyme integrase (IN). Since IN recognizes and cut the extremities of the proviral DNA, these tracts have to be generated with great accuracy. After the second strand transfer, both DNA strands are fully extended, generating the full provirus, which is longer than the original RNA genome, having at both ends two repetitive sequences called Long Terminal Repeats (LTR), which will be used during the integration process. In this way the integrated provirus will contain the entire genomic information of HIV-1, to be transcribed by the cellular RNA polymerase. The accuracy of the RNaseH cleavages is essential in precisely determining the U3 and U5 sequences at the extremities of the proviral DNA, as shown in **Figure 6.8b,g**. As mentioned above, the final proviral DNA contains a 90-nt flap at the boundary of the *pol* and *env* genes (**Figure 6.8h**). The presence of this flap has been proposed to play a role in facilitating the entry of the proviral DNA into the nucleus through the pore complex.[128] In the nucleus, this structure is likely recognized by the cellular Flap-specific endonuclease 1 (FEN-1), an enzyme normally involved in Okazaki fragment synthesis during DNA replication. FEN-1 removes the flap, generating a nick, which can be sealed by a cellular DNA ligase,[129] generating an intact proviral DNA, ready for the integration.

Recombination. HIV-1 virions contain two copies of the genome. This has very important consequences. In the presence of damaged (nicked, broken) RNA, if the minus-strand synthesis is blocked on one molecule, it can be transferred to the second RNA genome to be completed. Indeed, RT during minus-strand synthesis can shift back and forth between the two RNA strands. Since often virions are produced by cells which contain different integrated proviruses, they can bear RNA genomes with slightly different sequences. This will allow RT-mediated recombination during the retrotranscription process, setting the stage for the generation of viral recombinants. Thus, the presence of a diploid genome in HIV-1 is an important determinant of the intrinsic high genetic variability of this virus.[130] The basic mechanism of recombination is based on a classical "copy choice" model, whereby the extensive single-stranded DNA region generated by RNAse H activity at the end of the minus-strand DNA synthesis (**Figure 6.8e**), can anneal to the complementary region of a second copy of the RNA genome. This, in turn, can trigger the transfer of the nascent 3′-OH strand to the acceptor RNA resulting in template switching. The rate of template switching has been estimated in 2–3 events per genome/per replication cycle *in vivo*[131] and they are likely triggered by the presence of pausing sites and/or structured regions along the RNA template.

6.6.1.2 *Structural features of HIV-1 reverse transcriptase*

HIV-1 RT is an asymmetric dimer composed of two subunits, p66 and p51, both derived through proteolytic cleavage of the Gag-Pol polyprotein by the viral protease. The p51 subunit (440 aa) is derived from the p66 full length protein after removal by the viral protease of the last 120 aa (RNase H domain) at its C-terminus. Both the DNA polymerase and RNase H active sites are contained in the p66 subunit, while p51 plays a major structural role. RT is one of the most exploited antiviral targets, and several crystal structures have been obtained and published for this enzyme, either alone or in combination with its substrates.[132]

As shown in **Figure 6.9**, the DNA pol and RNase H domains of the p66 subunit are spatially separated by the so-called connector domain (aa 319–426). The polymerase domain adopts the classical polymerase folding into the three domains fingers (aa 1–85; 118–155), palm (aa 86–117; 156–236) and thumb (aa 237–318). The same domains can be recognized in the p51 subunit; however, their spatial arrangement relative to each other is different. The nucleic acid is positioned into a central cleft, with the connection and thumb domains forming its "floor". The α-helices H and I in the thumb of p66 contribute to the correct positioning of the nucleic acid, along with the β12–β13 hairpin (DNA primer grip), which constrains the 3-hydroxyl end of the primer within the polymerase active site. The active site is located in the thumb and comprises D110 along with the residues D185 and D186 of the highly conserved -Y*x*DD- motif (where *x* is M in HIV-1 RT and I, L or A in other retroviruses). These three carboxylates are involved in divalent metal coordination and phosphodiester bond formation

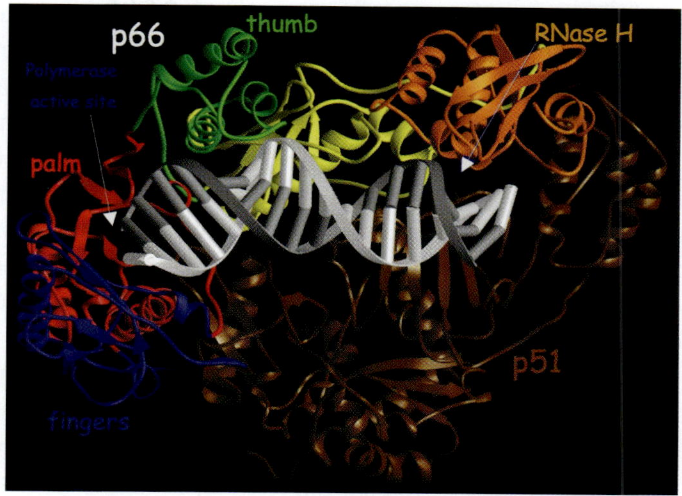

Figure 6.9 Structure of HIV-1 reverse transcriptase. The different domains of the p66 subunit are highlighted in different colors. The nucleic acid lattice is in grey. The p51 subunit is represented in brown. (Reproduced from Ref. 139 with permission.)

between the $3'$-OH primer and the $5'\alpha$-phosphate of the incoming nucleotide. The dNTP binding site is part of the active polymerization site and is lined by residues R72 and K65, which bind the β- and γ-phosphates of the nucleotide, Y115, which contacts the ribose ring, discriminating deoxy-versus ribo-nucleosides triphosphates and Q151 which interacts with the terminal $3'$-hydroxyl group of the primer strand. The RNase H domain of HIV-1 RT consists of 5β-strands and 4α-helices. The RNAse H active site contains four acidic residues (D443, E478, D498 and D549), which are involved in the coordination of divalent metal ions and are essential for catalysis. Similarly to the polymerase domain, also the RNAse H domain contains a "primer grip" consisting of residues K395 and E396 of p51, G359, A360 in the p66 polymerase domain, H361 in the p66 connection domain and T473, N474, Q475, K476, Y501 and I505 of the RNase H domain. This region makes specific contacts with the DNA strand which is base paired to the RNA positions -4 to -9 relative to the phoshodiester bond which will be cut. Since the RNAse H primer grip spans residues also in the polymerase domain, this element contributes to both DNA polymerization and RNA degradation efficiency.

6.6.1.3 *Enzymatic features of HIV-1 reverse transcriptase*

The overall catalytic cycle of HIV-1 RT has been extensively investigated both by steady state and pre-steady state techniques. Like the majority of DNA pols, HIV-1

RT follows an obligatory sequential ordered mechanism, whereby the nucleic acid binds first, followed by the nucleoside triphosphate. Binding of the nucleic acid to give the binary complex induces a first conformational change in the p66 subunit, exposing the nucleotide binding site. Upon ternary complex formation, a second, rate-limiting conformational change takes place, where the p66 fingers close down on the active site, properly positioning the $3'$-hydroxyl primer end and the $5'\alpha$-phosphate group of the incoming nucleoside triphosphate for nucleophilic attack. After chemical bond formation a fast conformational change causes the fingers to open again to release the pyrophosphate and to allow translocation of the enzyme along the nucleic acid lattice. The overall affinity of HIV-1 RT for both the nucleic acid and the nucleoside triphosphates are similar to replicative DNA pols (i.e. in the submicromolar and low micromolar range, respectively). The polymerization rate under single turnover conditions is $1-10\,s^{-1}$ for DNA/DNA substrates and $10-60\,s^{-1}$ for RNA/DNA hybrids, depending on the particular terminal base pair involved.[133,134] The nature of the nucleic acid (either RNA/DNA or DNA/DNA) can also influence the ground-state nucleotide binding affinity, with an higher efficiency for the DNA/DNA substrates. However, homopolymeric RNA/DNA hybrids such as poly(rA)/oligo(dT) or polyr(rG)/oligo(dC) are the preferred substrates for HIV-1 RT *in vitro* (**Table 6.3**) HIV-1 RT has the ability to incorporate $2',3'$-dideoxyribonucleoside triphosphates (ddNTPs), resulting in termination of the newly synthesized DNA strand. This property has been extensively exploited for the development of antiretroviral drugs (see Chapter 9). The ground-state binding affinity and incorporation rate for ddNTPs is similar to normal (i.e. $2'$-deoxy, $3'$-hydroxyl) dNTPs (**Table 6.3**). After incorporation of a ddNTP, further synthesis is blocked, due to the lack of an available $3'$-hydroxyl group, and the enzyme is forced to dissociate from the template. The dissociation from a ddNMP-terminated primer end occurs with slower kinetic than from a normal primer, causing the accumulation of a dead-end complex, stable enough to be resolved by electrophoretic shift mobility assays on polyacrylamide non-denaturing gels. The strong pausing of HIV-1RT at a ddNMP-terminated primer end allows an unusual reaction to take place, namely the condensation of the chain-terminating ddNMP (donor) with an incoming pyrophosphate molecule or a nucleoside triphosphate (acceptor) to generate a new ddNTP (in the case of pyrophosphate donor) or a di-nucleoside-tetraphosphate (in the case of a nucleotide donor).[135] Several donor deoxy- and ribonucleosides triphosphates can be used by HIV-1 RT (**Table 6.3**). This reaction, termed phosphorolysis, is the reversal of the polymerization step and generates a DNA product one residue shorter. However, the most important consequence of this reaction is that the terminal 3-hydroxyl group of the primer strand is freed from the chain-terminating ddNTP, thus becoming available for a new round of

Table 6.3 Kinetic parameters for substrate binding and catalysis by HIV-1 reverse transcriptase

Substrate	K_d, M[a]	k_{pol}, s^{-1}	K_m, M[b]	k_{cat}, s^{-1}[b]
poly(rA)/oligo(dT)	1×10^{-8}	n.a.		
poly(rC)/oligo(dG)	5×10^{-8}			
poly(dA)/oligo(dT)	8×10^{-8}			
DNA heteropolymeric p/t	6×10^{-8}			
ss RNA	5×10^{-8}			
ss DNA	9×10^{-8}			
dTTP	10×10^{-6}	20		
dGTP	7×10^{-6}	30	0.7×10^{-3}	0.1×10^{-3}
dATP	4×10^{-6}	30	1.6×10^{-3}	0.3×10^{-3}
dCTP	3×10^{-6}	15	3.9×10^{-3}	1.1×10^{-3}
ddTTP	5×10^{-6}	0.5		
AZTTP	5×10^{-6}	10		
ATP	0.7×10^{-3}	160×10^{-3}	1.6×10^{-3}	0.4×10^{-3}
CTP	1.4×10^{-3}	60×10^{-3}	3.1×10^{-3}	1.2×10^{-3}
GTP	7.2×10^{-3}	10×10^{-3}	1.5×10^{-3}	1.5×10^{-3}
UTP	1.75×10^{-3}	70×10^{-3}	2×10^{-3}	0.4×10^{-3}

a: K_d, ground-state binding affinity; p/t, primer/template; k_{pol}, polymerization rate; AZTTP, 2′, 3′-dideoxy, 3′-azido thymidine triphosphate; n.a., not applicable, binding of the nucleic acid substrate only does not lead to a catalytically competent complex.
b: Kinetic parameters for the phosphorolytic reaction.

elongation. This phosphorolytic mechanism is a major determinant of HIV-1 RT resistance to ddNTPs inhibition (see Chapter 9). HIV-1 RT is also able to bind and incorporate ribonucleoside triphosphates. Both the affinity and the incorporation rates are much reduced compared to dNTPs (**Table 6.3**), but similar to the values shown in the phosphorolytic reaction.[136]

6.6.1.4 *The RNase H activity of HIV-1 reverse transcriptase*

The RNase H of HIV-1 RT is an endonuclease which specifically hydrolyzes the RNA present in RNA/DNA hybrids generating a nick with 5′-phosphate and 3′-hydroxyl ends. This reaction can be carried out in three distinct types of cleavages: 3′-end directed, 5′-end directed and internal. In the 3′-end directed mode, HIV-1 RT acts on a DNA primer junction with a 3′-recessed end annealed to an RNA template. The RNase H domain then cleaves the RNA strand 15–20 nucleotides away from the 3′-recessed DNA end. In the 5′-end mode, HIV-1 RT is bound to the 5′-recessed end of an RNA template annealed to a longer DNA strand. The RNase H then cleaves the RNA strand 13–19 nucleotides away from the recessed 5′-RNA end. Thus, in both modes, the presence of a 3′-DNA or a 5′-RNA recessed end within an RNA/DNA hybrid duplex determines the position of the cleavage site on the RNA strand. In the internal cleavage mode, the position of the site to be cut by the RNAse H domain is strictly dependent on the neighboring sequences. The consensus sequence is A-A-(G/C)-C-(G/C)-A for positions -14, -12, -7, -4, -2 and $+1$, respectively, with respect to the cleavage site (referred as position 0).[137] This consensus is present, for example, in the PPT region of the HIV-1 genome which is degraded during the retrotranscription reaction (**Figure 6.8g**).

6.6.2 *Other Retroviral Reverse Transcriptases*

6.6.2.1 *HIV-2 reverse transcriptase*

HIV-2 RT has 60–70% homology with the corresponding HIV-1 enzyme and shows similar enzymatic properties including both RDDP and DDDP activities and RNase H function. A major difference, however, is that HIV-2 RT contains several amino acid changes in the vicinity of the active site[138] which renders it resistant to almost all the non-nucleoside RT inhibitors (NNRTIs, see Chapter 9). It is an heterodimer composed of a p68 and a p51 subunit, the latter derived from proteolytic cleavage of the p68 polypeptide by the viral protease, in a very similar fashion to HIV-1 RT. HIV-2 RT has a longer "connection" domain than HIV-1 RT between the polymerase and the RNase H domains,[139] which results in a 484-aa long p55 subunit (with respect to the 440 aa p51 of HIV-1 RT). The difference in length between HIV-1 and HIV-2 RT subunits can be ascribed to a different substrate specificity of their respective proteases. The catalytic activity is dependent on the presence of an intact p68 subunit in the heterodimer, since p55/p55 homodimers are virtually inactive.[140–142] The only available crystal structure for HIV-2 RT[143] reveals an overall folding very similar to the one of HIV-1 RT, but with some specific differences. In particular, the p55 subunit does not appear to contribute to nucleic acid and or nucleotide binding. The enzymatic properties of HIV-2 RT are also very similar to those of HIV-1 RT, in terms of metal ion, pH and template preferences as well as fidelity.[140–142] A major difference lies in the RNase H efficieny, which, at least *in vitro*, is 10-fold lower for the HIV-2 enzyme when compared to HIV-1, even though the RNA cleavage pattern between the two enzymes is substantially the same.[144] In addition, the initiation of plus-strand DNA synthesis was found to be less efficient in HIV-2 RT catalyzed reactions,[145] suggesting architectural differences in the substrate binding sites (palm and primer grip regions) of HIV-1 and HIV-2 RTs.

6.6.2.2 *Murine Leukemia Virus (MLV) reverse transcriptase*

MLV-RT is active as a monomeric protein of 671 aa showing polymerase and RNase H activites.[146] The polymerase and RNase H domains can be separately expressed as recombinant active forms, but the RNaseH domain needs to be complemented in *trans* with an active polymerase domain to show full efficiency and specificity.[147] The crystal structure of MLV-RT[148,149] indicated a folding very similar to HIV-1 RT, in spite of only 25% amino acid sequence identity. In particular, the highly invariant -YMDD- and -LPQG- motifs which define the nucleotide binding site in HIV-1 and HIV-2 RTs are also present in MLV-RT. Structural modeling and site-directed mutagenesis studies allowed also to identify other key residues involved in nucleotide binding and catalysis such as K103, R110, D153 and F115, which are equivalent to K65, R72, D113 and Y115 of HIV-1 RT.[150,151]

6.6.2.3 *Equine Infectious Anemia Virus (EIAV) reverse transcriptase*

EIAV-RT is a heterodimer formed by p68 and p51 subunits.[152] It has polymerase activity in the N-terminal domain and RNaseH activity in its C-terminal part, as seen in other RTs.[153] However, p51/p51 homodimers showed almost comparable DDDP activity and only slightly reduced RDDP activity than heterodimers,[154,155] in sharp contrast with HIV-1 or HIV-1 RTs. However, the p51/p51 homodimer was more distributive than the heterodimeric form, leading to incorporation of only a few nucleotides for each association event. Accordingly, kinetic studies showed a reduced affinity of the p51/p51 form with respect to the p66/p51 dimer, which was due to a substantial increase in the dissociation rate, in good agreement with the processivity data. The general mechanism of the retrotranscription reaction of EIAV-RT appears to be very similar to the general scheme as derived from HIV-1 RT studies.[153] The EIAV enzyme initiates minus-strand synthesis from a tRNA^{Lys} primer and catalyzes the obligatory strand-transfer steps in a manner similar to HIV-1 RT.

6.6.2.4 *Feline Immunodeficiency Virus (FIV) reverse transcriptase*

FIV-RT is a p66/p51 heterodimer, very similar to HIV-1 RT. Similar to EIAV-RT, the RDDP activity was comparable among all the possible isoforms (p66/p66, p51/p51 and p66/p51). However, contrary to EIAV-RT and more similar to HIV-1, the DDDP activity of FIV-RT homodimers was only about 1% of the DDDP activity of the p68/p51 isoform.[156] Studies with inter- and intra-molecular HIV-1/FIV chimeric constructs suggested that the p51 subunit was important for the activity and stability of the p66/p51 heterodimer.[157,158] FIV-RT has a high affinity for dNTPs, likely reflecting its ability to replicate in non-dividing cells, where the concentration of dNTPs is generally low.[159]

6.6.2.5 *Mouse Mammary Tumor Virus (MMTV) reverse transcriptase*

The MMTV-RT is a trans-frame protein, containing at its N-terminal domain 27 residues derived from the C-terminal part of the protease gene (*pro*) flanking the

RT gene (*pol*), but normally translated from a different reading frame. Contrary to most RTs, the MMTV enzyme is active as a monomeric protein of 603 aa, showing RDDP, DDDP and RNAse H activities.[160,161] Site-directed mutagenesis studies suggested that the nucleotide binding site of MMTV-RT is very similar to the one of HIV-1 RT.[162]

6.6.3 *Reverse Transcriptase Activity of Mobile Genetic Elements: The Retrotransposons*

Retrotransposons are genetic mobile elements which utilize reverse transcription to generate a DNA copy to be inserted into eukaryotic genomes.[163] Similarly to retroviruses, their genomes possess LTRs at both ends, as well as genes for RT, RNase H, integrase, protease and Gag-like proteins.[164] However, they lack *env* genes, so they do not generate virus-like particle and therefore do not spread ectopically among individuals, but rather are genetically inherited by the descendants if present in the chromosomes of germline cells. Enzymatic characterization of their RT activity has been performed for some of these mobile elements.

6.6.3.1 *Saccharomyces cerevisiae Ty reverse transcriptase*

The Ty yeast retrotransposons shows RT activity[165–167] and possess a 115-kDa protein constituted by the fused 55 kDa RT and 61 kDa integrase proteins.[168,169] The N-terminal integrase domain activates RT and is essential for the retrotranscription reaction.[169–171] Ty RT possesses relatively low fidelity as the retroviral enzymes.[172] Ty enzymes also possess the highly conserved -YxDD- motif in their active site.[173] Metal coordination and catalysis require a catalytic diad consisting of the first aspartate of the YxDD motif plus an additional aspartic acid. The predicted spatial arrangements of Ty RT are similar to the one of HIV-1 RT with thumb, fingers, palm and RNAse H domains. The overall reverse transcription reaction is similar to the one observed for retroviruses.[174]

6.6.3.2 *Schizosaccharomyces pombe Tf1 reverse transcriptase*

The Tf1 RT utilizes a unique mechanism for initiation of DNA synthesis.[175] Instead of using a tRNA annealed to the PBS, the reaction starts from an 11-nt RNA primer generated by self annealing of the 11 terminal bases of the RNA genome to a complementary sequence located within the 5′-LTR and subsequent cleavage of the loop by the RNAse H activity of Tf1 RT, to generate the 3′-hydroxyl end of the primer.[176] Another unique feature of this enzyme is its ability to act as a terminal transferase (see Chapter 4), adding one-two nucleotides to the 3′-terminal end of the DNA copy in a template independent fashion. This mechanism likely acts to protect the 3′ end of the DNA from degradation by cellular exonucleases.[177]

6.6.3.3 *Bombyx mori R2 reverse transcriptase*

The insect retrotransposon R2 encodes an RT enzyme with unsual enzymatic properties. It can use a 3′-hydroxyl end from any RNA or DNA molecule to prime its reaction without the need of any complementarity between the primer and the template. This property is likely correlated with its robust terminal transferase activity.[178,179] Thus, a free 3′-hydroxyl end is sufficient to initiate non-templated nucleotide addition until the added DNA is sufficient to enable priming of the reverse transcription on the RNA strand. Another unusual characteristic of Tf1 RT is its high rate of template switching.[178,179] This ability to jump from the 5′-end of an RNA template to the 3′-end of another template is likely correlated to the high efficiency of Tf1 RT in utilizing free 3′-ends to prime its reaction and to the terminal addition of non-templated nucleotides, generating 3′-overhangs which facilitate annealing to an acceptor RNA molecule. This activity results in the generation of a single DNA copy of multiple RNA templates.

6.6.4 *Hepadnavirus Reverse Transcriptase*

Hepadnaviruses are responsible for the hepatitis B disease in a number of mammalian and avian species, including humans. Even though their genomes are composed of DNA, they use an RNA intermediate in their replication process, thus requiring a reverse transcription step. The hepadnaviral genome is an open partially double-stranded circular DNA molecule of about 3 kb which, after duplication, is converted into a covalently closed circle which persists in an episomal state in infected cells. The plus (+) strand is shorter than the minus (−) strand and the circularity is maintained by overlapping 5′ ends which are annealed in close proximity of two direct repeat units DR1 and DR2. In addition, the (+) strand contains a 18-nt capped RNA at its 5′ end, which is important in the reverse transcription process.

A first critical step in genome duplication is the generation of a pregenomic mRNA, longer than a genome unit (3.5 kb) by the host RNA polymerase II, which is then exported into the cytoplasm where it serves as template for the reverse transcription process. Reverse transcription takes place upon complex formation of the pre-genomic mRNA with the viral polymerase and core proteins. The transcription by RNA polymerase II initiates 6 bp upstream of the DR1 region, proceeds through the entire DNA genome and terminates about 200 nt downstream the DR1 initiation site, generating a longer-than-genome RNA transcript having two DR1 units at its 3′- and 5′-ends, which is then capped and polyadenylated before nuclear export.

The reverse transcription process (**Figure 6.10**) uses two different priming mechanisms for each strand. Priming for the (−) strand synthesis is provided by the polymerase itself, through its N-terminal domain called terminal protein region. This protein-primed mechanism of synthesis is analogous to the one used by the AdDNA pol, and requires the 3′-hydroxyl group of a specific tyrosine residue.[180] This step leads to the synthesis of a four nucleotide product complementary to a specific sequence (-CAUU-) at the 5′-end of the pregenomic mRNA (**Figure 6.10a**).[181] The polymerase-bound -GTAA- DNA primer then jumps and reanneals to another CAUU sequence present in the other DR1 unit located at the 3′-end of the pregenomic mRNA (**Figure 6.10b**). RNA-dependent DNA synthesis then continues, while the RNA template which has been copied is degraded by the RNaseH activity of the viral polymerase (**Figure 6.10c**). Completion of (−) strand synthesis generates a terminally redundant cDNA copy of the original pre-genomic mRNA. The 5′-end terminal 18 nt of the RNA are not hydrolyzed by the RNase H activity and are used to prime the (+) strand synthesis. This 5′-capped terminal RNA is annealed in a second template jump to the complementary DR2 region present near the 5′ end of the (−) DNA strand. Upon reaching the 5′-end of the (−) DNA strand, (+) strand synthesis stops (**Figure 6.10d**). The terminal redundant sequences which are present at both ends of the (−) strand catalyze an intramolecular template exchange, generating an open circular DNA molecule. The (+) strand synthesis then resumes and continues for an additional tract, generating the open, partially double-stranded genome which will be encapsidated (**Figure 6.10e**).

Hepadnavirus DNA pol (also called P protein) is about 90 kDa in size and shows the presence of conserved reverse transcriptase and RNase H motifs, including the -YMDD- signature, in its C-terminal 550 nt.[182−186] The N-terminal 350 nt share no homology with other known proteins and encode for the terminal protein domain, containing the Y63 residue which provides the 3′-hydrxyl group for initiation of DNA synthesis. The fact that several nucleoside analogs designed to target HIV-1 RT are also active against human hepatitis B virus (HBV) polymerase (see Chapter 9), also point to a certain degree of structural similarity between the two proteins.[187] Unfortunately, no crystal structures of hepadnaviral pols are currently available, but molecular modeling and structural homology studies suggested that the nucleotide binding site is lined by residues M204 (of the -YMDD- motif), as well as A87, F88, P177 and L180, which constitute an hydrophobic pocket for the sugar of the incoming nucleotide.[188] HBV polymerase shows optimal activity in the presence of 0.25–1 mM Mn^{2+} or 2.5–5 mM Mg^{2+}, 400 mM KCl at pH 7.0–8.0. Remarkably, endogeneous HBV pol assayed in viral core particles showed apparent affinities (K_m) values for dNTPs ranging from 0.07 μM to 0.18 μM, thus much

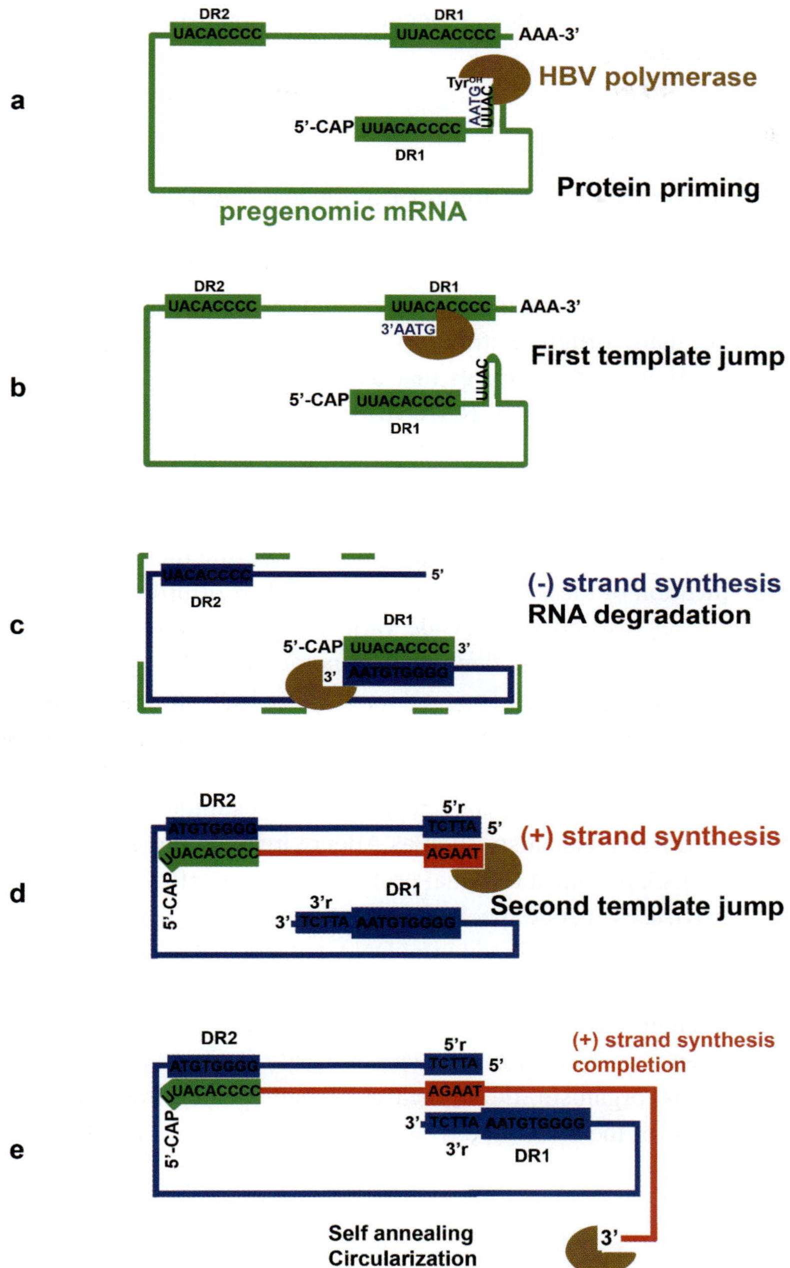

Figure 6.10   Scheme of the reverse transcription-dependent HBV DNA replication. For details see text.

lower than HIV-1 RT (see **Table 6.3** for comparison), indicating a very high affinity for nucleotide binding.[189]

6.7 Chapter Summary

Viruses are among the simplest forms of biological organization. They are intracellular parasites that cannibalize the cells of their hosts in order to duplicate their genetic information, often leading to cellular dysfunctions that, in turn, determine the insurgence of pathological manifestations in the infected organism. In the viral domain of life, many different strategies are exploited to store and duplicate the genetic information, so that viruses that use either RNA or DNA as the genetic material are known. In addition, RNA viruses exist, called retroviruses, that use DNA to insert their genetic information into the host cell genome. To further complicate the matter, the replicative strategies can be very different even among DNA or RNA viruses. Indeed, the kind of nucleic acid present in the viral particles, together with the mechanism of its duplication, constitute the most important tool for the classification of the different viral families. It is of little surprise, thus, that the viral replicases also possess a wide variety of different reaction mechanisms and catalytic properties. Due to their small genome size, viruses encode only for a limited set of essential proteins. The total number of virally encoded proteins can be very different from small- to large- genome sized viruses, but all viruses presented in Chapter 6 encode their own specialized replicases, indicating a strict co-evolution of replicative strategies and replicative enzymes. For example, those viruses which use a protein-primed mechanism such as Adenovirus and the ϕ29 bacteriophage, encode both a DNA pol and a terminal protein, that form a functional heterodimer. Similarly, retroviruses encode specialized RNA/DNA pols called reverse transcriptases. As the number of crystallized viral DNA pols increases, however, it becomes clear that these different functions have evolved starting from a common "core" of structural elements that can be found also in the DNA pols of prokaryotic and eukaryotic organisms. Since viral infections very often cause pathological conditions in the host organism, the replicases of pathogenic viruses represent very important targets for the antiviral chemotherapy, as will be outlined in Chapter 9.

References

1. Sinha NK, Morris CF, Alberts BM. 1980. *J Biol Chem* 255: 4290–3.
2. Alberts BM, Barry J, Bedinger P, Formosa T, Jongeneel CV, Kreuzer KN. 1983. *Cold Spring Harb Symp Quant Biol* 47 Pt 2: 655–68.
3. Formosa T, Burke RL, Alberts BM. 1983. *Proc Natl Acad Sci U S A* 80: 2442–6.
4. Chastain PD, 2nd, Makhov AM, Nossal NG, Griffith J. 2003. *J Biol Chem* 278: 21276–85.

5. Nossal NG, Makhov AM, Chastain PD, 2nd, Jones CE, Griffith JD. 2007. *J Biol Chem* 282: 1098–108.
6. Tsurimoto T, Stillman B. 1990. *Proc Natl Acad Sci U S A* 87: 1023–7.
7. Nossal NG. 1992. *Faseb J* 6: 871–8.
8. Young MC, Reddy MK, von Hippel PH. 1992. *Biochemistry* 31: 8675–90.
9. Trakselis MA, Mayer MU, Ishmael FT, Roccasecca RM, Benkovic SJ. 2001. *Trends Biochem Sci* 26: 566–72.
10. Benkovic SJ, Valentine AM, Salinas F. 2001. *Annu Rev Biochem* 70: 181–208.
11. Kodadek T. 2008. *Mol Biosyst* 4: 1043–5.
12. Spiering MM, Nelson SW, Benkovic SJ. 2008. *Mol Biosyst* 4: 1070–4.
13. Goulian M, Lucas ZJ, Kornberg A. 1968. *J Biol Chem* 243: 627–38.
14. De Waard A, Paul AV, Lehman IR. 1965. *Proc Natl Acad Sci U S A* 54: 1241–8.
15. Spicer EK, Rush J, Fung C, Reha-Krantz LJ, Karam JD, Konigsberg WH. 1988. *J Biol Chem* 263: 7478–86.
16. Delagoutte E, Von Hippel PH. 2003. *J Biol Chem* 278: 25435–47.
17. Fairfield FR, Newport JW, Dolejsi MK, von Hippel PH. 1983. *J Biomol Struct Dyn* 1: 715–27.
18. Capson TL, Peliska JA, Kaboord BF, Frey MW, Lively C, *et al.* 1992. *Biochemistry* 31: 10984–94.
19. Mace DC, Alberts BM. 1984. *J Mol Biol* 177: 313–27.
20. Alley SC, Jones AD, Soumillion P, Benkovic SJ. 1999. *J Biol Chem* 274: 24485–9.
21. Yang J, Zhuang Z, Roccasecca RM, Trakselis MA, Benkovic SJ. 2004. *Proc Natl Acad Sci U S A* 101: 8289–94.
22. Fidalgo da Silva E, Reha-Krantz LJ. 2007. *Nucleic Acids Res* 35: 5452–63.
23. Muzyczka N, Poland RL, Bessman MJ. 1972. *J Biol Chem* 247: 7116–22.
24. Bessman MJ, Muzyczka N, Goodman MF, Schnaar RL. 1974. *J Mol Biol* 88: 409–21.
25. Reha-Krantz LJ, Stocki S, Nonay RL, Dimayuga E, Goodrich LD, *et al.* 1991. *Proc Natl Acad Sci U S A* 88: 2417–21.
26. Berdis AJ. 2001. *Biochemistry* 40: 7180–91.
27. Kadyrov FA, Drake JW. 2002. *Nucleic Acids Res* 30: 4387–97.
28. Tanguy Le Gac N, Delagoutte E, Germain M, Villani G. 2004. *J Mol Biol* 336: 1023–34.
29. Villani G, Boiteux S, Radman M. 1978. *Proc Natl Acad Sci U S A* 75: 3037–41.
30. Khare V, Eckert KA. 2002. *Mutat Res* 510: 45–54.
31. Blanca G, Delagoutte E, Tanguy le Gac N, Johnson NP, Baldacci G, Villani G. 2007. *Biochem J* 402: 321–9.
32. Kadyrov FA, Drake JW. 2004. *J Biol Chem* 279: 35735–40.
33. Moarefi I, Jeruzalmi D, Turner J, O'Donnell M, Kuriyan J. 2000. *J Mol Biol* 296: 1215–23.
34. Mueser TC, Jones CE, Nossal NG, Hyde CC. 2000. *J Mol Biol* 296: 597–612.
35. Shamoo Y, Friedman AM, Parsons MR, Konigsberg WH, Steitz TA. 1995. *Nature* 376: 362–6.
36. Hamdan SM, Johnson DE, Tanner NA, Lee JB, Qimron U, *et al.* 2007. *Mol Cell* 27: 539–49.
37. Johnson DE, Takahashi M, Hamdan SM, Lee SJ, Richardson CC. 2007. *Proc Natl Acad Sci U S A* 104: 5312–7.
38. Hamdan SM, Richardson CC. 2009. *Annu Rev Biochem* 78: 205–43.
39. Grippo P, Richardson CC. 1971. *J Biol Chem* 246: 6867–73.
40. Studier FW. 1972. *Science* 176: 367–76.
41. Modrich P, Richardson CC. 1975. *J Biol Chem* 250: 5515–22.
42. Scherzinger E, Seiffert D. 1975. *Mol Gen Genet* 141: 213–32.
43. Adler S, Modrich P. 1979. *J Biol Chem* 254: 11605–14.
44. Modrich P, Richardson CC. 1975. *J Biol Chem* 250: 5508–14.
45. Huber HE, Russel M, Model P, Richardson CC. 1986. *J Biol Chem* 261: 15006–12.

46. Hori K, Mark DF, Richardson CC. 1979. *J Biol Chem* 254: 11598–604.
47. Hori K, Mark DF, Richardson CC. 1979. *J Biol Chem* 254: 11591–7.
48. Tabor S, Huber HE, Richardson CC. 1987. *J Biol Chem* 262: 16212–23.
49. Huber HE, Tabor S, Richardson CC. 1987. *J Biol Chem* 262: 16224–32.
50. Doublie S, Tabor S, Long AM, Richardson CC, Ellenberger T. 1998. *Nature* 391: 251–8.
51. McCulloch SD, Kunkel TA. 2006. *DNA Repair (Amst)* 5: 1373–83.
52. Li Y, Dutta S, Doublie S, Bdour HM, Taylor JS, Ellenberger T. 2004. *Nat Struct Mol Biol* 11: 784–90.
53. Brieba LG, Eichman BF, Kokoska RJ, Doublie S, Kunkel TA, Ellenberger T. 2004. *Embo J* 23: 3452–61.
54. Dutta S, Li Y, Johnson D, Dzantiev L, Richardson CC, *et al.* 2004. *Proc Natl Acad Sci U S A* 101: 16186–91.
55. Koffel-Schwartz N, Verdier JM, Bichara M, Freund AM, Daune MP, Fuchs RP. 1984. *J Mol Biol* 177: 33–51.
56. Lee JB, Hite RK, Hamdan SM, Xie XS, Richardson CC, van Oijen AM. 2006. *Nature* 439: 621–4.
57. Lee SJ, Marintcheva B, Hamdan SM, Richardson CC. 2006. *J Biol Chem* 281: 25841–9.
58. Dunn JJ, Studier FW. 1983. *J Mol Biol* 166: 477–535.
59. Matson SW, Richardson CC. 1985. *J Biol Chem* 260: 2281–7.
60. Matson SW, Tabor S, Richardson CC. 1983. *J Biol Chem* 258: 14017–24.
61. Crampton DJ, Ohi M, Qimron U, Walz T, Richardson CC. 2006. *J Mol Biol* 360: 667–77.
62. Guo S, Tabor S, Richardson CC. 1999. *J Biol Chem* 274: 30303–9.
63. Frick DN, Baradaran K, Richardson CC. 1998. *Proc Natl Acad Sci U S A* 95: 7957–62.
64. Kato M, Ito T, Wagner G, Richardson CC, Ellenberger T. 2003. *Mol Cell* 11: 1349–60.
65. Hamdan SM, Marintcheva B, Cook T, Lee SJ, Tabor S, Richardson CC. 2005. *Proc Natl Acad Sci U S A* 102: 5096–101.
66. Kim YT, Tabor S, Churchich JE, Richardson CC. 1992. *J Biol Chem* 267: 15032–40.
67. Kim YT, Richardson CC. 1994. *J Biol Chem* 269: 5270–8.
68. Kong D, Griffith JD, Richardson CC. 1997. *Proc Natl Acad Sci U S A* 94: 2987–92.
69. Kong D, Nossal NG, Richardson CC. 1997. *J Biol Chem* 272: 8380–7.
70. Kong D, Richardson CC. 1996. *Embo J* 15: 2010–9.
71. Yu M, Masker W. 2001. *J Bacteriol* 183: 1862–9.
72. Hollis T, Stattel JM, Walther DS, Richardson CC, Ellenberger T. 2001. *Proc Natl Acad Sci U S A* 98: 9557–62.
73. Muylaert I, Elias P. 2007. *J Biol Chem* 282: 10865–72.
74. Belnap DM, McDermott BM, Jr., Filman DJ, Cheng N, Trus BL, *et al.* 2000. *Proc Natl Acad Sci U S A* 97: 73–8.
75. Liu S, Knafels JD, Chang JS, Waszak GA, Baldwin ET, *et al.* 2006. *J Biol Chem* 281: 18193–200.
76. Komazin-Meredith G, Santos WL, Filman DJ, Hogle JM, Verdine GL, Coen DM. 2008. *J Biol Chem* 283: 6154–61.
77. Ramirez-Aguilar KA, Kuchta RD. 2004. *Biochemistry* 43: 9084–91.
78. Song L, Chaudhuri M, Knopf CW, Parris DS. 2004. *J Biol Chem* 279: 18535–43.
79. Challberg MD, Kelly TJ, Jr. 1979. *Proc Natl Acad Sci U S A* 76: 655–9.
80. de Jong RN, van der Vliet PC, Brenkman AB. 2003. *Curr Top Microbiol Immunol* 272: 187–211.
81. Joung I, Engler JA. 1992. *J Virol* 66: 5788–96.
82. Joung I, Horwitz MS, Engler JA. 1991. *Virology* 184: 235–41.
83. Roovers DJ, van der Lee FM, van der Wees J, Sussenbach JS. 1993. *J Virol* 67: 265–76.
84. Brenkman AB, Heideman MR, Truniger V, Salas M, van der Vliet PC. 2001. *J Biol Chem* 276: 29846–53.

85. Field J, Gronostajski RM, Hurwitz J. 1984. *J Biol Chem* 259: 9487–95.
86. King AJ, Teertstra WR, Blanco L, Salas M, van der Vliet PC. 1997. *Nucleic Acids Res* 25: 1745–52.
87. Brenkman AB, Breure EC, van der Vliet PC. 2002. *J Virol* 76: 8200–7.
88. Mysiak ME, Holthuizen PE, van der Vliet PC. 2004. *Nucleic Acids Res* 32: 3913–20.
89. Salas M, Blanco L, Prieto I, Garcia JA, Mellado RP, *et al.* 1984. *Adv Exp Med Biol* 179: 35–44.
90. Gutierrez C, Sogo JM, Salas M. 1991. *J Mol Biol* 222: 983–94.
91. Blasco MA, Esteban JA, Mendez J, Blanco L, Salas M. 1992. *Chromosoma* 102: S32–8.
92. Blanco L, Prieto I, Gutierrez J, Bernad A, Lazaro JM, *et al.* 1987. *J Virol* 61: 3983–91.
93. Blanco L, Salas M. 1984. *Proc Natl Acad Sci U S A* 81: 5325–9.
94. Hermoso JM, Mendez E, Soriano F, Salas M. 1985. *Nucleic Acids Res* 13: 7715–28.
95. Mendez J, Blanco L, Esteban JA, Bernad A, Salas M. 1992. *Proc Natl Acad Sci U S A* 89: 9579–83.
96. Kamtekar S, Berman AJ, Wang J, Lazaro JM, de Vega M, *et al.* 2006. *Embo J* 25: 1335–43.
97. Truniger V, Lazaro JM, Salas M, Blanco L. 1996. *Embo J* 15: 3430–41.
98. Oliveros M, Yanez RJ, Salas ML, Salas J, Vinuela E, Blanco L. 1997. *J Biol Chem* 272: 30899–910.
99. Garcia-Escudero R, Garcia-Diaz M, Salas ML, Blanco L, Salas J. 2003. *J Mol Biol* 326: 1403–12.
100. Kumar S, Bakhtina M, Tsai MD. 2008. *Biochemistry* 47: 7875–87.
101. Lamarche BJ, Kumar S, Tsai MD. 2006. *Biochemistry* 45: 14826–33.
102. Maciejewski MW, Shin R, Pan B, Marintchev A, Denninger A, *et al.* 2001. *Nat Struct Biol* 8: 936–41.
103. Showalter AK, Byeon IJ, Su MI, Tsai MD. 2001. *Nat Struct Biol* 8: 942–6.
104. Sampoli Benitez BA, Arora K, Balistreri L, Schlick T. 2008. *J Mol Biol* 384: 1086–97.
105. Lamarche BJ, Showalter AK, Tsai MD. 2005. *Biochemistry* 44: 8408–17.
106. Crick FH. 1958. *Symp Soc Exp Biol* 12: 138–63.
107. Watson JD, Crick FH. 1953. *Nature* 171: 964–7.
108. Watson JD, Crick FH. 1953. *Nature* 171: 737–8.
109. Franklin RE, Gosling RG. 1953. *Nature* 172: 156–7.
110. Temin HM, Mizutani S. 1970. *Nature* 226: 1211–3.
111. Baltimore D. 1970. *Nature* 226: 1209–11.
112. Barre-Sinoussi F, Chermann JC, Rey F, Nugeyre MT, Chamaret S, *et al.* 1983. *Science* 220: 868–71.
113. Gallo RC. 1983. *Jama* 250: 1015–21.
114. Gallo RC, Sarin PS, Gelmann EP, Robert-Guroff M, Richardson E, *et al.* 1983. *Science* 220: 865–7.
115. Jiang M, Mak J, Huang Y, Kleiman L. 1994. *Leukemia* 8 Suppl 1: S149–51.
116. Mak J, Jiang M, Wainberg MA, Hammarskjold ML, Rekosh D, Kleiman L. 1994. *J Virol* 68: 2065–72.
117. Cen S, Khorchid A, Gabor J, Rong L, Wainberg MA, Kleiman L. 2000. *J Virol* 74: 10796–800.
118. Khorchid A, Javanbakht H, Wise S, Halwani R, Parniak MA, et al. 2000. *J Mol Biol* 299: 17–26.
119. Beerens N, Groot F, Berkhout B. 2001. *J Biol Chem* 276: 31247–56.
120. Abbink TE, Berkhout B. 2008. *Virus Res* 134: 4–18.
121. Ramalanjaona N, de Rocquigny H, Millet A, Ficheux D, Darlix JL, Mely Y. 2007. *J Mol Biol* 374: 1041–53.
122. Basu VP, Song M, Gao L, Rigby ST, Hanson MN, Bambara RA. 2008. *Virus Res* 134: 19–38.
123. Yu H, Jetzt AE, Ron Y, Preston BD, Dougherty JP. 1998. *J Biol Chem* 273: 28384–91.
124. Hwang CK, Svarovskaia ES, Pathak VK. 2001. *Proc Natl Acad Sci U S A* 98: 12209–14.

125. Nikolenko GN, Svarovskaia ES, Delviks KA, Pathak VK. 2004. *J Virol* 78: 8761–70.
126. Auxilien S, Keith G, Le Grice SF, Darlix JL. 1999. *J Biol Chem* 274: 4412–20.
127. Muthuswami R, Chen J, Burnett BP, Thimmig RL, Janjic N, McHenry CS. 2002. *J Mol Biol* 315: 311–23.
128. Zennou V, Petit C, Guetard D, Nerhbass U, Montagnier L, Charneau P. 2000. *Cell* 101: 173–85.
129. Rumbaugh JA, Fuentes GM, Bambara RA. 1998. *J Biol Chem* 273: 28740–5.
130. Ramirez BC, Simon-Loriere E, Galetto R, Negroni M. 2008. *Virus Res* 134: 64–73.
131. Levy DN, Aldrovandi GM, Kutsch O, Shaw GM. 2004. *Proc Natl Acad Sci U S A* 101: 4204–9.
132. Sarafianos SG, Marchand B, Das K, Himmel DM, Parniak MA, et al. 2009. *J Mol Biol* 385: 693–713.
133. Kerr SG, Anderson KS. 1997. *Biochemistry* 36: 14064–70.
134. Kerr SG, Anderson KS. 1997. *Biochemistry* 36: 14056–63.
135. Meyer PR, Matsuura SE, So AG, Scott WA. 1998. *Proc Natl Acad Sci U S A* 95: 13471–6.
136. Sharma B, Crespan E, Villani G, Maga G. 2008. *Proteins* 71: 715–27.
137. Schultz SJ, Champoux JJ. 2008. *Virus Res* 134: 86–103.
138. De Clercq E. 2004. *Chem Biodivers* 1: 44–64.
139. Fan N, Rank KB, Leone JW, Heinrikson RL, Bannow CA, *et al.* 1995. *J Biol Chem* 270: 13573–9.
140. Hizi A, Tal R, Hughes SH. 1991. *Virology* 180: 339–46.
141. Hizi A, Tal R, Shaharabany M, Loya S. 1991. *J Biol Chem* 266: 6230–9.
142. Shaharabany M, Hizi A. 1991. *AIDS Res Hum Retroviruses* 7: 883–8.
143. Ren J, Bird LE, Chamberlain PP, Stewart-Jones GB, Stuart DI, Stammers DK. 2002. *Proc Natl Acad Sci U S A* 99: 14410–5.
144. Sevilya Z, Loya S, Hughes SH, Hizi A. 2001. *J Mol Biol* 311: 957–71.
145. Post K, Guo J, Howard KJ, Powell MD, Miller JT, *et al.* 2003. *J Virol* 77: 7623–34.
146. Kotewicz ML, D'Alessio JM, Driftmier KM, Blodgett KP, Gerard GF. 1985. *Gene* 35: 249–58.
147. Schultz SJ, Champoux JJ. 1996. *J Virol* 70: 8630–8.
148. Crowther RL, Remeta DP, Minetti CA, Das D, Montano SP, Georgiadis MM. 2004. *Proteins* 57: 15–26.
149. Das D, Georgiadis MM. 2004. *Structure* 12: 819–29.
150. Georgiadis MM, Jessen SM, Ogata CM, Telesnitsky A, Goff SP, Hendrickson WA. 1995. *Structure* 3: 879–92.
151. Gu Z, Arts EJ, Parniak MA, Wainberg MA. 1995. *Proc Natl Acad Sci U S A* 92: 2760–4.
152. Thomas DA, Furman PA. 1991. *Biochem Biophys Res Commun* 180: 1365–71.
153. Le Grice SF, Naas T, Wohlgensinger B, Schatz O. 1991. *Embo J* 10: 3905–11.
154. Wohrl BM, Howard KJ, Jacques PS, Le Grice SF. 1994. *J Biol Chem* 269: 8541–8.
155. Souquet M, Restle T, Krebs R, Le Grice SF, Goody RS, Wohrl BM. 1998. *Biochemistry* 37: 12144–52.
156. Amacker M, Hottiger M, Hubscher U. 1995. *J Virol* 69: 6273–9.
157. Amacker M, Hubscher U. 1998. *J Mol Biol* 278: 757–65.
158. Tasara T, Amacker M, Hubscher U. 1999. *Biochemistry* 38: 1633–42.
159. Operario DJ, Reynolds HM, Kim B. 2005. *Virology* 335: 106–21.
160. Taube R, Avidan O, Bakhanashvili M, Hizi A. 1998. *Eur J Biochem* 258: 1032–9.
161. Taube R, Loya S, Avidan O, Perach M, Hizi A. 1998. *Biochem J* 329 (Pt 3): 579–87.
162. Entin-Meer M, Sevilya Z, Hizi A. 2002. *Biochem J* 367: 381–91.
163. Beauregard A, Curcio MJ, Belfort M. 2008. *Annu Rev Genet* 42: 587–617.
164. Clare J, Farabaugh P. 1985. *Proc Natl Acad Sci U S A* 82: 2829–33.
165. Garfinkel DJ, Boeke JD, Fink GR. 1985. *Cell* 42: 507–17.
166. Mellor J, Malim MH, Gull K, Tuite MF, McCready S, *et al.* 1985. *Nature* 318: 583–6.

167. Wilhelm M, Boutabout M, Wilhelm FX. 2000. *Biochem J* 348 Pt 2: 337–42.
168. Eichinger DJ, Boeke JD. 1988. *Cell* 54: 955–66.
169. Kirchner J, Sandmeyer SB. 1996. *J Virol* 70: 4737–47.
170. Nymark-McMahon MH, Beliakova-Bethell NS, Darlix JL, Le Grice SF, Sandmeyer SB. 2002. *J Virol* 76: 2804–16.
171. Wilhelm M, Wilhelm FX. 2006. *Eukaryot Cell* 5: 1760–9.
172. Gabriel A, Mules EH. 1999. *Ann N Y Acad Sci* 870: 108–18.
173. Xiong Y, Eickbush TH. 1988. *Mol Biol Evol* 5: 675–90.
174. Heyman T, Agoutin B, Friant S, Wilhelm FX, Wilhelm ML. 1995. *J Mol Biol* 253: 291–303.
175. Le Grice SF. 2003. *Biochemistry* 42: 14349–55.
176. Hizi A. 2008. *J Virol* 82: 10906–10.
177. Kirshenboim N, Hayouka Z, Friedler A, Hizi A. 2007. *Virology* 366: 263–76.
178. Bibillo A, Eickbush TH. 2002. *J Biol Chem* 277: 34836–45.
179. Bibillo A, Eickbush TH. 2002. *J Mol Biol* 316: 459–73.
180. Lien JM, Petcu DJ, Aldrich CE, Mason WS. 1987. *J Virol* 61: 3832–40.
181. Staschke KA, Colacino JM. 1994. *J Virol* 68: 8265–9.
182. Chang LJ, Hirsch RC, Ganem D, Varmus HE. 1990. *J Virol* 64: 5553–8.
183. Howe AY, Elliott JF, Tyrrell DL. 1992. *Biochem Biophys Res Commun* 189: 1170–6.
184. Chen Y, Robinson WS, Marion PL. 1994. *J Virol* 68: 5232–8.
185. Sousa R. 1996. *Trends Biochem Sci* 21: 186–90.
186. Steitz TA. 2006. *Embo J* 25: 3458–68.
187. Beck J, Vogel M, Nassal M. 2002. *Nucleic Acids Res* 30: 1679–87.
188. Sharon A, Chu CK. 2008. *Antiviral Res* 80: 339–53.
189. Gaillard RK, Barnard J, Lopez V, Hodges P, Bourne E, *et al.* 2002. *Antimicrob Agents Chemother* 46: 1005–13.

CHAPTER 7

Synthetic Evolution of DNA Polymerases with Novel Properties

7.1 Why Design Enzymes with Novel Properties?

Most chemical reactions in a cell are catalyzed by enzymes. Enzymes possess a high substrate specificity that has evolved over billions of years. The versatility of the thousands of enzymes is enormous not only in their specificity, but also in their adaptive behavior to altered environmental and intracellular conditions. However, as indicated in a News and Views note in *Nature*,[1] our *ability of designing new enzymes on the basis of our knowledge of protein structure and reaction mechanism can charitably be described as primitive.* The first approaches to use enzymes came from industrial applications with the purpose to reduce high temperatures for catalytic chemical processes and to circumvent extremes of pH. Moreover, as reaction specificity can be increased, a more pure product can be obtained and, last but not least, the environment can be protected upon applying "mild" enzymatic conditions compared to the harsh chemical reactions with many side products. The growing use of industrial enzymes is dependent on innovation to improve performance and to reduce costs.[2]

As described in detail in Chapter 1, the polymerase chain reaction (PCR) has immediately brought the DNA pols into the very center of biotechnical applications. They include many novel purposes such as PCR itself for thousands of different applications, nested PCR, single-molecule amplification, inverse PCR, multiplex amplification, direct DNA sequencing, quantification of virus and bacterial load, single gamete typing, reverse transcriptase PCR (known as single enzyme RT-PCR), dUTP incorporation to prevent DNA carry-over, combinatorial libraries, SELEX to generate aptamers and ribozymes, amplification from ancient DNA, *in situ* PCR, sequence tagged sites, long PCR, accurate PCR by using a combination of DNA pol and $3' \rightarrow 5'$ exonuclease. Moreover, the research on thermophiles came into its golden age, since many DNA pols appeared with certain different properties. Even though we have learned a lot by structural studies as well as biochemical analyses on how DNA pols function, synthetic evolutionary approaches have given insight into the importance of certain amino acids and motifs in the active site of DNA pols.

In this Chapter we discuss:

(1) the tight active site of DNA pols to which the substrates fit,
(2) methods to evolve DNA pols with novel properties,
(3) applications of DNA pols with novel properties, and
(4) DNA pols with novel properties such as increased fidelity, decreased fidelity, being able to amplify from ancient and damaged DNA, creating RNA pols from DNA pols and evolving dNTP substrates for the expansion of the genetic code.

7.2 DNA Polymerases Have a Tight Active Site to Which the Substrates Fit

As already discussed in the previous Chapters 2–4, DNA pols have the form of a human right hand. The structures of many DNA pols have been solved in great details (reviewed in Refs. 3–5). On the other hand, DNA pols are complex enzymes that have to adopt in their active site a template, a primer and the four dNTPs. In view of this complexity one wonders how the active site of DNA pols is geometrically constructed to adopt all four dNTPs. It was argued that the Watson-Crick hydrogen bonds themselves are not necessary for efficient and accurate copy of a base pair by the DNA pol (reviewed in Ref. 6). Important issues in this connection appear to be the minor groove hydrogen bonding, base stacking, water solvation and steric effects. Many different approaches were tested to address this issue (see Ref. 6, and references therein). The first direction was the testing of nucleoside analogs as substrate probes for DNA pols. Such analogs included methylated bases, construction of bases with altered shapes and sizes that do not contain any hydrogen bonding, base analogs that still contain the hydrogen bonding but with an increased size of the base and sterically augmented sugars. E. Kool concluded that "among the noncovalent interaction in the active site, steric effects are the most likely candidates for the lion's share of fidelity during DNA synthesis. The steric effects depend on the complementarity of size and shape of the two bases being paired, and on the tightness of the polymerase active site under study. Watson-Crick hydrogen bonding groups may also contribute to fidelity, but may do so in part because of the steric effect of water bound to them".[6] Since it is known that different DNA pols have different active site geometries and thus different accuracies, one can expect that different modified bases have different effects on different DNA pols.

In general it is evident from **Figure 7.1**, that the four combinations of Watson-Crick base pairing have only little variation in the *overall* shape. If, however, the

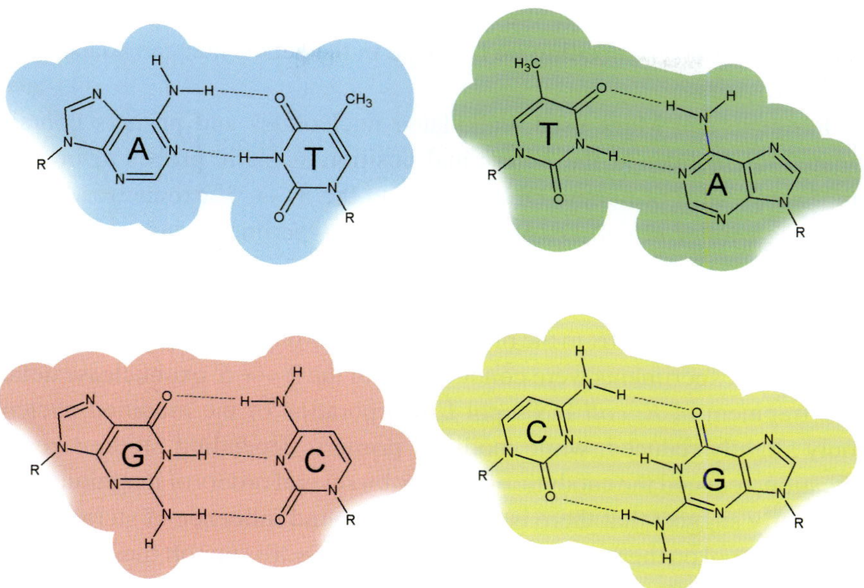

Figure 7.1   Space filling shapes of the four different naturally occurring base pairs. For details see text.

Figure 7.2   Overlay form of the four base pair shapes from Figure 7.1. Overlay of all four base pair combinations. Red: G-C; yellow: C-G; blue: A-T; and green: T-A.

four shapes are over-layed one can see from **Figure 7.2** that there is considerable variation in the minor grove and at the edges of the major grove. In **Figure 7.2** the overall shapes with the largest dimensions are shown. It was suggested that in the active site tightness hypothesis predicts to what degree the wall of this consensus pocket might be pushed outwards.[6] The size exclusion hypothesis in the active site of a DNA pol predicts that a base pair can only successfully be incorporated if the predicted shape fits in the active site. In summary this model clearly can predict how a particular DNA pol, depending on its shape in the active site, can adapt different types of base pairs.

Di Pasquale and Marx used 4′alkylated nucleotides and primers containing 4′alkylated nucleotides at the 3′-terminal position as steric probes to investigate different active sites of human DNA pol β and the 3′ → 5′ exonuclease deficient Klenow fragment of *E. coli* DNA pol I.[7] Kinetic experiments suggested that both DNA pols must vary considerably in the active site tightness. Small methyl and ethyl modification in nucleoside triphosphates compromised the activity of DNA pol β, while the extension from the modified primer is only slightly disturbed. When, however, these experiments were conducted with the 3′ → 5′ exonuclease deficient Klenow fragment of *E. coli* DNA pol I incorporation of the modified nucleotide was only slightly reduced, while modified primers, depending on the size of the modification, reduced the catalytic efficiency by several orders of magnitude. These results clearly suggest that there is indeed a profound influence of steric effects of different DNA pol on their accuracy of DNA synthesis. To support their hypothesis,[7] a graphic representation of the distances of the van der Waals radii between DNA and the two enzymes DNA pol β and the DNA pol from *Bacillus stearothermophilus* (Bst DNA pol) was made (**Figure 7.3**). Bst DNA pol is highly homologous to the the 3′ → 5′ exonuclease deficient Klenow fragment of *E. coli* DNA pol I,[8–10] of which there is not enough structural information available so far.[7]

From **Figure 7.3** it is evident that the shapes are different between DNA pol β and Bst DNA pol when either the incoming dNTP was compared to the coding

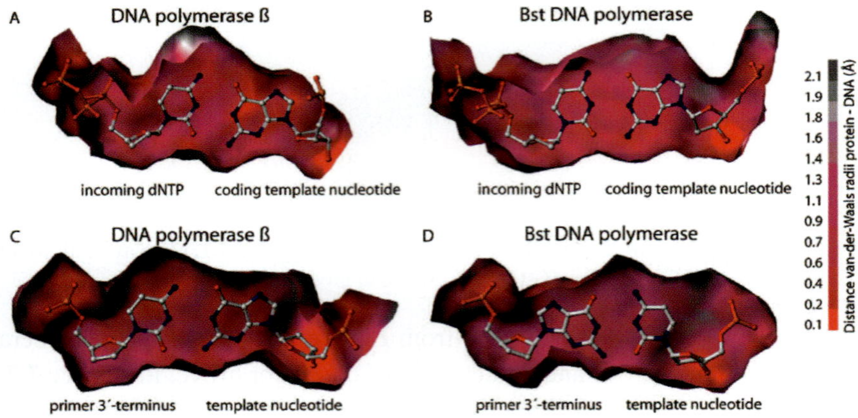

Figure 7.3   Comparison of human DNA polymerase β to *Bacillus stearothermophilus* DNA polymerase. The graphics represent the distances of the van der Waals radii between the DNA and the two DNA pols. For details see text. (Reproduced with permission from Ref. 7.)

template nucleotide (compare **Figure 7.3A and B**) or the 3′primer terminus was compared to the template nucleotide (compare **Figure 7.3C and D**). In summary, these experimental data supported nicely the size exclusion hypothesis.[6]

7.3 Methods to Evolve DNA Polymerases with Novel Properties

Several techniques have been developed to artificially evolve novel properties of DNA pols (reviewed in Ref. 11). Five of them are discussed here in more details.

7.3.1 Detection and Characterization of DNA Polymerases and Mutants Thereof by Functional Complementation in Escherichia coli

The first *in vivo* screens were developed in the laboratory of L. Loeb in 1993, in which a bacterial complementation system to identify and characterize mammalian DNA pol β mutants was designed and employed.[12] In this complementation system, DNA pol β was able to replace both the replicative and repair functions of DNA pol I in the *E. coli* recA718 polA12 double mutant, allowing isolation and evaluation of mutants of a DNA pol in *E. coli* and their characterization for a better understanding of their functions. Later, this technique was expanded to the Taq DNA pol I.[13] Expression of Taq DNA pol I in *E. coli* complemented the growth defect caused by a temperature-sensitive mutation in the host DNA pol I. The nucleotide sequence encoding the amino acids 659–671 of the O-helix of Taq DNA pol I, corresponding to the substrate binding site, was replaced with oligonucleotides containing random nucleotides. With this functional Taq, DNA pol I mutants were selected based on colony formation at the non-permissive temperature. Finally, catalytically active mutants of HIV RT were generated by random sequence mutagenesis and selected in *E. coli* for their ability to complement the temperature-sensitive phenotype of a DNA pol I mutant.[14] The mutations recovered included most of those associated with drug resistance as well as so far unidentified mutations.

Further studies with DNA pol I from *E. coli* revealed that several recently identified mutants with abilities to incorporate nucleotides with bulky adducts have mutations that are not located within conserved regions. Analysis of these mutants might be useful for our understanding of how DNA pols select bases with high fidelity.[15] Moreover, DNA pols with increased capacity to incorporate rNTPs[16] (see also Section 7.5.4) or with increased fidelity[17] were evolved. Mutations that produced an antimutator phenotype were found throughout the entire polymerase domain, with 12% clustered in the so called M-helix. A single mutation in the M-helix resulted in increased base discrimination suggesting that M-helix is a determinant of fidelity.

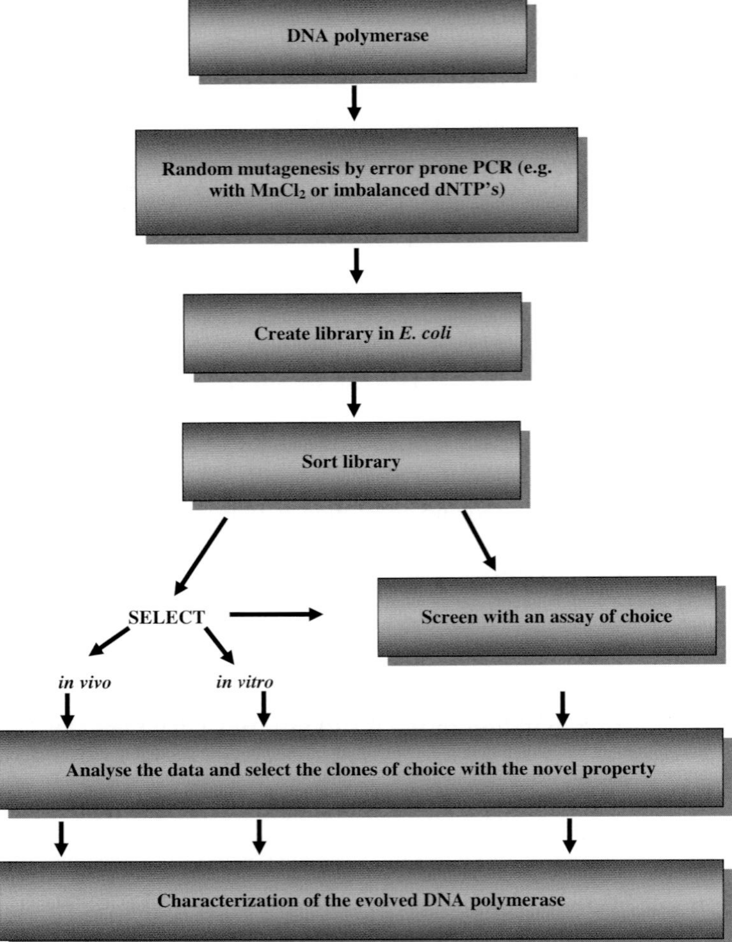

Figure 7.4   DNA polymerase evolution by random point mutagenesis. For details see text.

7.3.2 *DNA Polymerase Evolution by Random Point Mutagenesis*

This method includes the construction of a large library and creates random mutations in the DNA pol (reviewed in Ref. 18 and in **Figure 7.4**). For the DNA pol of choice primers are designed so the entire enzyme can bee PCR amplified. PCR is performed under error-prone conditions that have to be optimized in each case.

This can be achieved by varying the relative dNTP concentrations (imbalanced nucleotides) or by replacing Mg^{2+} with Mn^{2+}, which is known to render DNA pols less faithful. Next with these DNA molecules a library is constructed in *E. coli* and this library is sorted in microtiter plates (e.g. 384 wells). After growth of the bacteria, extracts are prepared and polymerase activities are measured with an assay of design. This is done to select for the property of choice (see Sections 7.4 and 7.5). These experiments can now be carried out by an automated robot system. As an example Sauter and Marx took Taq DNA pol I (Klentaq) performed error-prone PCR and screened the library clones for activity in the microtiter plates directly after heat inactivation of the host proteins (among others also the *E. coli* DNA pols I, II, III, IV and V). They performed the subsequent step in automated liquid handling. DNA pol activities were monitored by using SYBER green for the optical quantification of the different clones. With this a Taq DNA pol I was evolved that possesses reverse transcriptase activity.[19]

7.3.3 DNA Polymerase Evolution by Compartmentalized Self-Replication (CSR)

The compartmentalized self-replication (CSR) technology was pioneered by Holliger and Co-workers.[20] This method is based on the water in oil emulsion technology.[21] It is based on the self-replication onto a separate compartment that cannot exchange proteins and DNA anymore and ensures the linkage between phenotype and genotype. First, a library of clones is cloned and expressed in *E. coli* (**Figure 7.5**). Active Taq DNA pol I molecules were selected by incubating the bacteria in a solution containing flanking primers for the Taq DNA pol I and the dNTPs. Then the DNA pols are segregated into the aqueous solution to allow self-replication

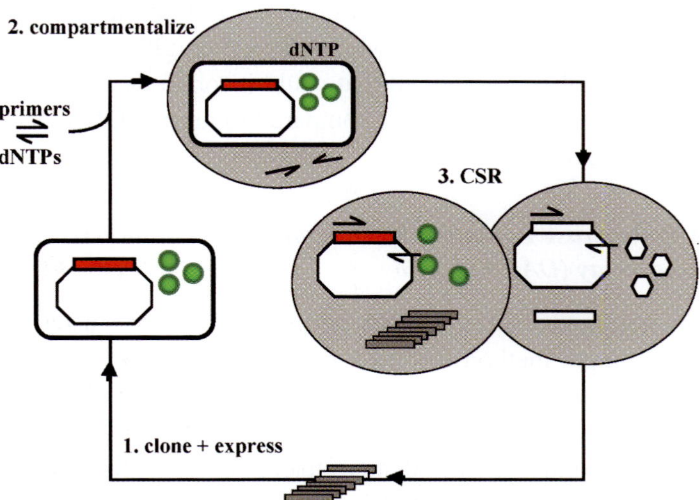

Figure 7.5   DNA polymerase evolution by compartmentalized self-replication (CSR). For details see text. Adapted from Ref. 20.

of the own polymerase gene by PCR. The thus created DNA pols are released and can be recloned for another round of CSF selection until the DNA pol molecules of choice are finally selected. With this technologies a Taq DNA pol I with 11 fold higher thermostability and a 130 fold higher resistance to the DNA pol inhibitor heparin was engineered.[20] Moreover, Taq DNA pol I molecules were constructed by this method that contain in addition to the DNA pol, also RNA pol and reverse transcriptase activities.

7.3.4 *DNA Polymerase Evolution by Phage Display*

Phage display is a powerful technique to covalently attach protein to a filamentous bacteriophage (reviewed in Ref. 22). DNA pols can now be covalently attached and expressed as a fusion protein. The DNA pols of choice are selected in several steps: first, a DNA substrate is cross-linked to the phage; second, the active DNA pols synthesize DNA and generate a product containing biotin, and third, the product is purified by affinity chromatography. This cycle can be repeated and the final DNA pol can then be characterized after recovery. This occurs after extensive washing, when the desired phage is recovered by treatment with DNase I and re-infection of *E. coli* by the DNA pol gene (**Figure 7.6**). The Stoffel fragment of Taq DNA pol I was made as a p3-Jun/Fos-Stoffel fragment fusion[23] and could thereby be evolved to be an efficient RNA polymerase.[24] Romesberg and Co-workers[25] created mutants Taq DNA pols that could efficiently synthesize an unnatural polymer from 2′-O-methyl ribonucleoside triphosphates. Their directed evolution resulted in relocating a critical side chain to a different position in the DNA pol, therefore re-engineering the overall active site while preserving critical protein-DNA interactions. One DNA pol was identified that could incorporate modified nucleotides with the same efficiency and accuracy compared to normal DNA as the wild-type DNA pol (see also Section 7.5.5).

7.3.5 *DNA Polymerase Evolution by Oligonucleotide Addressed Enzyme Assay (OAEA)*

A novel approach to characterize DNA pol mutants was taken by Kranaster and Marx by developing a method for DNA pol profiling on DNA arrays[26] and was named oligonucleotide addressed enzyme assay (OAEA). The principle consists of two spotting and an incubation step. In the first spotting step $5'NH_2$-$(CH)_6$ modified oligonucleotides are covalently attached to activated glass slides.[27] The olignucleotides are attached in defined spot rows (**Figure 7.7 A**). In the second spotting step DNA pol mutants to be characterized in the presence of dATP, dGTP, dCTP, biotin-dUTP and different DNA template to be tested will be applied to the

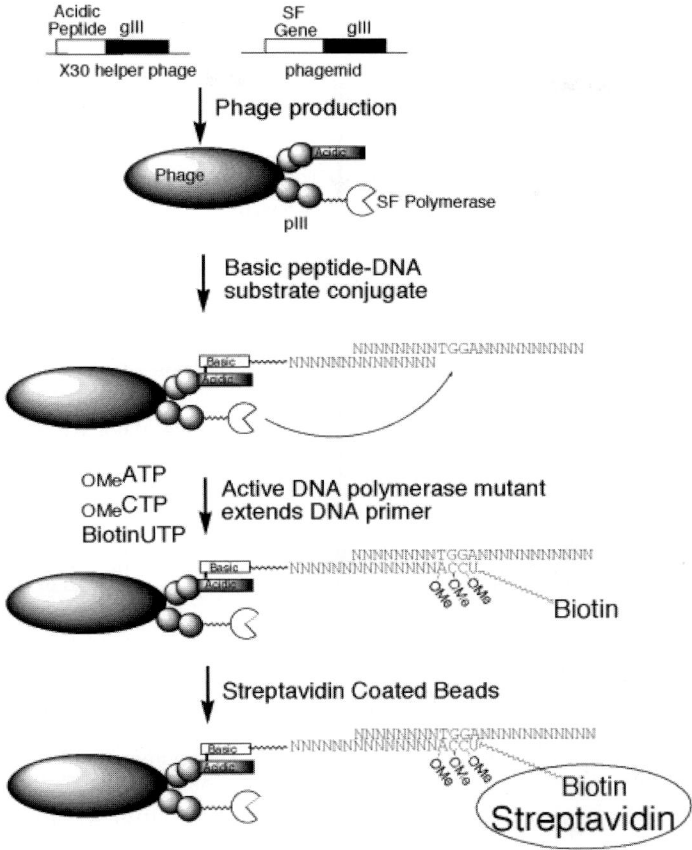

Figure 7.6   DNA polymerase evolution by phage display. For details see text. (Reproduced with permission from Ref. 25.)

oligonucleotides in nanoliters quantities. Interestingly, even though the nanodrops dried out, they can be rehydrated without loss of DNA pol function (exemplified here with Taq DNA pol I, Klentaq). The drops are then rehydrated in a humidity saturated chamber at $-50°C$. In case of DNA synthesis the incorporated biotin binds with an alexa546-streptavidine conjugate that will create a signal. In summary, this method is a nice microarray assay that can be used not only to characterize DNA pol mutants but can also be employed for the evolution of DNA pols with novel properties. Finally, this method, if combined in a single experiment with the already powerful microarrays containing several 100 000 of different oligonucleotides, might be extremely informative. However, one must bear in mind that evolving heat-labile DNA pols might need laborious optimization in the drying and rehydration steps, that might irreversible inactivate labile DNA pols.

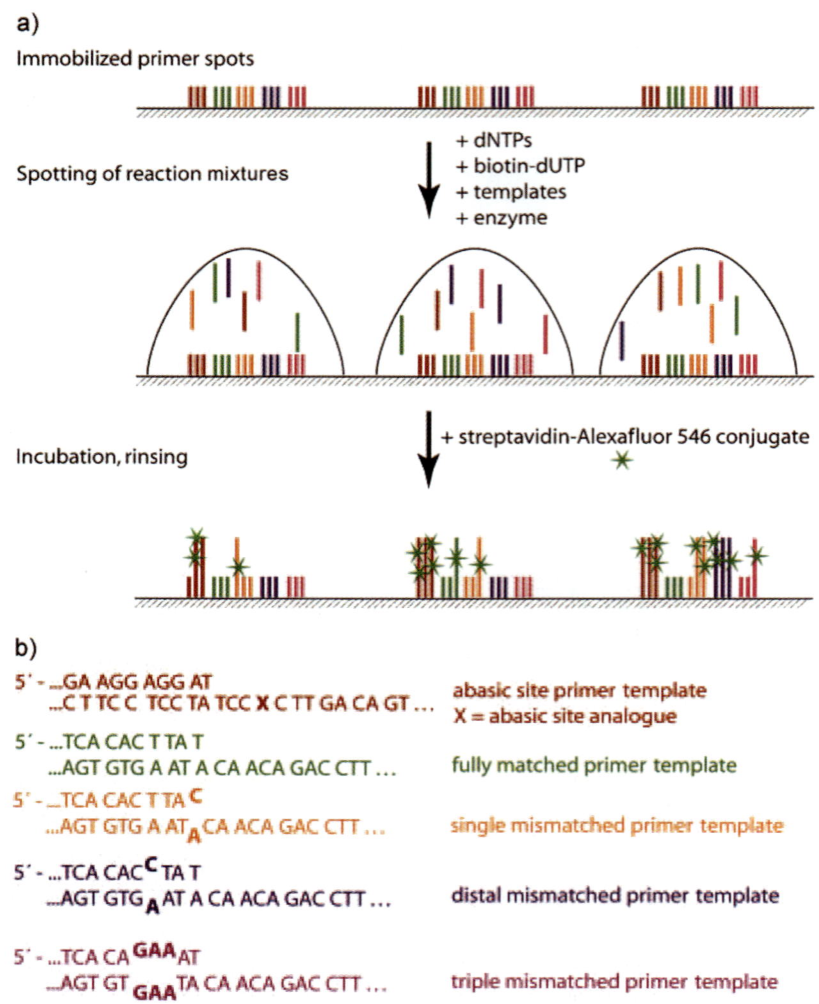

Figure 7.7   DNA polymerase evolution by the oligonucleotide addressed enzyme assay (OAEA). (a) Top: the short colored strips are immobilized primers; middle: the longer stripes represent DNA template that hybridize to the corresponding primers; bottom: after successful DNA synthesis the Alexa 546-streptavidin can bind to the incorporated biotin and this gives a green signal (stars). (Reproduced with permission from Ref. 26.)

7.4 Applications of DNA Polymerases with Novel Properties

DNA pols can be mutated, mostly in their active sites, to adopt novel properties (see e.g. Refs. 11 and 15) as reviewed. **Table 7.1**, which includes 26 citations, summarizes some novel properties of DNA pols and some of their applications. In Section 7.5 we will discuss in five subsections different novel properties that could

Table 7.1   Novel properties of DNA polymerases and their applications

DNA pol source	Mutation/ Deletion	Novel property/ application	Reference
Taq[a] DNA pol I, Stoffel fragment	$5' \rightarrow 3'$ exonuclease deficient	improved stability and fidelity	(28)
Two Thermus species	chimeric reverse transcriptase and with $3' \rightarrow 5'$ exonuclease activity	RT-PCR	(29)
Vent DNA pol	$3' \rightarrow 5'$ exonuclease mutants and Y412V	incorporates ribonucleotides at least 200-fold more efficiently	(30)
Vent and Taq DNA pol I	Vent (A488L) and Taq (F667Y)	increased incorporation of dideoxy terminators; DNA sequencing and genotype analysis	(31)
Taq DNA pol I	R660N	improvement in the accuracy of DNA sequencing	(32)
Taq DNA pol I	one to four amino acid substitutions	adds rNTPs into a growing polynucleotide chain	(16)
E. coli DNA pol I	E710A	adds rNTPs into a growing polynucleotide chain	(33)
Taq DNA pol I	S543N	reduced pausing, PCR amplification and sequencing of difficult templates, e.g. with a high GC content or strong secondary structure.	(34)
DNA pol β	N294Q	increase in fidelity	(35)
E. coli DNA pol I	insertion of the thioredoxin binding domain of the T3 DNA pol	increase in processivity	(36)
Taq DNA pol I	insertion of the thioredoxin binding domain of the T3 DNA pol	increase in fidelity and processivity	(37)
E. coli DNA I	mutations in the M-helix	increased fidelity	(17)
archaeal DNA pol 9 degrees N	two biotinylated peptide "legs" are inserted at positions flanking the DNA-binding cleft of the DNA pol	are active both in solution and when immobilized on a surface	(38)
E. coli DNA pol Klenow fragment, exo minus and Taq DNA pol I	879Q-880V-881H to PLQ, LVL and LVG	higher extension fidelity	(39)

(*Continued*)

Table 7.1   (*Continued*)

DNA pol source	Mutation/ Deletion	Novel property/ application	Reference
Taq DNA pol I (Klentaq)	M1: L322M, L459M, S515R,I638F, S739G and E773G M1: E789F	DNA pol with reverse transcriptase activity; RT-PCR	(19)
Taq DNA pol I	28 different mutants	DNA pol with reverse transcriptase activity; RT-PCR	(40)
Taq DNA pol I (Klentaq)	M747K	amplify highly UV-damaged DNA	(41)
Taq DNA pol I (Klentaq)	many randomized mutations in Q782-V783-H784	increased selectivity of mismatch extension	(42)
Taq DNA pol I (Stoffel fragment)	I614E615 to E614G615, further mutations in these two amino acids	synthesize an unnatural polymer from $2'$-O-methyl rNTP	(25)
Taq DNA pol I (Stoffel fragment)	F598I, I614F, and Q489H	replicates efficiently DNA containing a PICS self-pair	(43)
Taq DNA pol I	E602V, A608V, I614M, E615G	DNA polymerase, RNA polymerase and RT activities	(44)
Taq DNA pol I	F73S, R205K, K219E, M236T, E434D, A608V	increased thermostabiliy; PCR	(20)
Taq DNA pol I	K225, E388V, K540R, D578G, N583S, M747R	resistance to heparin; PCR	(20)
Taq DNA pol I	G84A, D144G, K314R, E520G, F598L, A608V, E742G	bypasses abasic sites, a thymidine dimer or the base analog 5-nitroindol	(45)
Several *Thermus* species (Taq, Tth, Tfl)[a]	many multiple mutations and combinations of DNA pols	synthesis over different damaged DNAs	(46)
Taq DNA pol I	cocktail of seven DNA pols with a mutation each	PCR amplification of a 47 000-year-old DNA from a cave bear (*Ursus spelaeus*)	(46)
Pfu1 DNA pol	hydrophobic substitution mutations at D514L and K593M	increase in selectivity	(47)
DNA pol λ	Y5050M	increase in selectivity over 6 mG	Parasuraman *et al.*, submitted, 2009

[a] Taq: *Thermus aquaticus*; Tth: *Thermus thermophilus*; Tfl: *Thermus flavus*; Pfu: *Pyrococcus furiosus*.

be evolved by the techniques mentioned above. They include increased fidelity, decreased fidelity, amplification from ancient and damaged DNA, RNA pols evolved from DNA pols and evolving dNTPs and expansion of the genetic code.

From **Table 7.1** a few general conclusions can be drawn:

(1) The most successful approaches were performed by using Taq DNA pol I. This is evident from its biotechnical applications mainly in PCR. Improvements were made in thermostability, fidelity, processivity, RT-PCR, decreased pausing at GC rich regions, resistance to heparin, an anticoagulant found in patient samples.

(2) Vent and Taq DNA pols I were generated that increase the incorporation of dideoxy chain terminators and they are used for improved DNA sequencing and genotyping.

(3) *E. coli* DNA pol I was evolved into RNA pols, with enhanced processivity, increased fidelity and higher extension fidelity.

(4) Several Thermus species (Taq, Tth and Tfl) were evolved to carry out PCR from different damaged DNA. This might be of great practical use to forensically analyze DNA from dead bodies that were decaying over a long time period.

(5) A cocktail of seven Taq pols with multiple mutations was able to PCR amplify a 47 000-year-old DNA form the cave bear *Ursus spelaeus*.

(6) A Pfu DNA pol with increase in base selectivity was rationally designed by substituting two amino acids in the active site into hydrophobic counterparts (D514L and K593M).

(7) Very little is known from heat-labile DNA pols from eukaryotes. A recent example is a mutation in human DNA pol λ Y505M, which incorporates opposite 6me-G the correct C, while the wild-type DNA pol λ incorporates an incorrect T. This mutant therefore has an increased fidelity over a lesion in the DNA.

7.5 DNA Polymerases with Novel Properties

7.5.1 *Increased Fidelity*

The accuracy of DNA pols is essential for their ability to copy genomes faithfully. The fidelity of those DNA pols that contain a $3' \rightarrow 5'$ exonuclease activity can be changed by the ratio of polymerase versus exonuclease activites. As we will discuss in Chapter 8 (DNA polymerases and diseases) mutations in the active site of DNA pol δ can lead to cancer. Moreover, with the techniques described above, DNA pols were now selected that contain an increased fidelity. A few examples are given here to illustrate the potential of the approaches.

Loeb's laboratory has, as already indicated, used a bacterial complementation system to identify and characterize mammalian DNA pol β mutants.[12] DNA pols of the family A are most well characterized in the motifs A and B, which regulate the fidelity of replication and the incorporation of nucleotide analogs such as dideoxyribonucleotides. Mutants were identified with abilities to incorporate nucleotides with bulky adducts and the analysis of these mutants will helped to understand how DNA pols select bases with high fidelity.[15] The tolerance of *E. coli* DNA pol I for amino acid substitutions in the active site and in different segments of DNA pol I were determined and the effects of these substitutions on the fidelity of DNA synthesis were tested.[17] The DNA pol I mutant library, with random substitutions throughout the polymerase domain, was evolved for activity. It was found that two-thirds of single base substitutions were tolerated without loss of activity, and this plasticity occurred at evolutionarily conserved regions. 408 mutants of the active library were screened for alterations in fidelity *in vivo* in *E. coli*. The mutation frequencies varied from 10^6-fold compared with wild type (e.g. 1000 fold lower up to 1000 fold higher fidelity). Mutations that produced a high fidelity and thus an antimutator phenotype were distributed throughout the polymerase domain, with 12% clustered in the M-helix[17] suggesting that this domain is a determinant of fidelity.

Another innovative approach was to create chimeric DNA pols. The thioredoxin binding domain (TBD) of the T3 DNA pol was inserted into thermostable Taq DNA pol I at an analogous position in the thumb domain.[37] This converts the Taq DNA pol from a low processive to a highly processive enzyme. The enhancement in processivity was 20–50-fold when compared with the wild-type Taq DNA pol I. The chimeric thioredoxin/Taq DNA pol I copied six to seven times more faithfully than Taq DNA pol I alone and synthesized 2–3-fold fewer AT $\rightarrow$ GC transition mutations.

Marx *et al.*, by randomizing a gene cassette of the *E. coli* DNA pol I Klenow fragment, that was $3' \rightarrow 5'$ exonuclease deficient, did a comparative automatic screening and identified mutants that had a significant higher extension fidelity.[39] These findings were then transferred to thermostable Taq DNA pol I as a tool to evolve a better PCR-based genotyping. Further work created a Taq DNA pol I that was randomly mutagenized in the motiv C at amino acids Q782/V783/H784 and these multiple mutants had an increase in selectivity of mismatch extension.[42] In view of the fact that the hydrogen bonding network between the DNA pol active site and its primer and template has an influence on the selectivity of DNA pol, a variant of Pfu DNA pol was evolved by hydrophobic substitution of the active site D541 to L and K593 to M.[47] These data further indicated that the complementary shape, as well as the polar interactions, have important influence on the selectivity of a DNA pol. By applying this principle to a hydrophobic substitution at the active site of human DNA pol λ (Y505 to M) a DNA pol was identified that better incorporated the correct nucleotide (dCTP) over an O-6-methylguanine (O-6-mG) template relative to the non complementary nucleotide (dTTP) while the wild-type DNA pol λ incorporated as expected preferentially the incorrect dTTP. This is the first report of a pol showing faithful bypass of the O-6-mG lesion and can help to elucidate the molecular basis for the mutagenic potential of this lesion (Parasuraman *et al.*, submitted, 2009).

7.5.2 *Decreased Fidelity*

Error rate of *E. coli* DNA pol I was increased by introducing point mutations in three structural domains that regulate its fidelity. This system of targeted mutagenesis in *E. coli* (see above for details of the methodology) had an important impact on enzyme-based applications in areas such as synthetic chemistry, gene therapy and molecular biology.[48]

7.5.3 *Amplification of Damaged and Ancient DNA*

A nice example where DNA pols with either higher or lower substrate specificity are needed is the amplification from damaged DNA. Evolving the M747 to K in Taq DNA pol I resulted in the property to amplify from damaged DNA such as abasic sites, from oxidative damages such as 8-oxo-G and 8-oxo-A.[41] Such enzymes are of great value in forensic medicine to identify individuals that were dead since a long time and whose DNA was severely damaged. Another work resulted in evolving a Taq DNA pol I that retain the high turnover, processivity and fidelity, but was able to replicate from various lesions and non-natural substrates.[45] These included abasic sites, CDP dimers, 7-deaza dGTP, aS-dNTPs, FITC labeled dATP, Biotin16-dUTP and the base analog 5-nitroindol, suggesting that this Taq DNA pol I, that contains seven mutations (see **Table 7.1**), could be of great help for many applications in biotechnology and diagnostic medicine.

Another application was the evolving of a DNA pol that can rescue and amplify from ancient DNA.[49] In particular to amplify highly degraded DNA is of great importance in science to study ancient DNA. Holliger's lab performed molecular evolution to created a mixture of DNA pols from different species (Taq, Tth, Tfl) each containing multiple mutations that could amplify damaged DNA.[46] Damaged DNA included abasic sites, 5-methyl-5-hydroxyhydantoin and 5-hydroxyhydantoin. Furthermore, extensions from multiple mismatches (up to four) and from misaligned DNA structures (e.g. hairpins after slippage) were possible with such multiple mutants. Finally a cocktail of the seven most promising Taq DNA pol I mutants was chosen to amplify DNA from a 47 000-year-old cave bear called *Ursus spelaeus*. When this blend was compared to wild type Taq DNA pol I the mutated cocktail amplified from two- to five-fold less DNA than the wild type. This suggested that the evolved DNA pol cocktail had a higher specificity with highly damaged ancient DNA.[46]

7.5.4 *A DNA Polymerase Becomes an RNA Polymerase*

We have already seen above that *Thermus* species DNA pols can be made into reverse transcriptase useful for RT-PCR.[29] Many publications exist that made a DNA pol into a RNA pol and two selected examples are discussed in more details here.

(1) By using genetic complementation, a library of approximately 8 000 active mutants Taq DNA pol I, was obtained and 350 were sequenced and analyzed resulting in a large collection of active polymerase mutants.[50] The high mutability of the polymerase active site *in vivo* and the ability to evolve altered enzymes may be required in nature for survival in environments that have increased mutagenesis. In a follow up work a large library of 200 000 mutants Taq DNA pol I was created containing random substitutions within a part of the dNTP binding site (amino acids 605–617 in motif A) and selected active Taq DNA pol I containing multiple mutations. Of these, 291 were screened for the ability to incorporate ribonucleotides.[16] This led to the identification of 23 mutants that added rNTPs into a growing polynucleotide chain and suggested that DNA and RNA pols could have evolved by divergent evolution from an ancestor that shared a common mechanism for polynucleotide synthesis.

(2) The second approach utilized the activity-based selection method to evolve DNA pol with RNA pol activity.[24] For this the Taq DNA pol I Stoffel fragment was displayed on a filamentous phage by fusing it to a pIII coat protein, and the substrate DNA template/primer duplexes were attached to other adjacent pIII coat proteins (for details see **Figure 7.6** and methodology in Section 7.3.4). The phage particles displaying Taq DNA pol I Stoffel fragments, were then selected to incorporate rNTPs and biotinylated UTP. When the Taq DNA pol I Stoffel fragment library was screened four times three mutants could be harvested that incorporate rNTPs with an efficiency as the wild-type enzyme can incorporate dNTPs. In summary, the phage method created an RNA pol.

7.5.5 *Evolving the dNTP Substrates and Expansion of the Genetic Code*

15 years ago, the Benner and Hübscher labs tested the ability of DNA pols from the B, X and RT families to catalyze the template-directed synthesis of duplex oligonucleotides containing non-standard Watson-Crick base pairs.[51] They included a nucleotide bearing a 5-(2,4-diaminopyrimidine) heterocycle (dκ) and a nucleotide

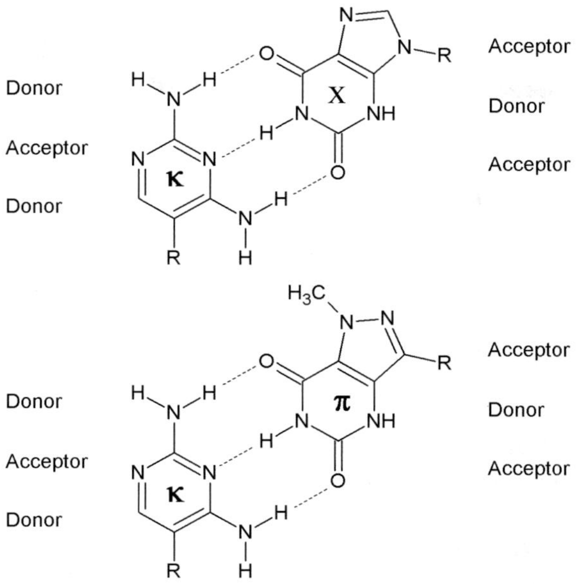

Figure 7.8   **Non-standard Watson-Crick base pairs.** Shown are κ, X and π including their donor hydrogens and their acceptor oxygens or nitrogens. (For details see text and Ref. 51.)

bearing either deoxyxanthosine (dX) or a N1-methyloxoformycin B (π). The dκ-X and dκ-π base pairs are joined by a hydrogen bonding pattern different from and exclusive of those joining the A-T and G-C base pairs (**Figure 7.8**). HIV-1 RT incorporated dXTP into an oligonucleotide opposite dκ in a template with good fidelity. With lower efficiency and fidelity, HIV-1 reverse transcriptase also incorporated dκTP opposite dX in the template. With dπ in the template, no incorporation of dκTP was observed with HIV reverse transcriptase. The Klenow fragment of DNA pol I from *E. coli* did not incorporate dκTP opposite dX in a template but did incorporate dXTP opposite dκ. Mammalian DNA pols α, β and ε incorporated neither dXTP opposite dκ nor dκTP opposite dπ. The replicative DNA pols α and ε (but not the repair DNA pol β) incorporated dκTP opposite dX in a template but discontinued elongation after incorporating a single additional base. These results were later discussed in the view of the ternary crystal structure of DNA pol β.[52]

The laboratory of Romesberg also made great efforts toward the expansion of the genetic alphabet. DNA pols were able to recognize a highly stable self-pairing hydrophobic base. For this a 7-propynyl isocarbostyril nucleoside was synthesized and converted into the nucleoside triphosphate.[53] This self-pairing base pair called PICS could be synthesized by *E. coli* DNA pol I and by Taq DNA pol I. However, the proteolytic fragment of Taq DNA pol I called the Stoffel fragment was unable to do this. The Stoffel fragment was evolved by using the phage display technology.[43]

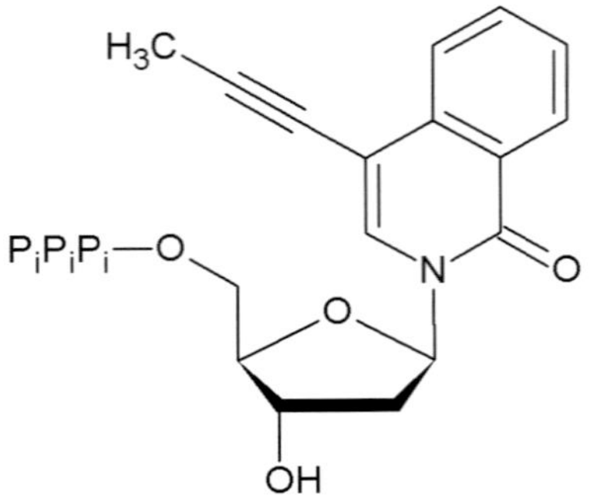

Figure 7.9   7-propynyl isocarbostyril nucleoside triphosphate. (For details see text and Refs. 25 and 53.)

The same laboratory also synthesized $2'$-O-methyl modified NTP and evolved with the same technology Stoffel fragment mutants that could efficiently synthesize an unnatural polymer from $2'$-O-methyl-rNTPs[25] and 7-propynyl iso carbostyril nucleoside triphosphate[53] (**Figure 7.9**). One evolved Stoffel fragment could incorporate the $2'$-O-methyl-rNTPs with an efficiency and fidelity equal to that of the wild-type Stoffel fragment with natural substrates. Such an enzyme is a great use for applications in biotechnology and therapeutic medicine.

7.6   Chapter Summary

In this Chapter we summarized the novel and exciting field of synthetic evolution of DNA pols with novel properties. Many biotechnological and forensic applications call for DNA pols with certain improved catalytic properties. First, the tight active site of DNA pols to which the substrates fit is discussed. These data can give prediction in which direction to evolve a DNA pol. Next five different methods are discussed: (i) detection and characterization of DNA pols and mutants thereof by functional complementation in *E. coli*, (ii) DNA pol evolution by random point mutagenesis and screening, (iii) DNA pol evolution by compartmentalized self-replication (CSR), (iv) DNA pol evolution by phage display, and (v) DNA pol evolution by oligonucleotide addressed enzyme assay (OAEA). Applications of DNA pols with novel properties include the construction of DNA pols for better PCR and sequencing properties, to DNA pols that are in addition RT's and RNA pols, to DNA pols that can amplify from heavily damaged and very old DNA. In particular DNA pols are discussed that have increased or decreased fidelity, can amplify from ancient and damaged DNA, create RNA pols from DNA pols and evolving the dNTP substrates of the genetic code. In summary, it can be foreseen that the synthetic evolution of DNA pols will have a great impact not only in biotechnology but also in biology and medicine, since eukaryotic DNA pols might be evolved with "better" properties to handle damaged DNA more faithfully than their wild-type counterparts.

References

1. Robertson MP, Scott WG. 2007. *Nature* 448: 757–8.
2. Cherry JR, Fidantsef AL. 2003. *Curr Opin Biotechnol* 14: 438–43.
3. Brautigam CA, Steitz TA. 1998. *Curr Opin Struct Biol* 8: 54–63.
4. Steitz TA. 1999. *J Biol Chem* 274: 17395–8.
5. Steitz TA. 2006. *Embo J* 25: 3458–68.
6. Kool ET. 2002. *Annu Rev Biochem* 71: 191–219.
7. Di Pasquale F, Fischer D, Grohmann D, Restle T, Geyer A, Marx A. 2008. *J Am Chem Soc* 130: 10748-57
8. Kiefer JR, Mao C, Braman JC, Beese LS. 1998. *Nature* 391: 304–7.
9. Johnson SJ, Taylor JS, Beese LS. 2003. *Proc Natl Acad Sci U S A* 100: 3895–900.
10. Johnson SJ, Beese LS. 2004. *Cell* 116: 803–16.
11. Henry AA, Romesberg FE. 2005. *Curr Opin Biotechnol* 16: 370–7.
12. Sweasy JB, Loeb LA. 1993. *Proc Natl Acad Sci U S A* 90: 4626–30.
13. Suzuki M, Baskin D, Hood L, Loeb LA. 1996. *Proc Natl Acad Sci U S A* 93: 9670–5.
14. Kim B, Hathaway TR, Loeb LA. 1996. *J Biol Chem* 271: 4872–8.
15. Loh E, Loeb LA. 2005. *DNA Repair (Amst)* 4: 1390–8.
16. Patel PH, Loeb LA. 2000. *J Biol Chem* 275: 40266–72.
17. Loh E, Choe J, Loeb LA. 2007. *J Biol Chem* 282: 12201–9.
18. Arnold F, Georgiou G. 2003. *Directed Evolution Library Creation*. Totowa: Humana Press.
19. Sauter KB, Marx A. 2006. *Angew Chem Int Ed Engl* 45: 7633–5.
20. Ghadessy FJ, Ong JL, Holliger P. 2001. *Proc Natl Acad Sci U S A* 98: 4552–7.
21. Tawfik DS, Griffiths AD. 1998. *Nat Biotechnol* 16: 652–6.
22. Russel M, Lowman H, Clackson T. 2004. In *A Practical Apporach*, ed. T Clackson, H Lowman, pp. 1–26: Oxford University Press.
23. Brunet E, Chauvin C, Choumet V, Jestin JL. 2002. *Nucleic Acids Res* 30: e40.
24. Xia G, Chen L, Sera T, Fa M, Schultz PG, Romesberg FE. 2002. *Proc Natl Acad Sci U S A* 99: 6597–602.
25. Fa M, Radeghieri A, Henry AA, Romesberg FE. 2004. *J Am Chem Soc* 126: 1748–54.
26. Kranaster R, Marx A. 2009. *Angew Chem Int Ed Engl* 48: 4625–8.
27. Kranaster R, Ketzer P, Marx A. 2008. *Chembiochem* 9: 694–7.
28. Lawyer FC, Stoffel S, Saiki RK, Chang SY, Landre PA, *et al.* 1993. *PCR Methods Appl* 2: 275–87.
29. Schonbrunner NJ, Fiss EH, Budker O, Stoffel S, Sigua CL, *et al.* 2006. *Biochemistry* 45: 12786–95.
30. Gardner AF, Jack WE. 1999. *Nucleic Acids Res* 27: 2545–53.
31. Gardner AF, Jack WE. 2002. *Nucleic Acids Res* 30: 605–13.
32. Li Y, Mitaxov V, Waksman G. 1999. *Proc Natl Acad Sci U S A* 96: 9491–6.
33. Astatke M, Ng K, Grindley ND, Joyce CM. 1998. *Proc Natl Acad Sci U S A* 95: 3402–7.
34. Ignatov KB, Bashirova AA, Miroshnikov AI, Kramarov VM. 1999. *FEBS Lett* 448: 145–8.
35. Kraynov VS, Showalter AK, Liu J, Zhong X, Tsai MD. 2000. *Biochemistry* 39: 16008–15.
36. Bedford E, Tabor S, Richardson CC. 1997. *Proc Natl Acad Sci U S A* 94: 479–84.

37. Davidson JF, Fox R, Harris DD, Lyons-Abbott S, Loeb LA. 2003. *Nucleic Acids Res* 31: 4702–9.
38. Williams JG, Steffens DL, Anderson JP, Urlacher TM, Lamb DT, *et al.* 2008. *Nucleic Acids Res* 36: e121.
39. Summerer D, Rudinger NZ, Detmer I, Marx A. 2005. *Angew Chem Int Ed Engl* 44: 4712–5.
40. Vichier-Guerre S, Ferris S, Auberger N, Mahiddine K, Jestin JL. 2006. *Angew Chem Int Ed Engl* 45: 6133–7
41. Gloeckner C, Sauter KB, Marx A. 2007. *Angew Chem Int Ed Engl* 46: 3115–7
42. Strerath M, Gloeckner C, Liu D, Schnur A, Marx A. 2007. *Chembiochem* 8: 395–401.
43. Leconte AM, Chen L, Romesberg FE. 2005. *J Am Chem Soc* 127: 12470–1.
44. Ong JL, Loakes D, Jaroslawski S, Too K, Holliger P. 2006. *J Mol Biol* 361: 537–50.
45. Ghadessy FJ, Ramsay N, Boudsocq F, Loakes D, Brown A, *et al.* 2004. *Nat Biotechnol* 22: 755–9.
46. d'Abbadie M, Hofreiter M, Vaisman A, Loakes D, Gasparutto D, *et al.* 2007. *Nat Biotechnol* 25: 939–43.
47. Rudinger NZ, Kranaster R, Marx A. 2007. *Chem Biol* 14: 185–94.
48. Camps M, Naukkarinen J, Johnson BP, Loeb LA. 2003. *Proc Natl Acad Sci U S A* 100: 9727–32.
49. Gilbert MT, Willerslev E. 2007. *Nat Biotechnol* 25: 872–4.
50. Patel PH, Loeb LA. 2000. *Proc Natl Acad Sci U S A* 97: 5095–100.
51. Horlacher J, Hottiger M, Podust VN, Hubscher U, Benner SA. 1995. *Proc Natl Acad Sci U S A* 92: 6329–33.
52. Pelletier H, Sawaya MR, Kumar A, Wilson SH, Kraut J. 1994. *Science* 264: 1891–903.
53. McMinn Dl, Ogawa AK, Wu Y, Lium J, Schultz P-G, Romesberg FE. 1999. *J Am Chem Soc* 121: 11585–6.

CHAPTER 8

DNA Polymerases and Diseases

8.1 Introduction

DNA replication is a highly organized process whose purpose is to transmit as faithfully as possible the genetic information from one generation to the next one. Many proteins participate to DNA replication, either directly at the polymerization step or indirectly by regulating the timing of DNA duplication and/or modifying the post-translational status of the proteins implicated (for more details see Chapters 2–5). Mutations in DNA pols or changing their expression levels can severely perturb the overall process resulting in genetic instability that can lead to diseases, such as cancer or degenerative disorders. In this Chapter we will summarize the evidences linking malfunction of DNA pols to diseases and discuss some of the most recent findings on (i) DNA pols and the maintenance of genetic stability, (ii) DNA polymerases and resistance to chemotherapy, and (iii) DNA pol γ and human diseases. Several reviews on this topic have been published in the last few years, suggesting that this issue is becoming of great importance in the near future.[1–3] **Table 8.1** presents an overview of DNA pols and diseases.

8.2 DNA Polymerases and Genetic Stability

As we have already indicated in previous chapters the main task of DNA pols is to correctly incorporate bases according to the Watson-Crick base pair rules. Misincorporations during DNA replication and in DNA repair processes can lead to alterations of the DNA sequence and consequently to altered function of proteins. Along this line Loeb's laboratory was the first to indicate the significance of multiple mutations in cancer.[4] In normal cells, most DNA damage is repaired without error leading to genetic stability. In tumor cells, on the other hand, this stability may be disturbed, leading to the accumulation of multiple mutations thus resulting in a mutator phenotype. During tumor development there is a selection for cells

261

Table 8.1   DNA polymerases and diseases[a]

DNA pol	Possible physiological functions	Knock out or mutation phenotype	Diseases related to DNA pol
DNA pol α	— initiation of Okazaki fragments (synthesis of short RNA primers in DNA synthesis) — chromosomal end replication	— unknown	— N-syndrome, a form of X chromosomal breakage and T-cell leukemia?
DNA pol β	— short and long patch BER	— embryos are not viable — cells (MEF) = hypersensitive to alkylating agents	— overexpressed in over 30% of tumors (prostate, breast, colon, ovarian, chronic myeloid leukemia) — mutations found in gastric carcinomas and prostate cancer
DNA pol δ	— DNA synthesis of the lagging strand — DNA repair (long patch BER, MMR, TLS)	— mouse pol δ exo $-/-$ = cancer susceptibility — yeast POL3 exo $-/-$ = mutator phenotype	— colon cancer, sporadic colorectal carcinomas — diverse cancers and reduced life-span if $3' \rightarrow 5'$ exonuclease deficient
DNA pol ε	— DNA synthesis of the leading strand — DNA repair (long patch BER and NER)	— Mouse pol ε exo $-/-$ = cancer susceptibility	— intestine tumors — sarcomas
TdT	— V(D)J recombination and somatic hypermutation	— unknown	— immunodeficiency, and overexpressed in several acute leukemia cells; overexpression is correlated with poor prognosis
DNA pol θ	— interstrand crosslinking repair and AP site translesion synthesis, short patch BER	— mutation in Drosophila showed sensitivity to DNA crosslinking agents, elevated frequency of	— chromosome instability with potential cancer susceptibility

(Continued)

Table 8.1   (*Continued*)

DNA pol	Possible physiological functions	Knock out or mutation phenotype	Diseases related to DNA pol
		chromosomal aberration and altered DNA metabolism	
DNA pol υ	— interstrand crosslinking repair	— unknown	— unknown
DNA pol ζ	— mitosis/meiosis (spindle checkpoint) — somatic hypermutation of IgG — TLS "mismatch extender" — mouse embryonic development	— *S. cerevisiae* −/− = sensitive to various DNA damage agents (UV-light, MMS, cisplatin and IR) — apoptosis, cell proliferation and control of cell-cycle — mice −/− dye during embryonic development	— might be involved in mammalian untargeted mutation, possibly leading to cancer
DNA pol η	— TLS insertion over TT dimers — NER — somatic hypermutation — homologous recombination — function in S-phase checkpoint	— mutation in human = xeroderma pigmentosum variant (XPV)	— skin cancer, squamous cell carcinoma
DNA pol κ	— TLS insertion and extension synthesis of various distorted lesions — NER	— overexpression causes DNA breaks and stimulates DNA exchange	— up-regulated in non-small cell lung carcinomas — reduced expression in mammary carcinoma — excess expression causes proliferation of solid tumors in immunodeficient mice
DNA pol ι	— TLS insertion	— overexpression in breast cancer leading to significant decrease in	— up- or down-regulated in different cancer cell types

(*Continued*)

Table 8.1   (*Continued*)

DNA pol	Possible physiological functions	Knock out or mutation phenotype	Diseases related to DNA pol
		DNA replication fidelity and consequently contributing to genetic instability	
REV1	— TLS insertion synthesis of AP site	— mice −/− are normal — cells −/− are sensitive to DNA damaging agents	— unknown
DNA pol λ	— DSBR (NHEJ) — long patch BER — TLS insertion and extension over 8-oxo-G — V(D)J recombination	— mice −/− = immotile cilia syndrome — cells −/− = sensitivity to hydroxyperoxide	— immotile cilia syndrome with situs inversus and hydrocephalus?
DNA pol μ	— V(D)J recombination and somatic hypermutation	— mice −/− = defect in light chain of IgG	— up-regulation might lead to non-Hodgkin's lymphoma in B lymphocytes
DNA pol σ	— sister chromatid cohesion	— cells with double mutant die and cannot complete S phase of the cell cycle	— unknown
telomerase	— chromosomal end replication		— overexpressed in 90% tumors — aging — Werner syndrome

[a] For detailed references see either reviews (Ref. 1–3) or in the text of Sections 8.2 to 8.4. DNA pol γ is not included here since it will be presented in **Table 8.3**.

containing mutations that can overcome adverse conditions that limit tumor growth. Moreover, Loeb *et al.* suggested that cancers must exhibit a mutator phenotype early during their development.[5] It was found first that thousands of random mutations in individual malignant cells exist, second that during tumor progression the selection of a mutator phenotype occurs and third that large numbers of mutations found in cancers might derive from "normal" replicative processes. The mutation frequency of cancer cells was estimated to be 210-fold higher than in normal cells.[6] In summary, one can suggest that mutation might occur via the action of "canonical" DNA pols during replication and repair as well as from the action of "specialized"

DNA pols that are either up-regulated or down-regulated. Moreover, mutations in the DNA pols might hamper their functional accuracy again leading to mutations and thus to genomic instability.

Not optimally regulated DNA polymerase levels might lead to genomic instability. In Chapter 5 we already touched the issue of the regulation of the fifteen known DNA pols. Even though very little is known so far about their global regulation a general statement is that too much of unfaithful DNA pols, e.g. from the X and Y families, might lead to genomic instability, while, on the other hand, reduced levels of faithful DNA pols can cause genomic instability as well. A nice study testing 68 different tumors measured the expression levels of a variety of DNA pols.[7] When the two X family DNA pols β and λ, the two Y family DNA pols ι and κ and the three replicative B family DNA pols α, δ and ϵ were tested, the data indicated that in more than 45% of these 68 tumors at least one of the X and/or Y DNA pols were over-expressed. Of particular interest was the fact that over 30% of all tumors tested had an over-expressed DNA pol β. DNA pols λ and ι were also over-expressed but less frequently than DNA pol β. Surprisingly DNA pol κ was under-expressed in many tumor samples. Finally, a few tumors also over-expressed the three replicative DNA pols α, δ and ϵ while in most cases they were under-expressed. All the 68 tumors were summarized in an over-expression scale that gave the following order: DNA pol β > DNA pol ι > DNA pol λ > DNA pol α > DNA pol δ > DNA pol κ > DNA pol ϵ. These data clearly indicated that the expression patterns of DNA pols in normal and cancer tissues are very complex and the expression levels of the DNA pols are changed in tumors and can vary from one type of a tumor to another. Another study suggested that up-regulation of DNA pol μ gene expression might contribute to the pathogenesis of a subset of non-Hodgkin's lymphoma and this might happen through DNA repair associated genomic instability.[8]

Individual DNA polymerases and their relationship to diseases. The involvement of DNA pol γ in diseases is outlined in Section 8.4.

DNA polymerase α. Even though no direct experimental evidences have implicated DNA pol α in human diseases so far, it is likely that even minor misregulations of its activity or response to checkpoint controls, might have a profound influence in, for example, the development of cancer. Indeed, altered catalytic properties have been reported for DNA pol α isolated from a hepatoma cell line. An early report suggested that DNA pol α is deficient in the N-syndrome, a form of X-chromosomal linked mental retardation associated with chromosomal breakage and a link with T-cell leukemia has been suggested but definitive studies were never published.[9]

DNA polymerase β. As indicated above over-expression of DNA pol β was found in about one third of all cancers tested.[7] DNA pol β −/− mouse embryos are not viable, suggesting its essential role in embryogenesis and development.[10] DNA pol β null (−/−) embryonic cells survive in culture but are severely compromised in their ability to carry out short patch BER resulting in hypersensitivity to alkylating agents.[11] Interestingly, high levels of DNA pol β have been detected at the transcriptional and protein levels in many cancer tissues, mostly solid tumors (prostate, breast, colon, ovarian) as well as in chronic myeloid leukemia.[12] Its up-regulation could contribute to enhancing chromosome instability and tumorigenesis even when over-expressed by only 2-fold in cells, suggesting that a rigorous regulation of its expression may be essential *in vivo*.[13] DNA pol β causes chromosomal instability probably either by competing with replicative DNA pols or by producing translesion synthesis over DNA lesions. Cells over-expressing DNA pol β are much more sensitive to IR treatments by increasing apoptosis and hypermutator phenotype in surviving cells. These data indicate that up regulation of DNA pol β strengthens both cell death and genetic changes associated with a malignant phenotype.[14] Also, over-expression of DNA pol β strengthens the mutagenicity of oxidative damages, concomitantly with a higher cellular sensitivity and increased apoptosis.[15] Regulation of DNA pol β over-expression could be fundamental in some cancer treatment, e.g. tumor cell resistant to cisplatin drugs (see Section. 8.3). Mutations in DNA pol β are often identified in gastric carcinomas and cell lines thereof (see Ref. 3 and references therein). Gastric cancer contains a Y265C mutation that renders DNA pol β a mutator polymerase. K289M and I260M mutations of DNA pol β might lead to colon cancer and prostate cancer, respectively.

DNA polymerase δ. Defective DNA pol δ proofreading causes cancer susceptibility in mice with a shorter life span.[16] Single point mutation inactivates the $3' \rightarrow 5'$ exonuclease of DNA pol δ and causes a mutator and cancer phenotype indicating that DNA pol δ proofreading suppresses spontaneous cancer development and suggesting that unrepaired DNA pol δ errors contribute to carcinogenesis.[17] Mutations in the catalytic subunit of DNA pol δ were reported in colon cancer,[18] in sporadic colorectal carcinoma species[19] and in Novikoff hepatoma cells.[20] These mutations likely render DNA pol δ a mutator polymerase. The structural basis for the high fidelity of DNA pol δ has recently been solved in *S. cerevisiae*.[21] The yeast structure of DNA pol δ is an excellent framework for understanding the effects of cancer causing mutations in human DNA pol δ.[21]

DNA polymerase ε. Mouse DNA pol ε and DNA pol δ proofreading can suppress discrete mutator and cancer phenotypes. Preston *et al.* found that inactivation of DNA pol ε proofreading elevates base substitution mutations and accelerates a unique spectrum of spontaneous cancers. Interestingly the types of tumors were completely different from those triggered by loss of DNA pol δ proofreading. These findings distinguish Pol ε and δ functions *in vivo* and reveal tissue-specific requirements for DNA replication fidelity.[22]

TdT. TdT is over-expressed in several acute leukemia cells, and its over-expression correlated with poor prognosis and low response to chemotherapy. It has been documented that the nucleoside analogue cordycepine in combination with the antitumor drug coformycine, inhibits TdT and displays selective toxicity against TdT$^+$ leukemic cells, suggesting an important functional role of TdT in the cancerous phenotype.[23]

DNA polymerase θ. A mutagenesis screen for chromosome instability in the mouse genome identified a mutation named *chaos1* (chromosome aberration occurring spontaneously 1) and mapped it to a region of chromosome 16 where DNA pol θ gene is located.[24] The identity of DNA pol θ and *chaos1* has been confirmed by direct disruption of DNA pol θ in the mouse and by correction of the phenotype with the DNA pol θ gene. This result provides evidence that DNA pol θ might potentially be involved in cancer susceptibility.

DNA polymerase υ. No experimental evidences of a direct involvement of DNA pol υ in human diseases have been reported so far.

DNA polymerase ζ. In mouse T-lymphoma cells, stress response induced by DNA damage agents (8-methoxy-psoralen or UV-A) leads to specific, delayed and untargeted mutations.[25] It has been found that a low concentration of N-methyl-N′-nitro-N-nitrosoguanidine (MNNG), a carcinogen which can induce gastric cancer, could induce mammalian untargeted mutations.[26] However, it is not clear which factor capable of inhibiting fidelity can be induced or activated. More recently, it was found that DNA pol ζ might be involved in the mammalian untargeted mutations induced by MNNG. The transcriptional level of REV3 gene, coding for the catalytic subunit of DNA pol ζ, is up-regulated when human cells are treated by low concentration MNNG.[27] Finally, human cells, in which the function of untargeted mutations is inhibited by antisense REV3 RNA, display characteristics of both untargeted mutations and targeted mutations.[28] The fact that DNA pol ζ is an essential gene and that it is the main TLS extender DNA pol makes it an attractive target for chemotherapy (see Chapter 9).

DNA polymerase η. DNA pol η is the prototype of a DNA pol which is correlated without any doubt to a disease. Xeroderma Pigmentosum Variant (XPV) cells were taken to isolate a DNA pol and this DNA pol was identified as a mutant form of DNA pol η.[29] Humans with mutations in the DNA pol η gene, suffer from a disease characterised, among other phenotypes, by an increased susceptibility to sunlight-induced skin cancer.[30] DNA pol η is the primary TLS DNA pol responsible in many organisms for error-free bypass of cis-syn cyclobutane dimers (CPDs) a major lesion resulting from UV irradiation. In fact CPDs block DNA synthesis by DNA pols like DNA pols α, δ, ϵ, and λ, but DNA pol η has been shown to bypass them with high accuracy and efficiency *in vitro*.[31] Consequently, mutations or deletions in DNA pol η result in a reduced efficiency to copy DNA containing CPDs.[32] Furthermore, DNA pol η independent CPD bypass, which likely involves other TLS DNA pols such as DNA pol ζ, ι and κ is more mutagenic, presumably leading to an increased frequency of cancer in XP-V patients.[33] DNA pol η in addition to CPDs has the capacity to bypass *in vitro* a number of other DNA lesions such as 8-oxo-G,[34] acetylaminofluorene-adducted guanine,[35] and adducts formed by cisplatin and oxaliplatin[36] (see also Section 8.3: DNA polymerases and resistance to chemotherapy). Moreover, among all DNA pols, DNA pol η is the one with the lowest fidelity on undamaged DNA *in vitro*.[37]

Analysis of DNA from lymphocytes of XP-V patients showed that DNA pol η generates hypermutation in the μ and γ switch regions of immunoglobulin genes.[38] In addition, it was shown *in vitro* that human DNA pol η and the two other Y-family enzymes DNA pols κ and ι, both of which are dispensable for somatic hypermutation, possess reverse transcriptase activity. It is thus possible that DNA pol η may act as both a RNA-dependent and a DNA-dependent pol in somatic hypermutation.[39]

DNA polymerase κ. There is evidence that DNA pol κ is also involved in tumorigenesis. It was shown that ectopic expression of DNA pol κ promoted DNA strand breaks, aneuploidy as well as tumorigenesis in nude mice.[40] Additionally, of eight non-small cell lung carcinoma biopsies overexpressing DNA pol κ, seven displayed losses of heterozygosity compared with adjacent non-tumoral tissues. Taken together, these data suggested that mis-regulation of DNA pol κ can promote the emergence of a large spectrum of genetic disorders associated with a malignant phenotype. However, conflicting results have been published about DNA pol κ, as much as was the case with DNA pol ι. A reduced expression of DNA pol κ in rat mammary carcinoma cell lines and primary mammary carcinomas was found in comparison to that of the normal tissues.[41] In addition, it was reported that DNA pol κ is up-regulated by p53 in human, as well as in murine cells.[42] The functional loss of p53 by mutation resulted in the up-regulation of DNA pol κ in human lung cancer tissues.[43] However, results from other investigators indicated that mouse but not human DNA pol κ was up-regulated in response to exposure to various DNA-damaging agents in a p53 dependent manner.[44] In sum, the available body of information suggests a complex regulation pattern of DNA pol κ.

DNA polymerase ι. Growing evidence points to the involvement of DNA pol ι in lung cancer progression. In one study, a systematic candidate gene analysis of the pulmonary adenoma resistance 2 locus was performed. Differential gene expression in lung tissues and nucleotide polymorphisms analysis revealed that the gene encoding DNA pol ι contained 25 nucleotide polymorphisms in its coding region between A/J and BALB/cJ mice, resulting in a total of ten amino acid changes. Purified BALB/cJ and A/J DNA pol ι were shown to differ in substrate discrimination *in vitro*. Moreover, altered expression of DNA pol ι and an amino acid–changing nucleotide polymorphism were observed in human lung cancer cells, suggesting a possible role in the development of lung cancer.[45] In another study, a single nucleotide polymorphism in DNA pol ι (Thr706Ala), which correlated with a significantly higher risk of lung adenocarcinoma and squamous cell carcinoma, was reported.[46] A third study provided a potential link between altered DNA pol ι expression and mutagenesis in breast cancer. DNA pol ι expression was elevated in breast cancer cells and this correlated with a significant decrease in DNA replication fidelity.[47] Interestingly, after immunodepletion of DNA pol ι from nuclear extracts of these cells, a reduction in mutation frequency was measured *in vitro*. These data suggested that DNA pol ι might play a role in the high mutation frequencies observed in breast cancer cells. Another study, however, questioned the existence of a link between overexpression and DNA pol ι and cancer. The transcripts of DNA pols κ, η, ι, and ζ were measured in 131 self-paired cancerous and non-tumor samples, including 23 lung cancers, 49 stomach cancers, and 59 colorectal cancers.[41] These results indicated that, except for DNA pol η in colorectal cancers, all other DNA pols tested were significantly down-regulated in human lung, stomach, and colorectal cancers. However, this study did not rule out the possibility that the variation in the expression levels might be a function of the tumor stage or is cell-type specific.

Rev1. Rev1-deficient chicken DT40 cells display reduced viability and are sensitive to a wide range of DNA-damaging agents.[48] "Knock down" of REV1 mRNA in human cells results in a hypomutable phenotype after UV treatment.[49] In addition to its catalytic domain, Rev1 possesses a so-called BRCA1 C-terminal (BRCT) domain. Mice containing a targeted deletion of this domain are healthy, fertile and display normal somatic hypermutation. *Rev1BRCT1* −/− cells display an elevated spontaneous frequency of intragenic deletions at HPRT locus. In addition, these cells were sensitized to exogenous DNA damages. UV-C light induced a delayed progression through late S and G2 phases of the cell cycle and many chromatid aberrations, specifically in a subset of mutant cells, but not enhanced sister chromatid exchanges. UV-C-induced mutagenesis was reduced and mutations at TT dimers were absent in *Rev1BRCT1* −/− cells, the opposite phenotype of UV-C-exposed cells from XP-V patients, lacking DNA pol η. This suggested that the enhanced UV-induced mutagenesis in XP-V patients might depend on error-prone Rev1-dependent TLS. These data indicate a regulatory role of the Rev1 BRCT domain in TLS of a limited spectrum of endogenous and exogenous nucleotide damages during a defined phase of the cell cycle[50] (see also Chapter 5: DNA polymerase switches due to PCNA ubquitination).

DNA polymerase λ. Upon treatment of mammalian cells with DNA damaging agents (UV, γ-irradiation and H_2O_2) mRNA of DNA pol λ is down-regulated[51] and mouse embryonic fibroblasts (MEF) depleted of DNA pol λ are sensitive to H_2O_2 but not to alkylating agents.[52] Knock out mouse DNA pol λ cells were obtained from two groups.[53] Kobayashi *et al.* showed no differences between MEF DNA pol $\lambda^{-/-}$ and DNA pol $\lambda^{+/+}$ cells to various DNA damaging agents. They were also successful in obtaining DNA pol $\lambda^{-/-}$ mice. The surviving males, but not the females were sterile and this as a result of spermatozoal immobility. The main phenotypes were hydrocephalus, situs inversus, and chronic sinusitis resulting in *the immotile cilia syndrome*. There is, however, suspicion regarding this strong phenotype of DNA pol $\lambda^{-/-}$ mice, and the current belief is that Kobayashi *et al.* might have knocked out another closer gene responsible for *the immotile cilia syndrome* (L. Blanco, personal communication). The group of Reynaud obtained knock out DNA pol $\lambda^{-/-}$ mice in which the males were fertile, and homozygous breeding has been performed up to the third generation without a noticeable problem,[54] suggesting that this enzyme is dispensable for mouse development. When DNA pol λ levels were tested in lung epithelium of lung cancer patients it was found that DNA pol λ expression in the bronchiolar epithelia was significantly correlated with the amount of habitual smoking in the respective individuals. On the other hand, DNA pol λ expression in cancer tissues themselves did not correlate with the smoking of the patients.[55]

DNA polymerase μ. The close association of DNA pol μ with cells of the germinal centres, its error-prone nature and its strong expression by postgerminal centres in non-Hodgkin's lymphoma B cells suggested that DNA pol μ is a candidate to be involved in this cancer. Moreover, a close association was established between DNA pol μ expression and B cells non-Hodgkin's lymphoma.[56]

DNA polymerase σ. No experimental evidences of a direct involvement of DNA pol σ in human diseases have been reported so far.

Single nucleotide polymorphism in DNA polymerases. Single nucleotide polymorphisms (SNPs) have been identified in most DNA pols (**Table 8.2**). So far over 120 SNPs were found in human populations in the different DNA pols and they were predicted to result in non-synonymous amino acid substitutions of DNA pols (reviewed in Ref. 3 and see also NCBI data base dbSNP). Very little is known about the functional consequences of these variations. These are examples where a deleterious effect led to haploinsufficiency. This was exemplified for DNA pol $\beta^{+/-}$ mice, which are tumor prone.[57] Furthermore DNA pol SNP variants might

Table 8.2   Human DNA polymerases, genes, chromosomal localizations and single nucleotides polymorphisms[a]

DNA pol	Gene of the catalytic subunit	Chromosomal localization[b]	SNPs[c]	Frequency (%)[d]
DNA pol α	POLA1	Xp22.1–21.3	0	0
DNA pol β	POLB	8p11.2	3	0.6–2.1
DNA pol γ	POLG1	15q25	9	3–5
DNA pol δ	POLD1	19q13.3	8	0.6–37
DNA pol ε	POLE	12q4.3	11	1–13
DNA pol θ	POLQ (chaos1)	3q13.33	14	0.6–48
DNA pol υ	POLN	4q16.3	8	2.3–50
DNA pol ζ	REV3L	6q21	25	0.6–50
DNA pol η	POLH	6q21.1	6	0.6–8.4
DNA pol κ	POLK (hDinB)	5q13	6	0.6–4.2
DNA pol ι	POLI	18q21.1	8	0.6–38
REV1	REV1L	2q11.1–11.2	16	0.6–40
DNA pol λ	POLL	10q23	2	2–14
DNA pol μ	POLM	7p13	5	5–10
DNA pol σ	POLS (Trf4)	5q15	2	6

a: Data from Ref. 3 where more details concerning SNPs are documented. b: Taken form the NIEHS Environmental Genome Project. c: Number of non-synonymous codon changes within the sequence of the POL gene. d: Allele frequencies in individuals.

also affect DNA pol gene expression[58] and as a consequence the deleterious effect could be the amount of a particular DNA pol, which is up- and/or down-regulated as a consequence of SNP variants (see also above and in Chapter 5).

8.3 DNA Polymerases and Resistance to Chemotherapy

Drug resistance is a major obstacle for the successful chemotherapy treatment of cancer. Although high response rates are often initially observed in patients, resistance frequently occurs, rendering subsequent therapy largely ineffective. An example of such a problem is the cellular resistance induced by the antitumor drug cisplatin, a platinum compound which is one of the most effective and broadly used anticancer drug so far, particularly useful for treatment of testicular cancer.[59] Cisplatin is believed to exert its cytotoxic effects by interacting with DNA where it inhibits both replication and transcription and induces programmed cell death.[60]

Cell lines exhibit different mechanisms that account for their acquired resistance to cisplatin. These mechanisms include (a) decreased platinum accumulation; (b) elevated levels of proteins such as glutathione transferase or metallothionein, which can sequester cisplatin before it reaches its pharmacological target;

(c) enhanced repair capacity to remove Pt-DNA lesions; (d) alteration in the type of Pt-DNA lesions. An additional mechanism of resistance is the consequence of an increased capacity of the cell to tolerate platinum-DNA lesions. This has been shown to be the primary mechanism that caused decreased cisplatin sensitivity in a series of cell lines.[61] The mechanisms of replicative bypass of cisplatin-DNA lesions that take place *in vivo* are still not completely understood but a number of eukaryotic DNA pols have been shown capable of mutagenic bypass *in vitro* of the major cisplatin-DNA adduct, the Pt-(GpG) intrastrand cross-link. Among these DNA pols are the X family DNA pol β[62,63] and DNA pol μ[64] and the Y family DNA pol η.[36] DNA pol β could be a particularly good candidate for participating to the *in vivo* TLS of cisplatin lesions because this DNA pol has been found to be over expressed in cisplatin resistant cell lines.[65] Moreover, it has been shown that the frequency of mutations induced by cisplatin increased significantly in CHO cells over expressing DNA pol β compared to control cells displaying a normal level of enzyme and the use of cell extracts supported the hypothesis that error-prone translesion replication was one of the key determinants of tolerance phenotype.[66] Therefore it can reasonably be postulated that over expression of DNA pol β contributes to replicative bypass of cisplatin-DNA lesions *in vivo*. Of course, in addition to DNA pol β, other DNA pols could participate, together with other proteins implicated in DNA replication, in TLS processes leading to cisplatin resistance.

Survival through S phase as a result of replicative bypass should give the cell an additional period of time to repair DNA adducts in arrested G2 phase, suggesting a link between TLS and resistance as a consequence of augmented capacity of the cell to repair DNA lesions. An additional link between replicative bypass and cisplatin resistance may be enforced by the finding that over expression of several proto-oncogenes is correlated with cisplatin resistance following drug exposure.[67] Although the molecular mechanisms that could relate proto-oncogene induction to cisplatin resistance have not yet been explored, *in vivo* mutagenic replication of a single-stranded vector bearing the major cisplatin lesion Pt-(GdG) placed on codon 13 within the human H-ras has been reported.[68] Interestingly, the mutations observed with the highest frequency in this system result in the amino acid substitution that is known to be key step in the activation of the H-ras proto-oncogene.

In conclusion, as far as resistance to cisplatin is concerned, an essential strategy would be to inhibit DNA pols involved in TLS in order to increase the replication inhibiting capacity of the drug. In the case of DNA pol β it should be noted that this enzyme is implicated in DNA repair so that inefficient DNA repair synthesis could strengthen the cytotoxicity of potentially lethal repair intermediates by, for instance, increasing the chances of formation of lethal double strand breaks.

8.4 DNA Polymerase γ and Human Diseases

Mitochondrial DNA polymerase. Mitochondrial DNA is replicated by the DNA pol γ encoded by the nuclear gene POLG. DNA pol γ is a heterotrimer composed of a 140-KDa catalytic subunit and two 55-kDa accessory subunits that form themselves a dimer.[69] Very recently the crystal structures of the human heterotrimeric DNA pol γ holoenzyme and of a variant of its processivity factor have been determined. The holoenzyme structure reveals an unexpected assembly of the mitochondrial DNA replicase where the catalytic subunit interacts with its processivity factor primarily via a domain absent in all other DNA pols. This domain provides a structural model for supporting both the intrinsic processivity of the catalytic subunit alone and the enhanced processivity of the holoenzyme. The DNA pol γ structure also provides a context for interpreting the phenotypes of diseases related mutations in the polymerase.[70,71] The catalytic subunit is synthesized as a precursor polypeptide with a 25 amino-terminal leader sequence that functions to target the polypeptide to the mitochondria. In addition to DNA pol γ, the Twinkle DNA helicase and the mitochondrial single-stranded DNA binding protein (mtSSB) are required for replication and repair of the mitochondrial genome. The catalytic subunit of DNA pol γ contains DNA pol, $3' \rightarrow 5'$ proofreading exonuclease and deoxyribosephosphate (dRP) lyase activities. The domain of the 140-kDa subunit comprises an exonuclease domain and a DNA pol domain separated by a linker region. The accessory subunits facilitate tighter dNTPs and double-stranded DNA binding and are required for highly processive DNA synthesis. DNA pol γ accurately replicates DNA *in vitro*, with a measured error frequency of 2 to 4×10^{-6} per nucleotide. For reviews on the role of DNA pol γ in mitochondrial DNA replication and repair see.[72,73]

Several human mitochondrial diseases have been associated with mutations in DNA pol γ and the most important ones are listed in **Table 8.3** and briefly discussed below. A comprehensive review of DNA pol γ forms harboring disease alterations has recently been published.[74]

Table 8.3   DNA polymerase γ and human diseases[a]

Physiological functions	Diseases related to DNA pol γ
Mitochondrial DNA replication	Progressive external ophthalmoplegia (PEO)
Mitochondrial DNA repair	Alpers syndrome
	Premature aging (?)
	Parkinson disease
	Ataxia-Neuropathy

[a] see text for details and references.

Progressive External Ophthalmoplegia (PEO). PEO is a mitochondrial disorder associated with mtDNA mutations and deletions and is usually transmitted in an autosomal dominant trait. PEO is characterized by progressive weakening of external eye muscle leading to blepharoptosis and ophtalmoparesis. The disease is often accompanied by cataract and hearing loss and may result in development of neuromuscular problems. PEO has been linked to four different chromosomal loci, including genes encoding for the mitochondrial DNA helicase Twinkle,[75] a gene that encodes adenine nucleotide translocator (ANTI),[76] the catalytic subunit of the mitochondrial DNA pol,[77] and, more recently, in the gene encoding the p55 accessory subunits.[78] All the dominant Pol mutations leading to PEO have been mapped to the polymerase domain[79] and four amino acids substitutions, G923D, R943H, Y955C and A957S, were analyzed biochemically.[80] All these substitutions showed low catalytic activity, ranging from 30% to 1% of the wt DNA pol and, for two of them, namely R943H and Y955C, also decreased processivity. The clinical severity of PEO correlated with the extent of reduction of DNA synthesis of the mutations is examined. The single mutation G451E found in the p55 subunit disrupts interaction between the accessory and the catalytic subunit, ultimately causing deletions in the mtDNA of this patient.[78]

Alpers syndrome. Alpers syndrome is a rare but severe, heritable autosomal recessive disease that affects young children. The disease may manifest within the first year of life and the patient show progressive cerebral degeneration leading to mental deterioration, cortical blindness and deafness eventually leading to death. To date 45 different point mutations in DNA pol γ have been associated with Alpers syndrome, but only two, A467T and W748S have been characterized so far. The A467T mutant enzyme possessed only 4% of the wt polymerase activity and failed to interact with the accessory subunit p55.[81] Similarly, W748S protein exhibited low DNA pol activity, low processivity and severe DNA binding defect.[82] Very recently sequence analysis of the C-terminal polymerase region of DNA pol γ revealed a cluster of four Alpers mutations at highly conserved residues in the thumb subdomain and two in the adjacent palm subdomain.[83] Biochemical characterization of purified recombinant mutant proteins revealed that, although retaining their binding capacity with the accessory subunit p55, the proteins mutated in the thumb domain had only 1% of the wt DNA pol activity while proteins mutated in the palm subdomain retained 50–70% of wt polymerase activity.

Premature aging. Two groups have reported that mice harboring genetic defects in the proofreading exonuclease activity of the mitochondrial DNA pol γ (PolG$^{mut/mut}$) displayed both increased mitochondrial DNA mutations and multiple symptoms of premature aging, providing a causative link between mtDNA mutations and ageing phenotypes in mammals.[84] However, recent work from another group, using an alternative method to measure mutations, came to the conclusion that mitochondrial single-base mutations do not limit the lifespan of mice.[85] By contrast, the frequency of deletion mutations has been found to correlate with the aging phenotype.[86]

However, these data still appear to be controversial, as documented by two recent reports. First, it has been reported that the absolute level of deletions in mutator mice is quite low, especially when compared with the level of point mutations in the same mice.[87] The authors argued that the available data are insufficient to conclude that mtDNA mutations drive premature aging in mtDNA mutator mice. Second, molecular analyses were carried out to determine the mechanism whereby the mtDNA mutations impair respiratory chain function.[88] These authors found that mitochondrial protein synthesis is unimpaired in mtDNA mutator mice, consistent with the observed minor alterations of steady-state levels of mitochondrial transcripts. Their findings consequently refute arguments that circular mtDNA molecules with large deletions are driving the premature aging phenotype. Clearly the question concerning a causative role for accumulating mtDNA mutations in aging needs further investigation.

Parkinson disease and Ataxia-Neuropathy. Parkinson disease is a frequent neurodegenerative disease and in 2004 a significant co-segregation of Parkinsonism with mutations in the POLG gene was described.[89] Mutations in POLG are associated with a form of ataxia-neuropathy called mitochondrial associated ataxia syndrome (MIRAS).[90] Symptoms of Ataxia involving mutations in POLG include peripheral neuropathy, involuntary movements and epileptic seizures. One mutation, the A467T, is located in the linker region of the catalytic subunit of DNA pol γ and has been also found in PEO and Alpers syndromes.[91,92] Biochemical analysis indicated that the A467T mutant had strongly reduced DNA pol activity but possessed almost normal proofreading exonuclease activity and failed to interact with the p55 subunit.[81]

8.5   Chapter Summary

In this chapter, we highlighted the possible connections between DNA pols, the main enzymes in the DNA metabolism, and diseases. In addition, we also attempted to critically evaluate those cases where the experimental data are not fully convincing.

In the last ten years, numerous novel DNA pols have been revealed, but their exact cellular functions still await clarification. **Table 8.1** summarizes the known eukaryotic DNA pols and their relationships with diseases. Since DNA pols are the working horses that faithfully copy the DNA in all DNA transactions such as DNA replication, DNA repair, DNA recombination and TLS, one is not surprised that slight changes in their behavior might lead to genomic instability and therefore to cancer. The literature suggests that mutations might occur via the action of "canonical" DNA pols during replication and repair as well as from "specialized" DNA pols, of which there is too much (up-regulated) or not enough (down-regulated) in the cell. Moreover, mutations in the DNA pols itself might also hamper their functional accuracy again leading to mutations and thus to genomic instability. So far over 120 single nucleotide polymorphisms (SNPs) were identified in human populations of different DNA pols and they were predicted to result in non-synonymous amino acid substitutions within their sequences, but very little is known about the functional consequences of these variations. Since DNA pols are targets for chemotherapy (for details see Chapter 9), it is important to discuss their potential role in chemotherapy resistance. DNA pol γ is the prototype of a DNA pol that has been directly linked to a variety of diseases other than cancer, such as Progressive External Ophthalmoplegia, Alpers syndrome, premature aging, Parkinson disease and Ataxia-Neuropathy.

Genetic or even epigenetic factors involved in misregulation of the DNA pols could represent novel prognostic, diagnostic and therapeutic tools in medicine. Understanding the molecular details of the processes linking DNA pols properties to diseases will be a big challenge for the future.

References

1. Loeb LA, Monnat RJ, Jr. 2008. *Nat Rev Genet* 9: 594–604.
2. Ramadan K, Maga G, Hübscher U. 2007. *DNA polymerases and Diseases.* Heidelberg Germany: Springer Verlag. pp. 69–102.
3. Sweasy JB, Lauper JM, Eckert KA. 2006. *Radiat Res* 166: 693–714.
4. Loeb KR, Loeb LA. 2000. *Carcinogenesis* 21: 379–85.
5. Loeb LA, Loeb KR, Anderson JP. 2003. *Proc Natl Acad Sci U S A* 100: 776–81.
6. Bielas JH, Loeb KR, Rubin BP, True LD, Loeb LA. 2006. *Proc Natl Acad Sci U S A* 103: 18238–42.
7. Albertella MR, Lau A, O'Connor M J. 2005. *DNA Repair (Amst)* 4: 583–93.
8. Chiu A, Pan L, Li Z, Ely S, Chadburn A, *et al.* 2002. *Am J Pathol* 161: 1349–55.
9. Floy KM, Hess RO, Meisner LF. 1990. *Am J Med Genet* 35: 301–5.
10. Gu H, Marth JD, Orban PC, Mossmann H, Rajewsky K. 1994. *Science* 265: 103–6.
11. Sobol RW, Horton JK, Kuhn R, Gu H, Singhal RK, *et al.* 1996. *Nature* 379: 183–6.
12. Louat T, Servant L, Rols MP, Bieth A, Teissie J, *et al.* 2001. *Mol Pharmacol* 60: 553–8.
13. Bergoglio V, Frechet M, Philippe M, Bieth A, Mercier P, *et al.* 2004. *FEBS Lett* 566: 147–50.
14. Frechet M, Canitrot Y, Bieth A, Dogliotti E, Cazaux C, Hoffmann JS. 2002. *Oncogene* 21: 2320–7.
15. Frechet M, Canitrot Y, Cazaux C, Hoffmann JS. 2001. *FEBS Lett* 505: 229–32.
16. Goldsby RE, Lawrence NA, Hays LE, Olmsted EA, Chen X, *et al.* 2001. *Nat Med* 7: 638–9.
17. Goldsby RE, Hays LE, Chen X, Olmsted EA, Slayton WB, *et al.* 2002. *Proc Natl Acad Sci U S A* 99: 15560–5.

18. da Costa LT, Liu B, el-Deiry W, Hamilton SR, Kinzler KW, *et al.* 1995. *Nat Genet* 9: 10–1.
19. Flohr T, Dai JC, Buttner J, Popanda O, Hagmuller E, Thielmann HW. 1999. *Int J Cancer* 80: 919–29.
20. Popanda O, Flohr T, Fox G, Thielmann HW. 1999. *J Cancer Res Clin Oncol* 125: 598–608.
21. Swan MK, Johnson RE, Prakash L, Prakash S, Aggarwal AK. 2009. *Nat Struct Mol Biol* 16: 979–86.
22. Albertson TM, Ogawa M, Bugni JM, Hays LE, Chen Y, *et al.* 2009. *Proc Natl Acad Sci U S A* 106: 17101–4.
23. Kodama EN, McCaffrey RP, Yusa K, Mitsuya H. 2000. *Biochem Pharmacol* 59: 273–81.
24. Shima N, Hartford SA, Duffy T, Wilson LA, Schimenti KJ, Schimenti JC. 2003. *Genetics* 163: 1031–40.
25. Boesen JJ, Stuivenberg S, Thyssens CH, Panneman H, Darroudi F, *et al.* 1992. *Mol Gen Genet* 234: 217–27.
26. Zhang X, Yu Y, Chen X. 1994. *Mutat Res* 323: 105–12.
27. Zhu F, Jin CX, Song T, Yang J, Guo L, Yu YN. 2003. *World J Gastroenterol* 9: 888–93.
28. Zhu F, Zhang M. 2003. *World J Gastroenterol* 9: 1165–9.
29. Masutani C, Kusumoto R, Yamada A, Dohmae N, Yokoi M, *et al.* 1999. *Nature* 399: 700–4.
30. Kunkel TA, Pavlov YI, Bebenek K, Rogozin IB, Matsuda T. 2003. *DNA Repair (Amst)* 2: 135–49.
31. Johnson RE, Prakash S, Prakash L. 1999. *Science* 283: 1001–4.
32. Johnson RE, Kondratick CM, Prakash S, Prakash L. 1999. *Science* 285: 263–5.
33. Wang Y, Woodgate R, McManus TP, Mead S, McCormick JJ, Maher VM. 2007. *Cancer Res* 67: 3018–26.
34. Maga G, Villani G, Crespan E, Wimmer U, Ferrari E, *et al.* 2007. *Nature* 447: 606–8.
35. Yuan Fea. 2000. *J Biol Chem* 275: 8233–9.
36. Vaisman A, Masutani C, Hanaoka F, Chaney SG. 2000. *Biochemistry* 39: 4575–80.
37. Prakash S, Johnson RE, Prakash L. 2005. *Annu Rev Biochem* 74: 317–53.
38. Zeng X, Negrete GA, Kasmer C, Yang WW, Gearhart PJ. 2004. *J Exp Med* 199: 917–24. Epub 2004 Mar 29.
39. Franklin A, Milburn PJ, Blanden RV, Steele EJ. 2004. *Immunol Cell Biol* 82: 219–25.
40. Bavoux C, Leopoldino AM, Bergoglio V, J OW, Ogi T, *et al.* 2005. *Cancer Res* 65: 325–30.
41. Pan Q, Fang Y, Xu Y, Zhang K, Hu X. 2005. *Cancer Lett* 217: 139–47.
42. Wang Y, Seimiya M, Kawamura K, Yu L, Ogi T, *et al.* 2004. *Int J Oncol* 25: 161–5.
43. O-Wang J, Kawamura K, Tada Y, Ohmori H, Kimura H, *et al.* 2001. *Cancer Res* 61: 5366–9.
44. Velasco-Miguel S, Richardson JA, Gerlach VL, Lai WC, Gao T, *et al.* 2003. *DNA Repair (Amst)* 2: 91–106.
45. Wang M, Devereux TR, Vikis HG, McCulloch SD, Holliday W, *et al.* 2004. *Cancer Res* 64: 1924–31.
46. Lee GH, Nishimori H, Sasaki Y, Matsushita H, Kitagawa T, Tokino T. 2003. *Oncogene* 22: 2374–82.
47. Yang J, Chen Z, Liu Y, Hickey RJ, Malkas LH. 2004. *Cancer Res* 64: 5597–607.
48. Simpson LJ, Sale JE. 2003. *EMBO J* 22: 1654–64.
49. Clark DR, Zacharias W, Panaitescu L, McGregor WG. 2003. *Nucleic Acids Res* 31: 4981–8.
50. Jansen JG, Tsaalbi-Shtylik A, Langerak P, Calleja F, Meijers CM, *et al.* 2005. *Nucleic Acids Res* 33: 356–65. Print 2005
51. Aoufouchi S, Flatter E, Dahan A, Faili A, Bertocci B, *et al.* 2000. *Nucleic Acids Res* 28: 3684–93.
52. Braithwaite EK, Kedar PS, Lan L, Polosina YY, Asagoshi K, *et al.* 2005. *J Biol Chem* 280: 31641–7.
53. Kobayashi Y, Watanabe M, Okada Y, Sawa H, Takai H, *et al.* 2002. *Mol Cell Biol* 22: 2769–76.

54. Bertocci B, De Smet A, Flatter E, Dahan A, Bories JC, *et al.* 2002. *J Immunol* 168: 3702–6.
55. Ohba T, Kometani T, Shoji F, Yano T, Ichiro Y, *et al.* 2009. *Mutat Res* 677: 66–71.
56. Chiu A, Pan L, Li Z, Ely S, Chadburn A, *et al.* 2002. *Am J Pathol* 161: 1349–55.
57. Cabelof DC, Ikeno Y, Nyska A, Busuttil RA, Anyangwe N, *et al.* 2006. *Cancer Res* 66: 7460–5.
58. Pastinen T, Ge B, Hudson TJ. 2006. *Hum Mol Genet* 15 Spec No 1: R9–16.
59. Wang D, Lippard SJ. 2005. *Nat Rev Drug Discov* 4: 307–20.
60. Kelland L. 2007. *Nat Rev Cancer* 7: 573–84.
61. Johnson SW, Laub PB, Beesley JS, Ozols RF, Hamilton TC. 1997. *Cancer Res* 57: 850–6.
62. Hoffmann JS, Pillaire MJ, Garcia-Estefania D, Lapalu S, Villani G. 1996. *J Biol Chem* 271: 15386–92.
63. Hoffmann JS, Pillaire MJ, Maga G, Podust V, Hubscher U, Villani G. 1995. *Proc Natl Acad Sci U S A* 92: 5356–60.
64. Havener JM, Nick McElhinny SA, Bassett E, Gauger M, Ramsden DA, Chaney SG. 2003. *Biochemistry* 42: 1777–88.
65. Ali-Osman F, Berger MS, Rairkar A, Stein DE. 1994. *J Cell Biochem* 54: 11–9.
66. Canitrot Y, Cazaux C, Frechet M, Bouayadi K, Lesca C, *et al.* 1998. *Proc Natl Acad Sci U S A* 95: 12586–90.
67. Scanlon KJ, Kashani-Sabet M, Tone T, Funato T. 1991. *Pharmacol Ther* 52: 385–406.
68. Pillaire MJ, Margot A, Villani G, Sarasin A, Defais M, Gentil A. 1994. *Nucleic Acids Res* 22: 2519–24.
69. Yakubovskaya E, Chen Z, Carrodeguas JA, Kisker C, Bogenhagen DF. 2006. *J Biol Chem* 281: 374–82.
70. Falkenberg M, Larsson NG. 2009. *Cell* 139: 231–3.
71. Lee YS, Kennedy WD, Yin YW. 2009. *Cell* 139: 312–24.
72. Kaguni LS. 2004. *Annu Rev Biochem* 73: 293–320.
73. Graziewicz MA, Longley MJ, Copeland WC. 2006. *Chem Rev* 106: 383–405.
74. Chan SS, Copeland WC. 2009. *Biochim Biophys Acta* 1787: 312–9.
75. Spelbrink JN, Li FY, Tiranti V, Nikali K, Yuan QP, *et al.* 2001. *Nat Genet* 28: 223–31.
76. Kaukonen J, Juselius JK, Tiranti V, Kyttala A, Zeviani M, *et al.* 2000. *Science* 289: 782–5.
77. Van Goethem G, Dermaut B, Lofgren A, Martin JJ, Van Broeckhoven C. 2001. *Nat Genet* 28: 211–2.
78. Longley MJ, Clark S, Yu Wai Man C, Hudson G, Durham SE, *et al.* 2006. *Am J Hum Genet* 78: 1026–34.
79. Longley MJ, Graziewicz MA, Bienstock RJ, Copeland WC. 2005. *Gene* 354: 125–31.
80. Graziewicz MA, Longley MJ, Bienstock RJ, Zeviani M, Copeland WC. 2004. *Nat Struct Mol Biol* 11: 770–6. Epub 2004 Jul 18
81. Chan SS, Longley MJ, Copeland WC. 2005. *J Biol Chem* 280: 31341–6.
82. Chan SS, Longley MJ, Copeland WC. 2006. *Hum Mol Genet* 15: 3473–83.
83. Kasiviswanathan R, Longley MJ, Chan SS, Copeland WC. 2009. *J Biol Chem* 284: 19501–10.
84. Trifunovic A, Wredenberg A, Falkenberg M, Spelbrink JN, Rovio AT, *et al.* 2004. *Nature* 429: 417–23.
85. Vermulst M, Bielas JH, Kujoth GC, Ladiges WC, Rabinovitch PS, *et al.* 2007. *Nat Genet* 39: 540–3.
86. Vermulst M, Wanagat J, Kujoth GC, Bielas JH, Rabinovitch PS, *et al.* 2008. *Nat Genet* 40: 392–4.
87. Kraytsberg Y, Simon DK, Turnbull DM, Khrapko K. 2009. *Aging Cell* 8: 502–6.
88. Edgar D, Shabalina I, Camara Y, Wredenberg A, Calvaruso MA, *et al.* 2009. *Cell Metab* 10: 131–8.
89. Luoma P, Melberg A, Rinne JO, Kaukonen JA, Nupponen NN, *et al.* 2004. *Lancet* 364: 875–82.

90. Hakonen AH, Heiskanen S, Juvonen V, Lappalainen I, Luoma PT, *et al.* 2005. *Am J Hum Genet* 77: 430–41.
91. Naviaux RK, Nguyen KV. 2004. *Ann Neurol* 55: 706–12.
92. Van Goethem G, Martin JJ, Dermaut B, Lofgren A, Wibail A, *et al.* 2003. *Neuromuscul Disord* 13: 133–42.

CHAPTER 9

DNA Polymerases and Chemotherapy

9.1 DNA Polymerases Are Important Chemotherapeutic Targets

Uncontrollable DNA replication certainly contributes to pathological states like cancer, autoimmune diseases, and many viral and bacterial infections. Therefore the DNA pols of human/animal cells, viruses and bacteria are potential targets for chemotherapy of these pathological states in humans and animals.

Antiviral compounds targeting DNA pols and reverse transcriptases have assumed unexpected importance only a few decades ago since before it appeared unconceivable to specifically interfere with viral replication without affecting normal cell metabolism.

Inhibitors have been developed that successfully target the DNA pols of human herpesviruses and the Hepatitis B virus, as well as the reverse transcriptase (RT) of HIV-1.

Albeit the plethora of different DNA pols existing in human cells and the fact that their differential expression in normal versus tumoral tissues offers a great deal of opportunities to develop novel agents, no anticancer drugs targeting human DNA pols have been approved so far. For this reason, in this chapter we will focus only on the antiviral compounds. In particular, our attention will be devoted to those inhibitors that have already yielded approved drugs for clinical use or that, for their mechanism of action, have represented or might represent prototype molecules for the development of novel useful drugs. In doing so, we will give particular emphasis to the rationale behind their design, the problems encountered in their administration, including metabolism and drug-resistance issues, and the possible approaches to overcome these problems.

9.2 Strategies and Problems for the Design of Inhibitors of DNA Polymerases

9.2.1 *Substrate Analogs*

The most straightforward approach to inhibit an enzymatic activity is to develop substrate analogs. These are molecules which retain chemical and structural features of the natural substrates that are important for their interaction with the enzyme's active site, but lack one or more reactive groups essential for the chemical step. Thus, substrate analogs (or "false substrates" as they are sometimes named) will compete with the natural substrates of the enzyme, resulting in a decrease of the overall efficiency of the enzymatic reaction.

The obvious limitation of such an approach is the need to know in detail the structure of the substrate, its mode of interaction with the enzyme and the nature and mechanism of the chemical reaction.

Since all these limitations have been overcome in the case of the DNA pols, as detailed in the previous Chapters, it is of no surprise that the first class of DNA pol inhibitors ever developed was based on analogs of one of the natural susbtrates of DNA pols: the nucleotides. Already in the 1970's it was evident that by using nucleotides which lacked the reactive 3′-hydroxyl group on the deoxyribose ring, essential for the polymerization of the growing DNA chain, it was possible to inhibit the reaction by DNA pols.[1,2] In fact, these analogs are incorporated in the growing DNA chain by DNA pols, then preventing further elongation (**Figure 9.1**). Due to their mechanism of action, compounds of this class, whose prototype are the 2′,3′-dideoxy-nucleotides, are called chain terminators and still represent the majority of the DNA pol and RT inhibitors used in antiviral chemotherapy.

However, one major obstacle in the use of nucleotide analogs is that they must carry three phosphate groups at the 5′-position of the sugar ring, in order to be functional. These deoxynucleoside triphosphates (dNTPs), however, do not easily traverse the cellular membrane, due to the negative charges of their phosphate groups. Thus, while *in vitro* reactions with purified enzymes can be easily inhibited by purified dNTPs, *in vivo* these analogs have to be administered in their nucleosidic precursor form (i.e. without phosphate groups). These prodrugs then need to be activated intracellularly by specific nucleoside kinases up to their triphosphate form. The need of additional activation steps for nucleoside analogs has several important implications. For example, modifications that might improve the binding of the dNTPs to the target enzyme can decrease their ability to interact with the activating kinases in their nucleoside form. Moreover, while deoxynucleosides triphosphates are selective substrates of DNA pols, unphosphorylated nucleosides are recognized by a number of different cellular enzymes, thus resulting in lower stability (due to

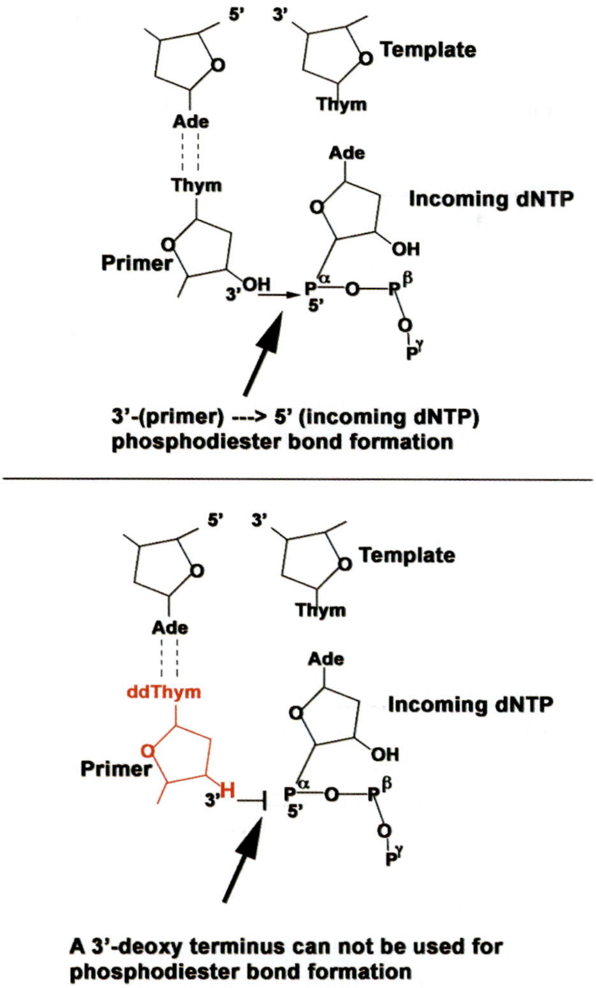

Figure 9.1   Mechanism of action of dideoxynucleotides. Upper panel: the hydroxyl group at the 3′ position of the deoxyribose carries on the nucleophylic attack to the α-phosphate group at the 5′-position of the incoming deoxynucleotide triphosphate to catalyze chain elongation. Bottom panel: when a dideoxynucleotide is incorporated, the resulting 3′ of the dideoxyribose lacks a suitable hydroxyl group to start the chemical bond formation, thus preventing chain elongation.

the action of nucleoside hydrolases) and/or in unwanted off-targeting effects and cytoxicity (as a result of their interference with other cellular metabolic pathways).

Thus, in designing nucleoside analogs, it is necessary to try not only to optimize their interaction with the target enzyme (DNA pols), but also their ability to be activated in the infected cell without being rapidly degraded and/or being toxic to the host organism. As will become clear later in this Chapter, in discussing the different analogs developed as antiviral agents, a number of strategies have been devised in order to balance efficacy and safety.

9.2.2 *Non-substrate Analogs*

As outlined in Chapters 2 and 4, we now have a detailed knowledge of the different steps of the DNA polymerization reaction and the thermodynamic and kinetic barriers responsible for the exact sequential order of molecular events leading to the incorporation of nucleotides into a nucleic acid lattice. In principle, any of these steps can be targeted by inhibitors that do not necessarily need to resemble natural substrates, but rather interact with amino acid residues of the enzyme which are important for the different conformational changes which drive the reaction (see also Chapter 4.2). The best known examples of such non-substrate inhibitors are probably the non-nucleoside inhibitors of HIV-1 RT (NNRTIs). As will be detailed later in this Chapter (see Section 9.5.2), these molecules bind to a hydrophobic pocket that is close to, but distinct from, the active site. Contrary to nucleoside analogs, they do not take any interaction with catalytic amino acids, but rather induce short- and long-range structural rearrangements that reduce the rate of the "open-to-close" conformational change of the enzyme, that is essential for catalysis. Thus, they act as a sort of allosteric inhibitors of RT. Clearly, the structural features which are important for the interaction of NNRTIs with their binding site are totally different from those governing the binding of the substrates to the enzyme and indeed NNRTIs have chemical structures completely unrelated to nucleosides. These non-substrate inhibitors have important advantages over nucleoside analogs: (i) they are generally very selective for the target enzyme and (ii) they do no require intracellular activation in the form of prodrugs. Indeed, NNRTIs have proven safer and more tolerated by patients than nucleoside analogs in clinical practice.

One major disadvantage of non-substrate analogs is that it is not intuitive, as it is for substrate analogs, to figure out which kind of chemical scaffold might be effective against a given target enzyme. Indeed, the first NNRTIs have been discovered in late 1980's by high-throughput random screening of hundreds of thousands different synthetic molecules. In recent years, however, more sophisticated technologies have been developed to allow the rational design of non-substrate inhibitors.

9.2.3 *Novel in silico Technologies for Designing Inhibitors of DNA Polymerases*

The availability of several crystal structures of proteins which are also relevant pharmaceutical targets, both in their unliganded state and in complex with substrates/inhibitors, coupled to the ever increasing power of modern computational methods, both in terms of hardware and software, opens new possibilities for drug design. *In silico* techniques allow to build pharmacophores, to screen virtual libraries and to identify specifically tailored inhibitors even in such cases when no previous inhibitors were known for that particular molecular target. Molecular dynamics, on the other hand, is starting to help to clarify the molecular details of ligand–protein interactions, hence providing a powerful tool for mechanism of action studies.

Virtual screening. Molecules specifically interacting with a given target can be identified through a combination of structural biology, screening of chemical libraries and rational drug design. The classical approach of experimental high-throughput screening (HTS) of large compound libraries is now being integrated, and in some instances replaced, by the more sophisticated knowledge-based computer-aided virtual screening (VS) approach (**Figure 9.2**).

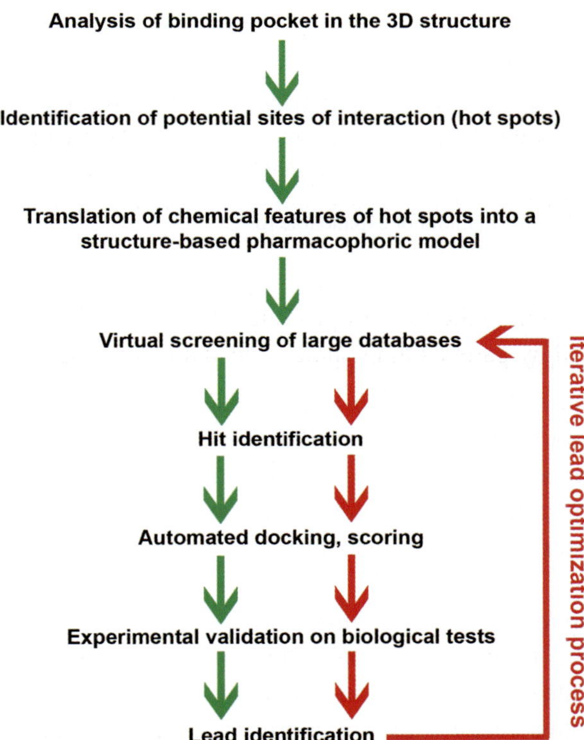

Figure 9.2   The virtual screening (VS) technique. After the first round (green arrows), the process is iterated to optimize the lead compound (red arrows).

Table 9.1   A comparison of HTS and VS approaches

	High-throughput screening (HTS)	**Virtual screening (VS)**
Main difference	Technology-driven	Knowledge-driven
Advantages	(1) Provides immediately the experimental confirmation for the hit compound (2) No structural information required for the target (3) No binding criteria required	(1) Low cost (2) No need for large-scale synthesis of compounds libraries (3) Provides mechanistic informations on ligand binding (4) Selectivity criteria of the desired stringency can be easily implemented (5) Generates criteria for compound optimization (6) Generates information that is easily expandable to other targets (7) Diverse chemistries can be exploited simultaneously
Disadvantages	(1) High costs (2) Requires large-scale synthesis of chemical libraries (3) No mechanistic information for ligand binding (4) Limited to the chemistry of the tested library (5) Information not expandable to novel targets	(1) Structural information required (2) Provides only virtual hits (need to be experimentally confirmed) (3) Highly dependent on the scoring/filter function (4) Requires precise binding criteria

The VS approach has several advantages over the classical HTS approach (**Table 9.1**). VS, also referred to as *in silico* screening, makes use of computational models to evaluate a specific biological activity of compounds in order to filter either existing databases or virtual libraries leading to the identification of molecules with activity against the target of interest.[3] By means of virtual screening of small molecules databases it is possible to identify new potential inhibitors against a target of interest.

Homology modeling and structure-based design. A pharmacophore is the description of the three-dimensional (3D) arrangements of essential features enabling a molecule to exert a particular biological effect. Structure-based pharmacophores constitute a very useful tool in drug design, both in leading to discovery and optimization processes.[4] The structure-based pharmacophore methodologies require the knowledge of the 3D-structure of the ligand-protein target complex. Briefly, the protocol involves the computation of points of minimum energy in the binding site of interest followed by the conversion of such points into pharmacophoric features. With this technology it is theoretically possible to determine all the physico-chemical features that a putative ligand (for example an inhibitor) should possess in order to selectively bind to any given target. Hence, specific pharmacophore models will result from this application.

Molecular dynamics. Another powerful methodology is certainly the molecular dynamics (MD).[5] MD is a computational tool that permits to explore the concerted movements of both small molecules (ligands) and large macromolecular (proteins) entities. MD simulations can be applied to selected ligand candidates in order to accurately calculate binding free energies and to optimize the ligand structure by taking into account the protein flexibility. MD is becoming an ideal companion tool to crystallography for the mechanism of action studies. In fact, while a crystal is a sort of snapshot of the most stable conformation of a protein-ligand complex, MD studies allow investigating the microscopic conformational rearrangements during the process of binding of the inhibitor to the target enzyme.

9.3 Inhibitors of Herpesvirus DNA Replication

Herpesvirus infections are among the most communicable diseases. Following primary or initial infection, herpesviruses persist for life in a latent form in peripheral sensory ganglia, from which they may periodically reactivate usually causing mild or, especially in immuno-compromised patients, severe diseases, often resulting in significant psychosocial distress. Through hematogenous or neuronal retrograde dissemination some herpesviruses are also among the most frequent causes of viral encephalitis.

A vaccine against varicella zoster virus (VZV) is available, but so far there are no effective vaccines against other human herpesviruses.

Although antivirals might in general exploit the host response, such as for example interferon-α, in the case of herpesviruses most antivirals were designed to inhibit the action of viral proteins/enzymes especially those involved in viral DNA replication.[6] And, among the first and still widely used drugs licensed for the treatment of HSV infections, most target the viral DNA pol and fall into two categories: nucleoside analogs and non-nucleoside inhibitors. The structures of the compounds discussed below are shown in **Figure 9.3**.

9.3.1 *Anti-Herpetic Nucleoside Analogs Require Activation by the Viral Thymidine Kinase (TK)*

As mentioned in Section 9.2.1, nucleoside analogs require activation by kinases to their triphosphorylated form. The herpesviruses encode for a thymidine kinase (TK),

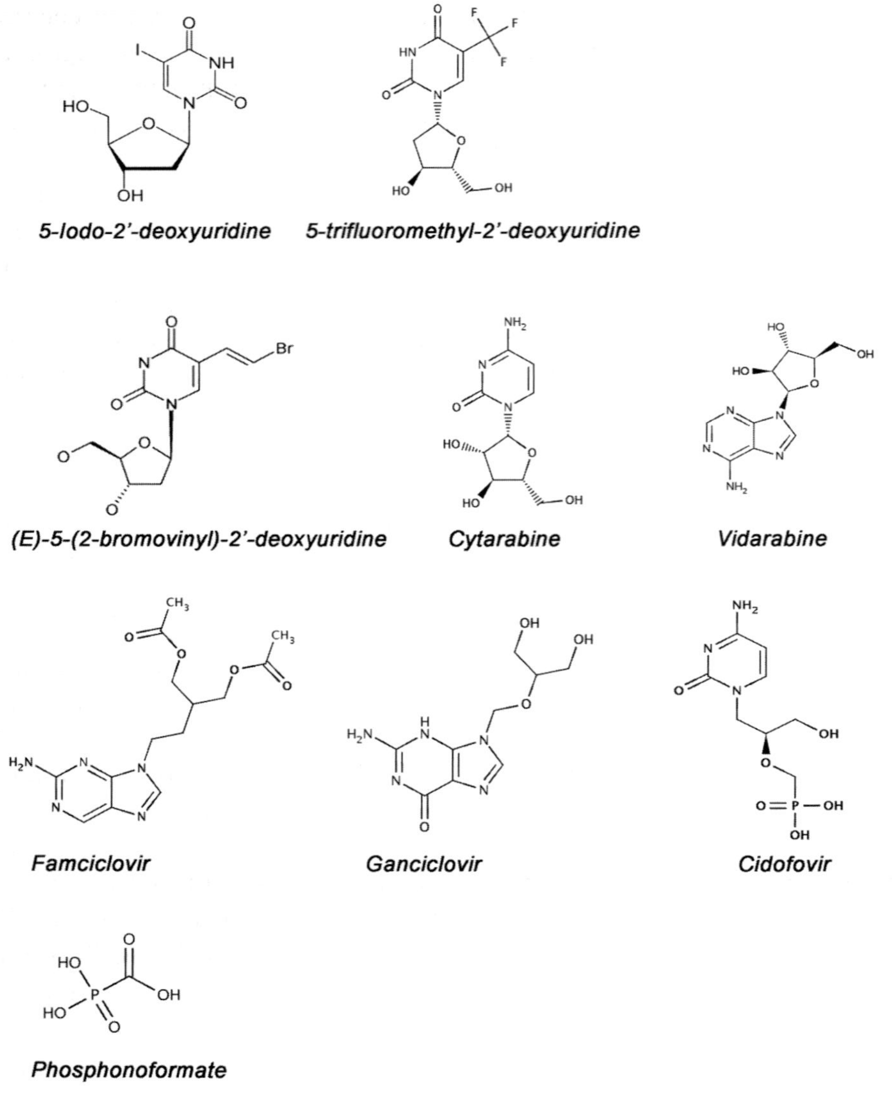

5-Iodo-2'-deoxyuridine   *5-trifluoromethyl-2'-deoxyuridine*

(E)-5-(2-bromovinyl)-2'-deoxyuridine   *Cytarabine*   *Vidarabine*

Famciclovir   *Ganciclovir*   *Cidofovir*

Phosphonoformate

Figure 9.3   Inhibitors of HSV DNA polymerase.

which has catalytic properties different from the corresponding cellular enzyme. This viral TK, in fact, efficiently catalyzes the phosphorylation not only of thymidine but also of cytidine as well as of a number of nucleoside analogs some of which are poor or no substrates for cellular TK. The HSV-1 TK, contrary to its cellular counterpart, also possesses a thymidylate (TMP) kinase activity, which converts TMP to thymidine 5'-diphosphate (TDP), as well as some other nucleoside

5′-monophosphate analogs. The diphosphate nucleoside analogs are then converted to their triphosphate forms by cellular kinases. In addition, human cytomegalovirus (HCMV) codes for a protein kinase which is capable also of phosphorylating the nucleoside analog ganciclovir, which explains its specificity for HCMV (see Section 9.3.3).

Thus, these differences were exploited in order to design nucleoside analogs, which could be selectively phosphorylated by the viral enzyme, which is present only in cells where the virus is replicating, thus limiting their toxicity towards uninfected cells. Chemically modified nucleosides that after phosphorylation retain the capacity to function as alternative substrate and/or to influence the rate of DNA synthesis by the viral DNA pol include those modified in the base and in the sugar.

9.3.2 *Nucleoside Analogs Modified in the Base Ring*

5-Iodo-2′-deoxyuridine. Also known as Idoxuridine (IdU, IUdR) or by trademark names such as *Herpid*®, *Idoxene*®, and *Stoxil*®, *Virudox*® may be considered as the prototype of the antiviral drugs and for over 40 years it has been applied in the topical treatment of HSV keratitis as eye drops or ophthalmic cream.[7] It is not very specific in its antiviral activity, since it is phosphorylated both in infected and uninfected cells being a substrate of cellular TK as well. The main target for the antiviral action of IdU is the DNA itself; the compound is readily incorporated into DNA, and its incorporation may disturb DNA replication and transcription.[8,9] Both the viral and the cellular thymidine kinases can phosphorylate IdU, but infected cells have 10-fold higher TK activity, and furthermore, viral TK has a higher affinity for IdU than human TK. Thus the infected cells will trap more IdU by phosphorylation than the adjacent uninfected cells. After conversion to monophosphate, IdU is phosphorylated to di- and tri-phosphate (IdUTP) and is then preferentially incorporated into the viral DNA. Furthermore, the viral DNA pol has a greater affinity for nucleoside triphosphates (including IdUTP) than cellular DNA pols. The high toxicity, however, made this drug useless as an antiviral agent for systemic treatment of herpesvirus infections.

In addition to IdU, several other substituted 2′deoxyuridines are incorporated into DNA, for example 5-vinyl-, 5-ethyl-, 5-propyl-, and 5-hydroxymethyl-2′-deoxyuridine. Their antiviral specificity can be improved by modifications that minimize phosphorylation by cellular TKs or increase their phosphorylation by viral-induced TK and consequently their incorporation into DNA of virus-infected cells. Thus the 5′-amino analog of IdU (AIdU) showed antiviral activity but lower cytotoxicity to uninfected cells. Its selectivity depends on preferential activation by herpes-encoded TK and subsequent incorporation into the viral DNA by a phosphoramidate linkage. The incorporation of AIdUMP induces double- and single-stranded breaks in the viral DNA, thus affecting the transcription of viral genes. The 5′-amino analog of IdU is useful in the topical therapy of experimental herpetic keratitis in rabbits, but it is not as potent as the parent drug, IdU.

5-trifluoromethyl-2′-deoxyuridine (CF₃dU) or trifluorothymidine (TFT). Known as *Viroptic®*, like IdU, is in clinical use only for topical therapy of herpes keratitis.[10] It is phosphorylated by cellular as well as by viral TKs to its monophosphate. Subsequent phosphorylation by cellular enzymes gives the active triphosphate form CF_3dUTP, which is both an inhibitor of dTTP incorporation and a partial substrate for HSV DNA pol and mammalian DNA pol α. The K_i values for the viral and the mammalian enzymes are 4 and 17 μM, respectively. It is incorporated into DNA by the viral DNA pol 20-fold more efficiently than by the mammalian enzyme. 5-trifluoromethyl-2′-deoxyuridine triphosphate incorporation into viral DNA is correlated with the observed loss of viral infectivity.

(E)-5-(2-bromovinyl)-2′-deoxyuridine, or bromovinyldeoxyuridine (brivudin, BVDU). Known under trademark names such as *Zostex®*, *Zonavir®*, *Zerpex®* and *Brivirac®* and *(E)-(2-halogenovinyl)-dU* derivatives in general differ from IdU and CF_3dU in that they are more specifically phosphorylated by the herpesvirus TK. E-5-(2-bromovinyl) 2′-deoxyuridine (where E stands for *Entgegen* or *trans*, referring to the positions of the hydrogens at C_1 and C_2 of the vinyl group) is a very potent and highly selective inhibitor of VZV, HSV-1 and HSV-2 replication with minimum inhibitory concentration (MIC) values of 0.002, 0.01, and 1 pg/ml, respectively.[11,12] Phosphorylation by the virus-induced TK can be considered as the first step in the intracellular metabolism of BVdU. The HSV-1 and VZV TKs phosphorylate efficiently BVdUMP to its diphosphate, BVdUDP, whereas the HSV-2 enzyme does not. This fact explains the lower sensitivity of HSV-2 to BVdU. The active metabolite of BVdU is its triphosphate form BVdUTP, which interferes with the viral DNA polymerase as either substrate or inhibitor. When acting as inhibitor, BVdUTP shuts off viral DNA synthesis. It has been shown that BVdUTP inhibits herpetic DNA pols more strongly than cellular DNA pols α or β. For example, in an *in vitro* system, at 1 μM BVdUTP the HSV-1 DNA pol was inhibited by 60%, while DNA pols α and β were inhibited only 6% and 3%, respectively.

Depending on the concentration at which BVdU is added to infected cells, it is incorporated into both viral and cellular DNA in place of thymidine, and the amount of BVdU incorporated into viral DNA is closely correlated with the reduction in virus yield. It has been shown that BVdU-substituted DNA is more labile than normal viral DNA, as demonstrated by a dose-dependent increase in single-stranded breaks in alkaline medium. However, BVdU showed relevant cytotoxicity on several cultured cell lines. When exponentially growing Vero cells were exposed for 72 h to different concentrations of BVdU, a reduction of 50% in cell number was observed at 0.8 μM. With Hep-2 and HEF cells, the half-maximal inhibitory dosage values for BVdU were 0.15 and 0.2 μM, respectively.

Another problem with BVdU is the conversion to its free base bromovinyluracil (BVU) by thymidine phosphorylase. BVU then inhibits dihydropyrimidine dehydrogenase (DPD) the enzyme that degrades thymine, uracil and 5-fluorouracil as well. Thus BVdU may enhance the toxicity of 5-fluorouracil in patients under treatment with this drug.

From a therapeutic point of view, BVdU has been first pursued in a number of experimental HSV-1 and VZV model infections in mice, guinea pigs, rabbits, or monkeys and for several years its use in humans was restricted to eye drops or cream for herpetic cheratitis and herpes labialis, respectively. More recently in a number of European countries it has also been licensed as a systemic drug for the treatment of herpes zoster especially in immuno-compromised patients where oral BVDU appeared superior to either intravenous or oral acyclovir.

9.3.3 *Nucleoside Analogs Modified in the Sugar Moiety*

Several modifications have been exploited for the design of novel antiherpetic drugs, including (1) the replacement of the deoxyribose moiety with an arabinose sugar or with cyclopentyl or a cyclobutyl ring to minimize the degradation of pyrimidine nucleoside analogs by pyrimidine phosphorylases and their phosphorylation to monophosphate nucleoside by cellular kinases; (2) the cleavage of the sugar ring leading to acyclic nucleosides more selectively phosphorylated by the herpes thymidine kinase and usually with a longer intracellular half-life; (3) the inverted configuration of hydroxyl groups or their elimination leading to dideoxy- or didehydro-nucleosides; (4) the replacement of the endocyclic oxygen with a methylene group or a sulfur atom; and (5) the transposition of the endocyclic oxygen or the insertion of an additional heteroatom.

9.3.3.1 *Arabinonucleosides*

Replacement of the deoxyribose moiety with an arabinose sugar led to the development of a series of nucleoside analogs with antitumor and antiviral activity that again is partly ascribable to their conversion into triphosphate forms. Because of therapeutic interest of Ara-C (cytarabine, *Cytosar-U*®, *Tarabine PFS*®, *Depocyt*®) to treat cancer as well as Ara-A (vidarabine, *Vira-A*®) and Ara-T as antiviral agents, their corresponding triphosphates have received most attention and have been shown to inhibit all eukaryotic and viral DNA pols.[13] Once more the anticancer and antiviral specificity depend on the selective phosphorylation of the nucleoside forms by cellular or viral induced nucleoside kinases. For instance, Ara-C is converted into the active triphosphate form by the cellular deoxycytidine kinase. Competing with this kinase in some cells and tissues is cytidine deaminase, which destroys the usefulness of Ara-C as a drug.

The inhibition of DNA pols by AraNTPs is in general competitive with the base-analogous dNTP, and K_i values are of the order of magnitude of K_m of the competitor triphosphate.[14] If the Ara analog is incorporated by the sensitive DNA pol, it distorts the primer template, and the Ara-3-OH primer is poorly extensible and completely blocked when two or more residues must be incorporated into a DNA sequence.

9-β-D-Arabinofuranosyladenine. Known as Ara-A, vidarabine or *Vira-A®*, it interacts with different enzymes including cellular DNA pols. It is phosphorylated to its triphosphate form, Ara-ATP, by cellular kinases rather than by virus-induced TK and the Ara-ATP then acts as a substrate for both viral and cellular DNA pols, but it is incorporated preferentially into viral DNA.[15] *In vitro*, Ara-ATP inhibits HSV-DNA pol competing with dATP with a K_i of 0.14 µM under conditions where cellular DNA pol α was inhibited with a K_i of 7.4 µM. Incorporation of a single molecule of AraAMP slows DNA elongation, thereby behaving as a pseudoterminator of DNA pol. The isolation of resistant herpes mutants with mutations linked to the DNA pol gene[16] is a proof that HSV DNA pol is at least one of the selective target. However other enzymes such as the virus-induced ribonucleotide reductase are possible targets. From a clinical view-point, Ara-A is sufficiently nontoxic, in contrast with IdU and CF_3dU, to be useful in the treatment of HSV encephalitis and neonatal infections in immunocompromised patients and it was the first antiherpes drug to be given systemically, i.e. intravenously. It has been approved also for the topical treatment of herpes keratitis. Being phosphorylated by cellular kinases as well, Ara-A is also active against TK⁻ mutants of HSV and VZV that are resistant to acyclovir.

1-β-D-Arabinofuranosylthymine. Indicated as Ara-T, it is a naturally occurring nucleoside analog. Several other 5-substituted arabinofuranosyl-uracil derivatives have been shown to be reasonably selective anti herpetic drugs. l-β-D-Arabinofuranosylthymine and derivatives are in fact more efficiently phosphorylated by herpes-encoded TK than by cellular kinases, and the triphosphate form, Ara-TTP, inhibits the viral DNA pol competing with dTTP. However, using various concentrations of Ara-T, phosphorylation also occurs in mammalian uninfected cells due to cellular TKs, and Ara-TMP is incorporated into cellular DNA.[17] Obviously, the phosphorylation of Ara-T as well as of other nucleosides mainly depends on two factors: the maximum velocity (Vmax) of the enzyme and the affinity of the nucleoside for the binding site of the enzyme. In case of low nucleoside concentrations, the binding is rate limiting; at high concentrations, the Vmax of the enzyme becomes more important, and significant phosphorylation of nucleoside with low affinity may occur. *In vitro* studies showed that cellular DNA pols α and δ do not accept Ara-TTP as substrate. Incorporation of Ara-TTP was observed *in vitro* by DNA pol β,[18] but measurement of unscheduled DNA synthesis in cultured cells in presence of labeled Ara-T after ultraviolet irradiation did not show any increase of Ara-TMP incorporation. 1-β-D-Arabinofuranosylthymine has proven to be effective against HSV infections and reactivation in animal models,[19] but the systemic application has been limited by its neurotoxicity and the local application is limited by its poor solubility in water and inability to penetrate lipophilic membranes to reach therapeutic concentrations in the target tissues.

9.3.3.2 *Acyclic nucleoside analogs*

A variety of nucleoside analogs with an acyclic moiety in place of the sugar ring have been synthesized and some of them are potent inhibitors of herpes simplex viruses.

9-(2-hydroxyethoxymethyl)guanine. Commonly referred to as acyclovir, ACV or *Zovirax*®, it is the best known and the first specific antiherpetic agent licensed for clinical use for therapy of herpes keratitis, herpes encephalitis, varicella zoster, neonatal herpes, genital herpes in normal and immunocompromised patients, and as a prophylactic repressor of herpesvirus infections in patients receiving immunosuppressive drugs. Acyclovir is quite effective and selective as an antiherpetic drug because it is phosphorylated to acyclovir monophosphate (ACVMP) in infected cells by herpes-encoded TK but very poorly by the host cell kinases.[20] Acyclovir monophosphate is further phosphorylated by cellular kinases to acyclovir triphosphate (ACVTP), the active form of the drug that inhibits the herpes-encoded DNA pol by a novel mechanism.[21] Acyclovir triphosphate is in fact a potent inhibitor of HSV, EBV, and cytomegalovirus DNA pols, and it is weakly active against mammalian DNA pol α and other cellular DNA pols. It is readily incorporated into viral DNA, where it inhibits DNA elongation. However, incorporation of ACVMP at the 3′-end of the growing chain does not lead *per se* to chain termination. In fact, inhibition of the viral DNA pol occurs by the formation of a dead end complex upon binding of the next nucleotide triphosphate coded by the template, subsequent to the incorporation of ACVMP at the 3′-end of the primer.[22] Thus inhibition involves formation of an inactive but reversible "dead end" complex between DNA pol, drug-terminated template/primer, and a dNTP substrate molecule. When HSV-1 is propagated under the selective pressure of ACV, emergence of resistant viruses occurs.[23] Genetic analysis has shown that resistance to the drug can be conferred by two independent loci. The first locus is the gene for the viral TK[24,25] and mutations that decrease TK activity also render the virus resistant to ACV while a second level of resistance corresponds to the decrease in TK activity. The second locus is the gene for the viral DNA pol[26]: HSV-1 mutants with altered DNA pol also showed marked resistance to ACV. After clinical use of ACV, mutant viruses with altered DNA pol are however rare. Most variants have arisen in immuno-compromised hosts and are lacking TK. They are therefore resistant to antivirals that act by similar mechanisms to ACV but may be susceptible to agents such as Ara-A or phosphonoformate.

Other limitations of oral ACV are its low oral bioavailability (approximately 15%) and a short half-life thus requiring multiple administrations per day. They are now overcome by other analogs such as Valaciclovir and Famciclovir.

L-valine ester of acyclovir. Valaciclovir (VACV, *Zelitrex®*, *Valtrex®*) serves as oral prodrug of acyclovir and then acts as acyclovir. It is absorbed 3 to 5 times better than ACV and achieves higher serum concentrations after oral administration.

9-(4-hydroxy-3-hydroxymethyl-but-1-yl)guanine. Penciclovir (PCV, *Denavir®*, *Vectavir®*) is used topically as cream against recurrent herpes labialis and its mechanism of action is similar to that of acyclovir. Unlike ACVTP, penciclovir triphosphate has a 3′-hydroxyl-like moiety and is not an obligate chain terminator.

Diacetyl ester of 9-(4-hydroxy-3-hydroxymethyl-but-1-yl)-6-deoxy guanine. Famciclovir (FCV, *Famvir®*) is the oral prodrug of pencyclovir and then acts as pencyclovir against HSV-1, HSV-2 and VZV infections. It is better adsorbed and can be given less often than ACV.

9-(1,3-dihydroxy-2-propoxymethyl)guanine. Also called DHPG, ganciclovir, GCV, it is used in different commercial formulations (*Cymevene®* — for intravenous infusion, *Cytovene®* — systemic gamgiclovir, *Vitrasert®* — ophthalmic ganciglovir). It is another acyclonucleoside analog which is selectively phosphorylated by HSV-encoded TK. Contrary to ACV, DHPG is also a potent inhibitor of HCMV proliferation. It is in fact selectively phosphorylated to its triphosphate form DHPGTP in HCMV-infected cells to levels at least 100-fold higher than the levels found in uninfected cells. There is no evidence for the coding of herpesvirus-like TK by HCMV, but HCMV codes for a protein kinase capable of phosphorylating ganciclovir which explains its specificity for HCMV. A four-amino acid deletion in a conserved region of such protein caused impaired ganciclovir phosphorylation and resistance of HCMV to the drug in the clinic is ascribable to impaired drug phosphorylation presumably as a result of a mutation in the above protein. DHPGTP inhibits both HSV and HCMV DNA pols with similar K_i values (0.022 µM), but it is a poor inhibitor of cellular DNA pol α ($K_i = 23$ µM).[27] Of a racemic mixture of DHPG, only the (S)-enantiomer is efficiently phosphorylated up to DHPGTP. (S)-ganciclovir triphosphate is then efficiently incorporated at G sites by HSV and HCMV DNA pols, but not by cellular DNA pols. Deoxyribonucleic acid containing DHPGMP residues at primer termini is a potent inhibitor of viral DNA pol with a Ki value of 0.015 µM.

Ganciclovir is poorly absorbed following oral administration and is principally used in clinical practice for the treatment of HCMV infections, particularly HCMV retinitis, in immunocompromised patients.

L-valine ester of ganciclovir. Valganciclovir (VGCV, *Valcyte®*) is an oral prodrug of GCV and then acts as described for GCV, with principal indication for HCMV infections.

(R)-9-(3,4-dihydroxybutyl)guanine. Bucyclovir (BCV) is selectively phosphorylated by HSV TK, and its triphosphate form BCVTP potently inhibits viral DNA pol ($K_i = 0.6\,\mu M$) but not cellular DNA pol α (estimated $K_i > 100\,\mu M$).

9.3.3.3 *Carbocyclic nucleoside analogs*

The antiviral activity of nucleoside analogs is often limited because of their rapid degradation by catabolic enzymes. For instance, BVdU and (E)-5-(2-iodovinyl)-2'-deoxyuridine (IVdU) are excellent substrates for pyrimidine nucleoside phosphorylases that cleave the glycosidic bond between the pyrimidine ring and the sugar moiety. In an attempt to avoid the degradation of pyrimidine nucleoside analogs by pyrimidine phosphorylases, carbocyclic derivatives of several uridine analogs with antiviral activity, in which the sugar moiety has been replaced by a cyclopentyl or a cyclobutyl ring, have been synthesized and found more resistant to nucleoside phosphorylases.[28–30] Like their parent compounds, C-IdU, C-BVdU, and C-IVdU, the carbocyclic analogs of IdU, BVdU and IVdU, have been shown to markedly affect HSV replication, and their antiviral selectivity also depends on their phosphorylation by the HSV-encoded TK.[31] None of the compounds proved to be inhibitory to TK negative HSV strains. It has also been shown that C-IVdU, as well as IVdU, is incorporated into both viral and cellular DNA of HSV-I infected cells, and there is a close correlation between the inhibition of viral DNA synthesis and the antiviral activity. Virus replication was completely suppressed by IVdU and C-IVdU at 2 and 5 μM, respectively. Cellular DNA synthesis was affected only at concentrations l0-fold (IVdU) and 40-fold (C-IVdU) higher than those required for inhibition of viral DNA synthesis. Incorporation of IdU, but not C-IVdU, resulted in extensive double-stranded DNA breaks. These results suggested that carbocyclic analogs behave differently from the parent compounds in their rate of incorporation into viral DNA and effect on virus DNA integrity. The inhibitory effects of IdU, IVdU, and BVdU on viral DNA synthesis may be a consequence of the single- and double-stranded DNA breakage resulting from the incorporation of the compounds into viral DNA, whereas it seems that incorporation of carbocyclic derivatives alters the functioning of DNA as template for both replication and transcription, as demonstrated for C-BVdU incorporation into synthetic templates, without affecting DNA integrity.

Other analogs like cyclic [1-α-(E), 2-β-3-α]-1-[2,3-bis(hydroxy-methyl) cyclobutyl]guanine and [1-α-(E), 2-β-3-α]-1-[2,3-bis (hydroxymethyl) cyclobutyl]-5-(2-halovinyl) uracil analogs have been synthesized. The guanine analog is a potent inhibitor of several herpesviruses including HSV-1 and HSV-2, VZV, and HCMV. In HSV-infected cells, it is phosphorylated by HSV-encoded TK to diphosphate, and the diphosphate form is readily converted to the triphosphate, which is a potent inhibitor of HSV DNA pol. Among the 5-halovinyl uracil analogs, the bromovinyl, iodovinyl, and chlorovinyl derivatives are all potent inhibitors of VZV replication but are less inhibitory to the replication of HCMV and HSV. Varicella zoster virus TK readily converts these compounds to their monophosphate but not to their corresponding diphosphates.

Despite their potential no carbocyclic nucleoside analog has so far been licensed as antiherpetic drug.

9.3.3.4 *Phosphonate nucleoside analogs*

For their mechanism of selective activation by herpesvirus TK, some of the above nucleoside analogs may obviously be fully inactive against viral mutants with altered resistant TK or against viruses which do not code for their own TK. A way to circumvent this problem was the development and intracellular delivery of prodrug monophosphorylated nucleoside analogs (such as nucleoside monophosphonates) that avoid the first kinase dependent phosphorylation step needed to start their activation to a triphosphate form.

Among them a variety of acyclic nucleoside phosphonates have been synthesized and found active against various viruses. They all can actually be regarded as derivatives of phosphonoacetic or phosphonic acid. The phosphorous atom is in fact attached to the side chain of the purine or pyrimidine base via a P-C bond thus forming a phosphonate linkage which, contrary to the physiological P-O bond, is not susceptible to hydrolysis by esterases. Thus, whereas P-O nucleotides are first hydrolyzed to nucleosides to enter the cell, these compounds, resistant to esterases, are apparently taken up as such by cells via a process resembling endocytosis and exhibit a broad spectrum of activities against a variety of RNA and DNA viruses, including TK⁻ strains, and retroviruses. Introduction of a phosphonate group, as in phosphonylmethoxyethyladenine (PMEA), obviates the need of the first phosphorylation step, which in the cell is reasonably selective as well as a rate-limiting step in the activation process of, for instance, 2′,3′-dideoxynucleosides. It has also been shown that some of these analogs can undergo intracellular phosphorylation to "triphosphates" in a single step by transfer of the pyrophosphate group of phosphoribosyl pyrophosphate (PRPP) mediated by a PRPP synthetase. The triphosphate forms are then inhibitors of viral polymerases or reverse transcriptase, and if incorporated they also act as chain terminators.

(S)-1-(3-hydroxy-2-phosphonylmethoxypropyl)   cytosine. Cidofovir (CDV, HPMPC, *Vistide®*, *Forvade®*), is active not only against herpesviruses such as HSV1, HSV2, VZV. CMV) but also against papilloma-, polyoma-, adeno- and pox-viruses because it mimics dCMP and it is phosphorylated by cellular kinases.[32] Following two intracellular phosphorylations to a triphosphate-like form it is incorporated at the 3′-end of the viral DNA chain where it acts as chain terminator. In the case of CMV DNA synthesis two consecutive residues must be incorporated to functions as chain terminator. Licensed to treat CMV retinitis in AIDS patients, it is also effective in the treatment of acyclovir-resistant (viral TK-deficient) HSV infections because the intracellular phophorylation is independent from the HSV- or VZV thymidine kinase or from CMV-encoded protein kinase. It is used also to treat recurrent genital herpes infections. It can be administered either topically as gel or cream (*Forvade®*) or intravenously (*Vistide®*).

9.3.4   *Active-Site Directed Non-nucleoside Inhibitors of Herpesvirus DNA Polymerases*

Relatively few compounds are known to bind directly and to inhibit DNA pols, but some of them have been introduced in clinical medicine or are being exploited for their clinical potential. Phosphonoacetic and phosphonoformic acid are non-nucleoside derivatives and therefore do not require an activation step to act at the target DNA pol, and despite their simple structures, they are effective antiviral drugs.[33] They operate as analogs of inorganic pyrophosphate and interact with DNA pols at the pyrophosphate binding site, binding with K_i near $1\,\mu$M that is 1000-fold tighter than pyrophosphate thus preventing normal pyrophosphate release.[34] Whereas the toxicity of phosphonoacetate precludes its clinical use, phosphonofor-mate (*Foscavir®*) has significant therapeutic value, representing a valid second line treatment againsts HSV-1 and -2, VZV and HCMV viruses that have lost thymidine kinase activity and became resistant to acyclovir or ganciclovir.[35] The viral DNA pols are >100 times more sensitive than the human DNA pols. It is not orally available and is administered intravenously.

9.4   The Lack of Enantioselectivity of Viral and Human Enzymes and the L-Enantiomers of Nucleosides: The Dawn of a New Generation of Antiviral Drugs

For a long time enzymes were assumed to be sterically specific when they act upon or form substances containing chiral centers and biological stereospecificity was regarded as an absolute and inviolate feature of enzymatic reaction. Therefore it was assumed that only nucleoside analogs having the natural D-configuration would exhibit biological activity.

Thus although the first L-nucleoside was synthesized in 1964, L-nucleoside enantiomers were not evaluated as potential antiviral agents until early 1990s when, among enzymes involved in the synthesis of nucleotides and DNA, some exceptions were found to the universal rule that enzymes act only on one enantiomer of a chiral substrate and that only one of the enantiomeric forms of chiral molecules may bind efficiently at the catalytic site displaying biological activity.[36–44] In particular, it was shown that these exceptions include: herpesvirus thymidine kinases, cellular deoxycytidine kinase and deoxynucleoside mono- and diphosphate kinases, cellular and viral DNA pols such as DNA pol α, terminal transferase and HIV RT. The ability of these enzymes to utilize unnatural L-β-nucleoside or nucleotides as substrate was thus exploited from a chemotherapeutic point of view.

9.4.1 *Herpesvirus Thymidine Kinase Has Low Enantioselectivity*

Spadari and co-workers demonstrated for the first time that the herpetic TK, an enzyme known to have a more relaxed substrate specificity for nucleoside analogs than cellular TKs, is poorly or not stereospecific.[44] The affinity of HSV-1 TK for L-thymidine is very close to that for the natural substrate D-thymidine whose K_m (2.8 µM) is comparable with the K_i (2 µM) of the analog. *In vivo*, L-thymidine selectively inhibited viral proliferation without detectable toxic effect on non-infected cells.

Similar results were obtained also with the pseudorabies virus (PRV) TK,[42] HSV-2 TK and varicella zoster TK suggesting that the lack of stereospecificity for the substrate was possibly a common feature of the TKs encoded by herpesviruses.

The stereospecificity of human TK and the lack of stereospecificity of herpesviruses TK with respect to thymidine enantiomers was absolute in the sense that the L-enantiomer did not inhibit the utilization of the substrate by human TK and the viral TKs exhibit a lack of stereospecificity towards both enantiomers utilized as substrate.

Interestingly, Spadari and coworkers have also demonstrated that human cytosolic TK and HSV-1 TK retain their stereospecific and non-stereospecific behaviour, respectively, also towards enantiomers of thymidine analogs such as L-IdU and L-BVdU, whose corresponding D-enantiomers are among the most potent anti-herpetic drugs. As mentioned above (see Section 9.3.2), their therapeutic use is limited by their cellular toxicity. In contrast, although efficiently phosphorylated by viral TK, L-IdU and L-BVdU are approximately 1000-fold less cytotoxic than their corresponding D-enantiomers on HeLa TK$^-$/HSV-1 TK$^+$ cells because of their lack of affinity for cytosolic TK and a 20-fold lower affinity than the corresponding D-enantiomers for the mitochondrial TK. Moreover, L-IdUMP and L-BVdUMP, contrary to their D-enantiomers, do not inhibit cellular thymidylate synthase, an enzyme shown to be higly sensitive to the inverted configuration of the sugar ring in the L-stereoisomers.

When evaluated for their activity against HSV-1 it was found that, despite their much lower cytotoxicity, L-IdU was only 6-30 times less potent than D-IdU and L-BVdU was 30-200 times less potent than D-BVdU (whose antiviral activity is clearly potentiated by its high cytotoxicity) with a potency comparable to that of acyclovir. The described high tolerance of HSV-1 TK to the overall sugar configuration was confirmed by Balzarini *et al.*,[45] who demonstrated that the (D) and (L) enantiomers of the carbocyclic analogs of IdU and BVdU, where the sugar moiety is replaced by a cyclopentenyl ring, show similar affinity for HSV-1 TK. The antiviral effect of L-Thymidine, L-IdU and L-BVdU, as in the case of the D-enantiomers, appears finally due to their corresponding triphosphates. If TK is clearly responsible for the phosphorylation of L-Thymidine, L-IdU and L-BVdU to L-TMP, L-IdUMP and L-BVdUMP respectively, the subsequent phosphorylation to di- and triphosphates is carried out by the cellular deoxyribonucleotide kinases which are no or poorly stereospecific enzymes.

9.4.2 *The Discovery of a Relaxed Enantioselectivity of Human and Viral DNA Polymerases*

Spadari and co-workers[40] showed that L-TTP inhibits HSV-1 DNA pol, HIV-1 RT, human DNA pols α and γ and terminal transferase by competing with D-TTP. DNA pol δ was inhibited in a non-competitive way when D-TTP or DNA were the variable substrates and in a mixed-type when PCNA was the variable substrate. DNA pol ε, like δ, is inhibited in a non-competitive way when D-TTP is the variable substrate, but in a competitive way when template-primer is the variable substrate. DNA pol β was the only tested pol found resistant to L-TTP. When the capability of these pols to utilize L-TTP as substrate was studied, it was found that DNA pols β, δ and ε were unable to use L-TTP as substrate, whereas DNA pol α, HIV-1 RT and terminal transferase clearly incorporate one L-TMP residue. DNA pol α and HIV-1 RT were also able to further elongate the DNA chain by catalyzing the phosphodiester bond between the incorporated L-TMP and an incoming L-TTP. The potent 3$'$ $\rightarrow$ 5$'$ exonuclease activity of HSV-1 DNA pol did not allow verifying with a primer extension assay, whether or not L-TTP is a substrate for HSV-1 DNA pol. However, even though L-TTP would not be a substrate, it interestingly exerted a very strong inhibition, probably contributing to the antiviral effect of L-thymidine and of other L-nucleoside analogs. Similarly to HIV-1RT, HBV DNA pol has been also shown to be capable to bind and incorporate L-nucleoside triphosphates.

9.4.3 *Lack of Enantiospecificity of Human 2′-Deoxycytidine Kinase: Relevance for the Activation of L-Deoxycytidine Analogs*

Mammalian cells possess kinases for all deoxyribonucleosides and their role is more relevant in cells (e.g. lymphocytes), tissues (e.g. thymus and spleen), or neoplasias (e.g. lymphocytic leukemia, low-grade lymphomas, hairy cell leukemia) with more active nucleotide salvage pathways. Of particular interest among cellular deoxyribonucleoside kinases is deoxycytidine kinase (dCK), which supplies dCTP also for the biosynthesis of membrane lipids. The dCK has been isolated and purified from many sources and, contrary to cellular TK that tolerates only 3′ modifications of the sugar ring and some nonbulky 5′-susbstitutions on the pyrimidine base, is characterized by a considerably lower selectivity for both the substrate and the phosphate donor. It can phosphorylate not only dCyd but also dAdo, dGuo, and several pyrimidine and purine deoxyribonucleoside analogs modified in both the base and the sugar ring (e.g. ddCyd, 1-β-D-arabinofuranosylcytosine, and 2-chloro-2′-oxyadenosine), using various nucleoside triphosphates as phosphate donors. For these properties, dCK greatly differs from the other cellular deoxyribonucleoside kinases and resembles the herpesvirus TK, which, as we have seen, efficiently catalyzes the phosphorylation not only of thymidine but also of dCyd as well as of a number of pyrimidine and purine analogs that are poor substrates or are not substrates for the cellular TK.

Interestingly, the comparison of the amino-acid sequence of 15 higher vertebrate herpesvirus viral TKs with the human dCK, has revealed that four or five of six conserved sites of the herpesvirus TKs are related to human dCK, suggesting that herpesvirus TKs evolved from a captured cellular dCK gene. Because of the possible evolution of herpesvirus TKs from cellular dCK, and its potential relevance in the activation of L-deoxycytidine analogs to potential antineoplastic and antiviral agents, the enantioselectivity of human dCK was investigated. Surprisingly, human dCK was shown to phosphorylate deoxycytidine (D-dCyd), the natural substrate, and L-deoxycytidine (L-dCyd), its enantiomer, with the same efficiency.[39] Kinetic studies showed that L-dCyd is a competitive inhibitor of the phosphorylation of D-dCyd with a K_i value of 0.12 μM which is lower than the K_m value for D-dCyd (1.2 μM). Moreover, it was shown that L-dCyd was resistant to cytidine deaminase and competed in cell cultures with the natural D-dCyd as substrate for dCK, thus reducing the incorporation of exogenous [3H]dCyd into DNA. L-dCyd had no effect on the pool of dTTP deriving from the salvage or from the *de novo* synthesis, nor did it inhibit short term RNA and protein syntheses, showing little or no cytotoxicity.

Consistently with these findings, it was reported that both enantiomers of 2′,3′-dideoxy-3′-thiacytidine and of its 5′-fluoro derivative served as substrates for human dCK with a surprising preference for the L-enantiomers. The discovery of a cellular nucleoside kinase endowed with low enantioselectivity provides a mean for designing selective L-nucleoside analogs against viruses lacking their own kinases, but having not-enantioselective polymerases such as HIV-1 and HBV.

9.5 Inhibitors of HIV-1 Reverse Transcriptase

One of the major advances in the recent history of the treatment of HIV infections has been the development of different classes of effective antiretroviral drugs. In particular, the reverse transcriptase (RT) inhibitors still represent the majority of the clinically used anti-HIV drugs and constitute the main backbone of currently employed combinatorial regimens.[46] They fall into two classes: the nucleoside inhibitors (NRTIs) and the non-nucleoside inhibitors (NNRTIs). The clinically approved drugs of both classes are listed in **Table 9.2**. The corresponding structures are shown in **Figure 9.4**.

9.5.1 *Nucleoside Reverse Transcriptase Inhibitors*

The NRTIs were the first class of compounds to be used in anti-HIV-1 therapy and are a cornerstone in antiretroviral therapy.[47] All approved NRTIs

Table 9.2   Current clinically approved HIV-1 reverse transcriptase inhibitors

Drug	Commercial name	Company	Licensed
NRTIs			
Zidovudine, AZT	Retrovir	Glaxo-SK	1987
Didanosine, ddI	Videx	Bristol-Meyer Squibb	1991
Zalcitabine, ddC	HIVID	Roche	1992
Stavudine, d4T	Zerit	Bristol-Meyer Squibb	1995
Lamivudine, 3TC	Epivir	Glaxo-SK	1998
Abacavir, ABC	Ziagen	Glaxo-SK	1999
Tenofovir-DF	Viread	Gilead	2001
Emtricitabine	Emtriva	Gilead	2003
NNRTIs			
Nevirapine, NVP	Viramune	Boehringer Ingelheim	1996
Efavirenz, EFV	Sustiva	Bristol-Meyer Squibb, Merck	1998
Delavirdine, DLV	Rescriptor	Pharmacia Upjohn, Agouron, Pfizer	1999

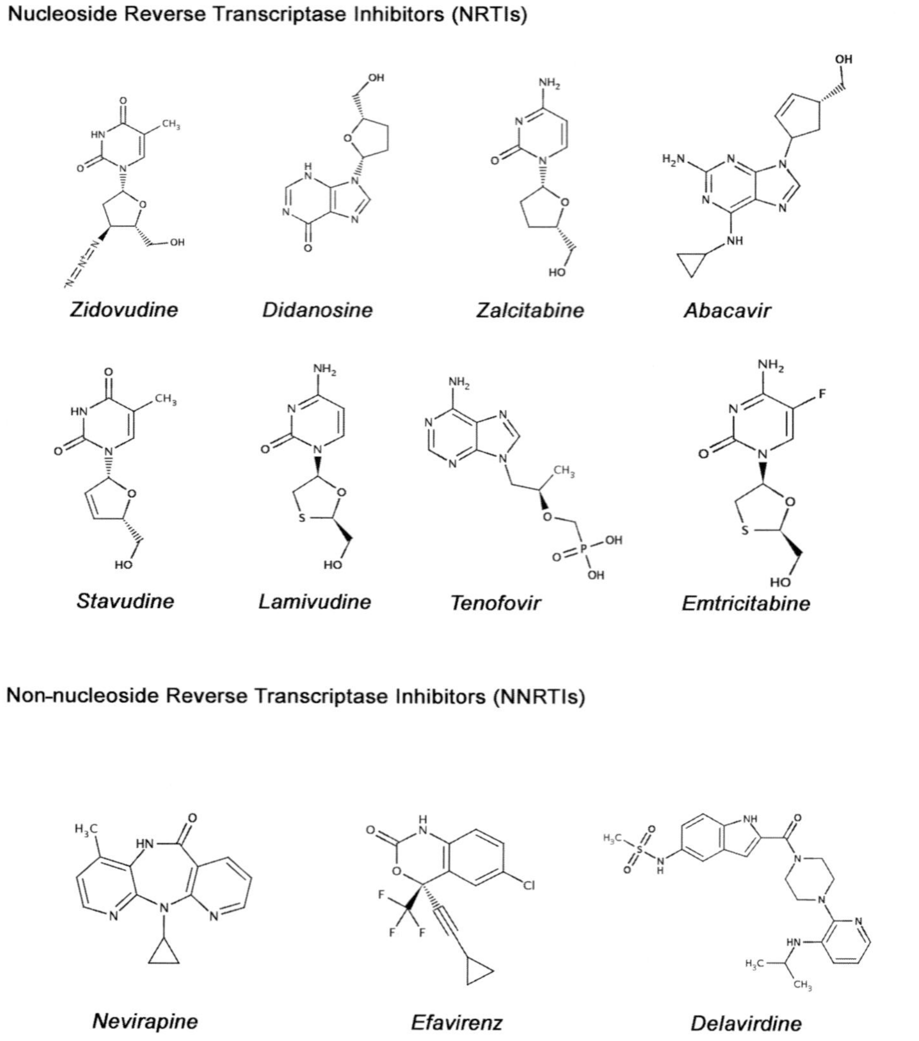

Nucleoside Reverse Transcriptase Inhibitors (NRTIs)

Non-nucleoside Reverse Transcriptase Inhibitors (NNRTIs)

Figure 9.4   HIV-1 nucleoside (top) and non-nucleoside (bottom) reverse transcriptase inhibitors.

are 2′,3′-dideoxy-derivatives of the natural nucleotide substrates. Following intracellular conversion to their 5′-triphosphate derivatives (the active form of the drug), they compete with natural nucleotides for binding to RT and, subsequently, cause chain termination through incorporation into the nascent DNA strand. Chain termination is caused by the lack of a hydroxyl moiety at the 3′ carbon of the pentose ring that is necessary to form a 3′–5′ phosphodiester bond with the next incoming nucleotide substrate during DNA chain elongation. The efficacy of a nucleoside analog is dependent on several factors, including its oral bioavailability, cellular uptake, the intracellular anabolism to its triphosphate derivative, the ability to compete with natural nucleotides as a substrate for RT and the degree of drug resistance developed by the virus.

Currently approved NRTIs (**Table 9.2**) are the thymidine analogs AZT (AZT), the first approved drug for the treatment of HIV-1 infections, and d4T (stavudine); the dideoxy-derivatives of inosine, ddI (didanosine), and of cytidine, ddC (zalcitabine); the hetero-substituted L-(β)-enantiomer of 2′-deoxy-3′-thiacytidine, 3TC (epivir or lamivudine), the carbocyclic analog abacavir (1592U89) and the recently approved tenofovir disoproxil (PMPA-DF) and emtricitabine (FTC).

As outlined in Section 9.2.1, these compounds are administered as nucleoside prodrugs, which necessitate phosphorylation up to their triphosphate forms to be active. The intracellular metabolism of these drugs is complex and the intracellular concentration of the active (i.e. triphosphate) drug is always the result of the balance of anabolic and catabolic pathways.

9.5.1.1 *Intracellular anabolism of nucleoside reverse transcriptase inhibitors*

Figure 9.5 summarizes the anabolic and catabolic pathways acting upon NRTIs.

AZT. The first step in the phosphorylation of AZT, in which a monophosphate (MP) derivative is produced, is catalyzed by cytosolic thymidine kinase. This step is not rate limiting for AZT, since thymidine kinase has similar affinities for its natural substrate thymidine and for AZT. The second phosphorylation step, conversion of AZTMP to its diphosphate (DP) form, is catalyzed by thymidylate kinase and is rate limiting for AZT activation.[48] Given that the rate-limiting step in the metabolism of AZT is conversion of AZTMP to AZTDP, it would be expected that, at usual clinical exposures, the pathway is saturated due to limited efficiency of thymidylate kinase. Indeed, *in vitro* studies showed that a linear increase in AZT concentrations over a constant incubation time was accompanied by a linear increase in intracellular AZTMP concentrations, whereas AZTDP and AZTTP concentrations reached a steady-state maximum. The clinical studies conducted with AZT yielded considerable variability about the drug's intracellular metabolism over time.[49,50] In asymptomatic patients, the total phosphorylated AZTTP was similar to a control group of healthy volunteers, however, largely elevated AZTMP concentrations were found in HIV infected patients.[51] No difference was found in AZTTP concentrations of 11 patients receiving long-term (>18 months) and 10 receiving short-term (<2 months) AZT. Preliminary data on total phosphorylated AZT and AZTTP concentrations from 23 patients (12 naive and 11 experienced) who were receiving AZT as part of a combination therapy regimen over 12 months indicated

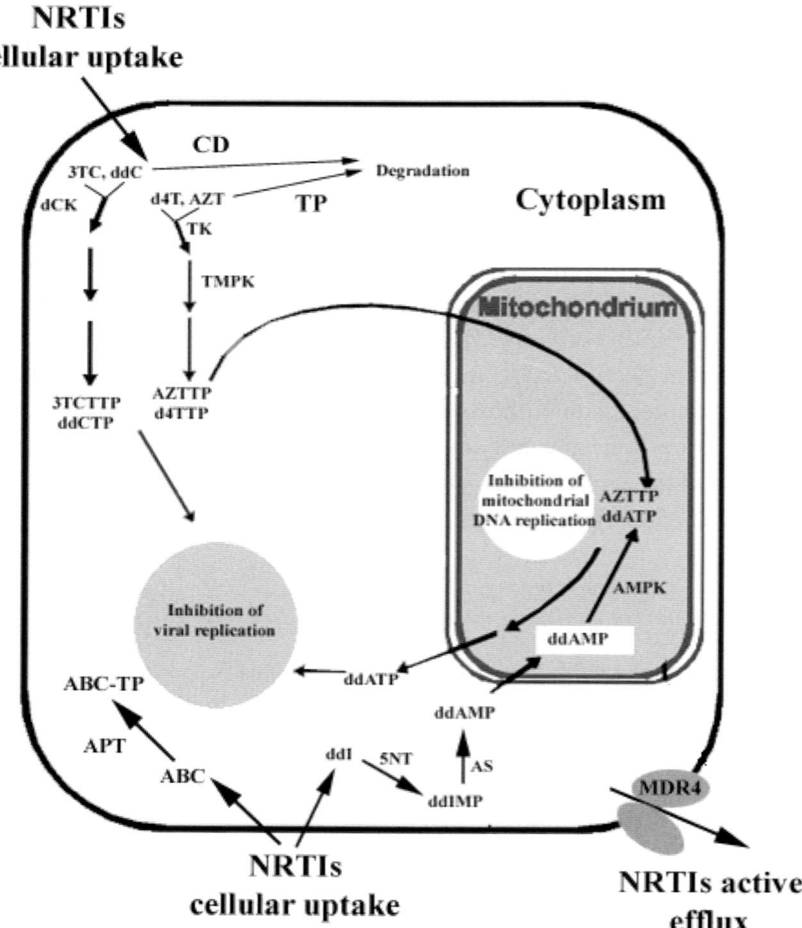

Figure 9.5   Overview of intracellular metabolism of NRTIs. Abbrevations are: dCK, deoxy-cytidine kinase; TK, thymidine kinase; TMPK, thymidylate kinase; CD, cytidine deaminase; TP, thymidine phosphorylase; APT, adenosine phosphotransferase; 5NT, 5′-nucleotidase; AS, adenylo-succynate synthetase; AMPK, mitochondrial adenylate kinase.

no systematic decrease or trends in intracellular phosphorylation over time and no difference in phosphorylation in naive versus experienced patients.[52]

d4T. The first step in phosphorylation of d4T, in which a monophosphate derivative is produced, is also catalyzed by the cytosolic thymidine kinase.[53] This step is rate limiting for d4T, since it is a poor substrate with up to a 700-fold decrease in affinity for the enzyme with respect to deoxythymidine.[53] Consistent with the rate-limited formation of d4TMP, the majority of intracellular d4T is not phosphorylated.[54] In several experiments, depending on the cell lines used, 82- to

230-fold more d4T than d4TTP was measured. In lymphoblastoid and CD4+ T cells, d4TTP concentrations peaked between 3 and 3.9 hours; d4TMP and d4TDP concentrations peaked between 3.5 and 7 hours after exposure. Increasing extracellular d4T concentrations by 200- to 400-fold resulted in nonlinear increases in intracellular concentrations of d4T, d4TMP, d4TDP, and d4TTP in these cell lines. However, the majority of intracellular drug remained as unphosphorylated d4T.

ddC. The first phosphorylation step for ddC is catalyzed by the cellular enzyme deoxycytidine kinase. Conversion of ddC to ddCMP by deoxycytidine kinase is the slowest step in accumulation of ddCTP, since ddC is a weak substrate for deoxycytidine kinase compared with the natural substrate deoxycytidine.[55]

3TC. The first phosphorylation step for 3TC is also catalyzed by the cellular deoxycytidine kinase, however, the predominant metabolite of 3TC phosphorylation is 3TCDP, both *in vitro* and *in vivo*, which suggests that formation of 3TCTP from 3TCDP is primarily the rate-limiting step.[56] The intracellular activation of 3TC was somehow surprising, given the fact that its sugar ring has the unnatural L-conformation, enantiomeric to the normal D-ribose of natural dNTPs. As already mentioned, it was shown that, among cellular nucleoside kinases, deoxycytidine kinase has the unique property of being poorly enantioselective with respect to its nucleoside substrates, thus being able to phosphorylate both D- and L-enantiomeric forms of nucleosides, a feature previously recognized only in viral enzymes (see also Section 9.4.3). Unlike d4T and AZT, 3TC shows less evidence of phosphorylation saturation *in vitro* at usual clinical exposures. In PBMCs and promonocytic cells, over a range of extracellular 3TC exposures (0.1–10 μM), there are proportional increases in intracellular 3TC and 3TCDP, slightly greater than dosage-proportional increases in 3TCMP, and less than dosage-proportional increases in 3TCTP. In *in vitro* studies at all exposures, 3TCMP is lower than 3TCDP and 3TCTP, consistently with the suggestion that the rate-limiting step is at 3TCTP formation. Additional studies on 3TC indicate saturation of 3TCTP as evidenced by lack of significant increase in 3TCTP concentration with increasing dosages in patients receiving 3TC in combination with AZT. Together, these findings are consistent with results of randomized, controlled trials indicating no additional benefits of administration of high dosages 3TC.[57]

ddI/ddA. Both ddI and ddA diffuse into the cell without active transport. Phosphorylation of ddI to ddIMP is carried out by the cellular 5′-nucleotidase, ddIMP is then converted to ddAMP by the adenylosuccinate synthetase. Further activation of ddAMP to ddADP requires the action of the mitochondrial adenylate kinase. Studies of ddI and ddA in various cell lines (promonocytic, CD4+T cells) consistently show 2–3 times greater ddATP than ddADP. However, the majority of drug is present as ddIMP and ddAMP.[58,59] This implies either a block in the formation of ddATP, although a specific rate-limiting site has not been identified, or a rate of intracellular degradation that exceeds its phosphorylation rate. The ddATP concentrations appear to plateau by 6 hours when measured up to 24 hours.

Abacavir. Abacavir is an adenosine analog that is metabolized to a 2′,3′ dideoxyguanosine analog, without an apparent rate-limiting step, by a unique set of enzymatic steps including a novel enzyme, adenosine phosphotransferase (**Figure 9.5**).[60] The triphosphate form of Abacavir is the only ddNTP to compete with deoxyguanosine triphosphate. The lack of significant rate-limiting steps in the phosphorylation pathway, is reinforced by *in vitro* data that show a linear relationship between intracellular abacavir triphosphate and extracellular abacavir concentration over a 1000-fold exposure range.[61]

Thus, stepwise intracellular phosphorylation to an active triphosphate form creates a pool of dideoxynucleoside analog triphosphates (ddNTPs) that compete with the cell's endogenous deoxynucleotide triphosphates (dNTPs) for binding to RT. The affinity for ddNTPs and the ratio of ddNTPs to endogenous dNTPs, all influence the ability of RT to incorporate the triphosphate form of the nucleoside analog into nascent proviral DNA, resulting in chain termination. Several intracellular metabolic parameters in addition to extracellular pharmacokinetics determine the efficacy and safety of nucleoside analogs. One or more steps in sequential phosphorylation of NRTIs from initial monophosphate to diphosphate and/or to triphosphate formation can be rate limiting for the accumulation of the active drug. For some agents, metabolism can vary by cell type. *In vitro* and *ex vivo* studies with NRTIs indicated that differential phosphorylation kinetics and amount of phosphorylated products in different cell types are related to the expression and efficiency of cellular kinases, intracellular dNTPs concentration and intracellular ddNTP:dNTP ratio. For example, differential ddNTP:dNTP ratios in different activation states together with other findings suggest that AZT and d4T (both phosphorylated by thymidine kinase, whose expression is dependent upon the cell proliferative state) are more efficiently phosphorylated and therefore have higher concentrations of phosphorylated metabolites in activated cells than in resting cells, whereas ddC, ddI, and 3TC (which depend for their activation upon cell proliferation-independent enzymes) are preferentially phosphorylated in resting cells.[62]

9.5.1.2 *Catabolism of nucleoside reverse transcriptase inhibitors*

Beside anabolic activation pathways, catabolic degradation is also important in determining the intracellular concentrations of NRTIs. There are two main catabolic pathways that act upon NRTIs, lowering their intracellular concentration: degradation and active transport.[63]

Two key cellular enzymes are responsible for the intracellular degradation of the majority of nucleosides and nucleoside analogs: thymidine phosphorylase (TP) and cytosine deaminase (CD).[64] TP acts by breaking the linkage between the ribose moiety and the base, leaving ribose-5′-phosphate and thymine (or uracil) as reaction products. CD, instead, catalyses the deamination of cytidine, deoxycytdine and several analogs. Interestingly, in one of the authors' laboratory it has been shown that CD is strictly stereospecific, being unable to act upon L-ddC.[65] Thus, presumably, the nucleoside analog 3TC, by virtue of the L-conformation of its sugar ring, is more resistant to intracellular degradation. The human multidrug resistance-associated protein (MRP) family, has seven members.[66] The ability of several of these membrane proteins to transport a wide range of anticancer drugs out of cells and their presence in tumors make them likely candidates in cellular drug resistance. Interestingly, overexpression of a member of this family, MRP4, was associated with high-level resistance to the NRTIs 9-(2-phosphonylmethoxyethyl) adenine and AZT.[67] Another transporter which seems to be implicated in cellular drug resistance is the P-glycoprotein, a 170-kd glycoprotein encoded by the MDR 1 gene, which is a member of a highly conserved superfamily of ATP-binding cassette (ABC) transport proteins. P-glycoprotein acts as an energy-dependent efflux pump that appears to transport structurally diverse agents ranging from ions to peptides. P-glycoprotein has been implicated as playing a role in multidrug resistance in cancer and possibly in resistance to anti-HIV NRTIs.[68]

9.5.2 *Non-nucleoside Reverse Transcriptase Inhibitors*

Non-nucleoside reverse transcriptase inhibitors (NNRTIs) are important drugs for the treatment of HIV, that are used in combination with other anti- HIV drugs such as NRTIs.[46,69] These inhibitors were discovered either from large screening programs for specific HIV-reverse transcriptase inhibitors or as the result of comprehensive evaluation of compounds with anti-HIV activity in infected cells. Both approaches were based on random screening of different molecules, with subsequent selection of the most active compounds as potential leads. Nevirapine and Delavirdine are the only first-generation NNRTIs in clinical use (**Figure 9.4**). In spite of their very different structures, these compounds are absolutely specific for the HIV-1 RT, being not active against HIV-2. The molecular basis for their inhibitory activity as well as for this absolute preference for the HIV-1 RT, has been fully elucidated only after the resolution of the crystallographic structures of different NNRTIs/RT complexes.[70,71] All the known NNRTIs bind to a common allosteric site of RT, close to, but distinct from the active site, causing a conformational change in the enzyme which impairs catalysis (see also below). The small differences in the amino acid residues of this site between HIV-1 and HIV-2 RTs are responsible for the lack of inhibition of the latter enzyme. However, the first generation of NNRTIs suffered from the rapid development of resistance, through the accumulation in the RT gene of drug resistance mutations which are readily selected during drug administration, due to the high viral turnover and mutation rate. The fast emergence of resistance mutations, coupled to the high level of cross-resistance, highlighted the urgent need of novel NNRTIs, with a broader spectrum of activity. Two approaches have been exploited: the improvement of already known classes of drugs and the screening in search of novel lead structures. In both cases, all the available information about the resistance mutations (their location in the RT structure, their mechanism of action, their interactions with the drug) has been incorporated into the drug-screening programs. This implemented knowledge-based approach, allowed more rational structure-activity relationship studies, aimed to select compounds not only highly active against the wild type virus, but also less sensitive to some of the known drug resistance mutations.[46] As a result of these efforts, the second-generation NNRTI Efavirenz has been introduced in clinical use.[72,73]

9.5.2.1 *Metabolism of non-nucleoside reverse transcriptase inhibitors*

Contrary to NRTIs, the NNRTIs do not require metabolic activation, thus their potency can be directly correlated to the intracellular levels of the drug. The high specificity of these compounds for the HIV-1 RT limits their interaction with cellular enzymes, reducing their cytotoxicity.

Nevirapine. The first NNRTI to be licensed for clinical use was the dipyridodiazepinone derivative BI-RG-587, Nevirapine. It is still the most widely used NNRTI and it is currently employed in combinatorial regimens with NRTIs.[74,75] Nevirapine is readily absorbed after oral administration. Peak plasma concentrations were attained 4 hours after a single 200-mg dose. Nevirapine is highly lipophilic and essentially nonionized at physiological pH. It readily crosses the placenta and is found in breast milk. Nevirapine is extensively biotransformed via cytochrome P450 metabolism to form several hydroxylated metabolites, which are primarily eliminated via urinary excretion. FDA approved Nevirapine in 1996 for use in combination with approved NRTIs for treatment of HIV-infected adults who have experienced clinical and/or immunologic deterioration. No dosage adjustments are required when Nevirapine is taken with NRTIs such as AZT, ddI, or ddC. Also no dosage adjustments are needed when Nevirapine is taken with the PI ritonavir. Co-administration of Nevirapine and the PIs indinavir and saquinavir led to decrease in availability of these drugs. The most frequently reported adverse events related to therapy were rash, fever, nausea, headache, and abnormal liver function tests. The majority of severe rashes occurred within the first 28 days of treatment. Monitoring of liver function should be considered during drug therapy due to frequent asymptomatic elevations in GGT levels.

Delavirdine. The FDA approved in 1997 another first-generation NNRTI, Delavirdine (U90152S), for use in combination with appropriate antiretrovirals for the treatment of HIV infection.[76] Delavirdine is distributed primarily into blood plasma. In the body, Delavirdine is converted to several inactive antimetabolites, primarily through the enzyme cytochrome P450 3A (CYP3A). The metabolic pathways for Delavirdine are N-desalkylation and pyridine hydroxylation. The mean elimination time from plasma is 5.8 hours. In clinical studies it has been used in combination with AZT/3TC or didanosine, for initial management. Clinical trial ACTG 359 showed that adding Delavirdine to dual protease inhibitor salvage therapy decreases viral load in about 30 to 40% of patients. Delavirdine is metabolized by and inhibits the 3A isoenzyme of the cytochrome P450 system. Since this system is the pathway for the metabolism of many common drugs, Delavirdine is likely to have interactions with such drugs. Co-administration with the NRTI ddI lowers the plasma levels of both drugs. Coadministration with the PI indinavir increases indinavir plasma levels, thus reduced levels of that drug should be taken. Co-administration of the PI saquinavir causes a fivefold increase in saquinavir plasma levels. As with other NNRTIs, the most significant side effect of Delavirdine is rash, which occurs in about 18% of the patients. Other side effects reported for Delavirdine are blisters, conjunctivitis, fever, joint aches, muscle aches, oral lesions, and swelling.

Efavirenz. The benzoxazinone DMP266 (efavirenz), has been approved for clinical use in 1998.[72] This compound showed good pharmacokinetics profile, with peak plasma levels achieved 5 hours after single-dose oral administration and steady-state levels after 6 to 10 days. Moreover, Efavirenz was found to cross the blood-brain barrier, achieving CSF concentrations of 0.2 to 1.2% of the corresponding plasma levels. These concentrations exceed the ID_{95} for most wild type and clinical strains of HIV-1, with a daily dose of 200 mg. Indeed, one of the major advantages of Efavirenz is that it can be administered once daily due to its long plasma half-life (40 hours). Efavirenz bound to human plasma proteins, especially albumin and it is metabolised in the liver by the cytochrome P450(CYP)3A4 and 2B6 isozymes, into inactive hydroxylated metabolites. Due to its interaction with the cytochrome system, Efavirenz reduced the levels of co-administered PIs indinavir and saquinavir, while did not show any unfavorable interactions in combination with NRTIs. Observed side effects included nervous system symptoms (headache, insomnia) and dermatological effects (rash). Several clinical trials have been designed to evaluate the efficacy of Efavirenz in combination with other anti-HIV drugs, including PIs (Indinavir, Nelfinavir) and NRTIs (3TC, AZT). In most cases, Efavirenz reduced plasma HIV RNA levels below detection. In a clinical study[77] two regimens containing efavirenz, one with a protease inhibitor and the other with two NRTIs, were compared with a standard three-drug regimen. 450 patients who had not previously been treated with lamivudine or any NNRTI or PI were randomly assigned to one of three regimens: efavirenz (600 mg daily) plus zidovudine (300 mg twice daily) and lamivudine (150 mg twice daily); the protease inhibitor indinavir (800 mg every eight hours) plus zidovudine and lamivudine; or efavirenz plus indinavir (1000 mg every eight hours). Suppression of plasma HIV-1 RNA to undetectable levels was achieved in more patients in the group given efavirenz plus nucleoside reverse-transcriptase inhibitors than in the group given indinavir plus nucleoside reverse-transcriptase inhibitors (70 percent vs. 48 percent, $P < 0.001$). More patients discontinued treatment because of adverse events in the group given indinavir and two NRTIs than in the group given efavirenz and two NRTIs (43 percent vs. 27 percent, $P = 0.005$). Thus, as antiretroviral therapy in HIV-1-infected adults, the combination of efavirenz, zidovudine, and lamivudine has greater antiviral activity and is better tolerated than the combination of indinavir, zidovudine, and lamivudine. Promising results have been obtained also for the indinavir/efavirenz combination in NRTIs-experienced HIV patients.[78]

9.5.3 *Combined Toxicities of Reverse Transcriptase Inhibitors*

The frequency, severity, and tolerability of side effects related to administration of antiretroviral agents, represent a major obstacle to initiation and continuation of therapy. The search to characterize and understand these problems has turned up a number of abnormalities in individuals with HIV infection that may or may not be related to specific drugs or drug classes.[79] Moreover, as many patients are co-infected with other infectious agents (in particular mycobacterium, herpesviruses and the hepatitis B and/or C viruses) or require treatment for concomitant medical problems, the management of HIV infection is increasingly complicated by problems that may not be related to HIV infection, antiretroviral therapy, or immune reconstitution. The situation is further complicated by the interference of certain drugs with other, a phenomenon called drug-drug interaction. **Table 9.3** summarizes the known side effects and drug interaction for NRTIs and NNRTIs.

It is clear that combination of antiretroviral drugs does not always result in beneficial effects and several examples have been reported of antagonistic activity, increased toxicity, metabolic interference as well as increased drug resistance and mutation rates for certain drug associations.

Table 9.3   Common adverse effects and drug interactions of reverse transcriptase inhibitors

Drug (commercial name)	Adverse Effects	Drug Interactions
	Nucleoside RT inhibitors	
AZT (Retrovir)	Malaise, headache, nausea, insomnia, myalgias, anemia, granulocytopenia, thrombocytopenia, lactic acidosis, mitochondrial toxicity, hepatomegaly.	Documented interactions with other myelosuppressive drugs (i.e. trimethoprim-sulfamethoxazole, ganciclovir); coadministration with ddC increases pancreatic toxicity.
ddI (Videx)	Pancreatitis, peripheral neuropathy, nausea, rash, hyperglycemia, headache, insomnia, elevated triglycerides, thrombocytopenia, lactic acidosis, mitochondrial toxicity, aminotransferases elevation.	Adverse interactions with neurotoxic drugs (ddC, ganciclovir); can reduce adsorption of indinavir[a], delavirdine, ritonavir[a], tetracyclines.
ddC (Hivid)	Peripheral neuropathy, stomatitis, rash, aphtous ulcers, pancreatitis, esophageal ulcerations, lactic acidosis, cardiomyopathy, mitochondrial toxicity, aminotransferases elevation.	Adverse interactions with other neurotoxic drugs (ddI, isoniazid, ganciclovir).
d4T (Zerit)	Peripheral neuropathy, anemia, macrocytosis, insomnia, lactic acidosis, mitochondrial toxicity, aminotransferases elevation.	Antagonistic effect with concomitant therapy with AZT.
Abacavir (Ziagen)	Nausea, headache, malaise, abdominal pain, rash, hepatomegaly, lactic acidosis, mitochondrial toxicity, hypersensitivity syndrome (fever, dyspnea, cough, pharyngitis, nausea, vomiting, aminotransferases elevation).	In case of hypersensitivity, do not re-challenge after interruption, due to possible life-threatening anaphylactic reactions; reduces 3TC levels in blood.
3TC (Epivir)	Headache, fatigue, insomnia, peripheral neuropathy, muscle aches, rash, nausea, thrombocytopenia, lactic acidosis, mitochondrial toxicity.	Indinavir decreases 3TC levels in blood.

(*Continued*)

Table 9.3   (*Continued*)

Drug (commercial name)	Adverse Effects	Drug Interactions
	Non-nucleoside RT inhibitors (NNRTIs)	
Nevirapine (Viramune)	Rash, fever, nausea, headache, hepatotoxicity.	No reported interactions with NRTIs; decreases levels of co-administered indinavir and saquinavir.[a]
Delavirdine (Rescriptor)	Rash, blisters, conjunctivitis, fever, muscle aches, oral lesions.	Lowers levels of ddI; increases levels of indinavir and saquinavir.
Efavirenz (Sustiva)	Rash, headache, insomnia.	No interactions with NRTIs; lowers levels of indinavir and saquinavir.

[a]Indinavir, Ritonavir, Saquinavir are inhibitors of HIV-1 protease used in therapy.

9.5.4 *Molecular Interactions of HIV-1 Reverse Transcriptase with Nucleoside- and Non-nucleoside Inhibitors: The Problem of Drug Resistance*

Appearance of drug resistant viruses is an inevitable consequence of prolonged exposure of HIV-1 to antiretroviral therapy.[46] This is believed to be caused by the combination of the elevated turnover of HIV-1 in patients with the high mutation rate of the virus, due to the low fidelity of RT. **Table 9.4** lists the most frequent mutations selected during anti-HIV-1 therapy. In **Figure 9.6**, the structure of the NNRTIs binding site is shown in complex with the drug Nevirapine. The most frequently mutated amino acids in drug-resistant strains are represented. As can

Table 9.4   **Resistance mutations selected by combinations of anti-RT drugs during therapy**

Mutations	Cross-resistance	Fold resistance
R211K	AZT, 3TC	2–5
L214F	AZT, 3TC	2–5
Q151M	AZT, ddC, ddI, d4T	5–20
V75I/F77L/F116Y/Q151M	AZT, ddI, ddC, d4T	10–50
A62V/V75I/F77L/F116Y/Q151M	AZT, ddI, ddC, d4T	10–65
T69S-(S-S/G)-K70	AZT, ddC, ddI	10
M41L/T215Y	AZT, ddI, d4T	4
M41L/M184V/T215Y/G333E	AZT, 3TC	7.5
M41L/T215Y/44A/118I	AZT, 3TC	4–50
K103N/Y181C	All NNRTIs	5–100
L100I/V106A/Y181C	Nevirapine, Delaviridine	20–100

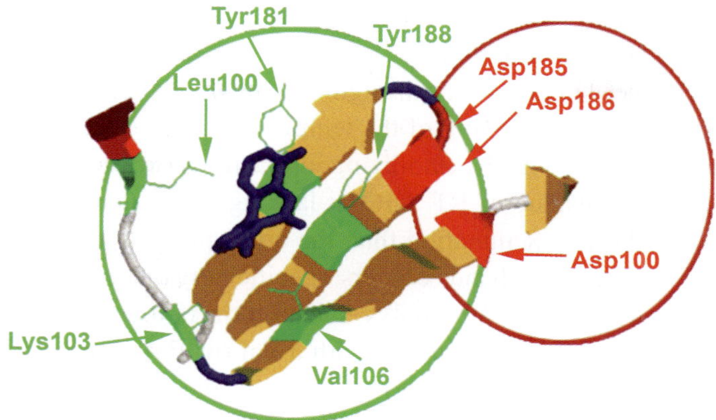

Figure 9.6   The NNRTIs binding site. The structural elements of the NNRTIs binding site (circled in green) are shown, along with the side chains (green) of the most commonly mutated amino acids in drug resistant viral strains. In red are shown the three Asp residues constituting the catalytic site (circled in red). In blue is the drug Nevirapine.

be seen, all the mutations fall into amino acid side chains which are taking direct contact with the drug.

9.5.4.1 *NRTIs drug resistance*

Drug resistant viruses have been well documented in patients undergoing either monotherapy or combination therapy with two drugs that include AZT, ddI, ddC and 3TC.[80] A single amino acid substitution within the RT enzyme is sufficient to cause resistance *in vitro*, although a combination of mutations is required to confer high-level resistance to AZT. Of particular concern is the fact that mutant viruses are often cross-resistant to several NRTIs (**Table 9.4**) and these mutated viruses might be transmitted during *de novo* infections.[81] The molecular mechanisms which are responsible for the observed resistance have been fully clarified by the detailed analysis of the enzymatic reaction catalyzed by HIV RT and by the resolution of the crystal structures of HIV-1 RT either alone or in complex with its substrates.[82–84] To date, three different mechanisms have been identified that account for the observed resistance to NRTIs[85]: (i) changes in discrimination between NRTIs and the corresponding dNTPs, due to alterations of the binding/incorporation characteristics of the analog by HIV-1 RT; (ii) alterations in the interactions between RT and the template-primer and (iii) pyrophosphorolytic removal of the chain-terminating residue from the 3′ end of the growing DNA strand. Different mutations are responsible for these mechanisms; moreover, different mutations can induce a similar phenotype through alternative molecular mechanisms. For example, both the K65R and the M184V mutations induce resistance at the discrimination level. However, in the case of K65R, substitution of an Arg for a Lys alters the orientation of the phosphate backbone, leading to a misalignment of the base moiety of the NRTI with the template. This weakening of the template — base interaction is compensated in the case of dNTPs by the stabilizing interactions between the ribose 3'-OH group and the Q151 residue of RT. NRTIs, lacking the 3'-OH function, are thus more sensitive to the K65R mutation than the corresponding dNTPs. The substitution M184V, on the other hand, confers specific resistance to 3TC, an analog with sulfur in its sugar ring. This bulky substitution is tolerated when a Met is present at position 184, but substitution with a Val (or Ile) at this position causes steric hindrances which, by a matter of facts, exclude the NRTI from the nucleotide binding pocket. The natural dNTPs or other NRTIs without bulky substitutions are not affected by this mutation. Template-primer repositioning effects have been proposed to account for the resistant phenotype induced by the L74V and M184I mutations. Crystal structures of HIV-1 RT in complex with a DNA substrate showed that L74 contacts the template nucleotide which is base paired to the incoming nucleotide. This led to the proposal that substitution at this position might induce partial displacement of the template strand. Direct evidences that some mutation can induce a repositioning of the template-primer have been derived from comparisons between the structures of the wild type and M184I mutant RT bound to DNA, showing alterations of the template-primer poisition.[86] High-level of AZT resistance requires multiple mutations, contrary to the other NRTIs, thus the resistance mechanism might be different. The AZT-specific mutations are clustered around the nucleotide binding pocket, suggesting a direct effect on nucleotide binding. Indeed, the crystal structure of the quadruple mutant D67N/K70R/T215F/K219Q suggested that substitutions at the 215 and 219 positions induced long-range conformational alterations of the catalytic site. However, enzymological analysis failed to detect any difference in AZTTP (the triphosphate form of AZT) binding and/or incorporation. The most likely explanation for the induced resistance comes from recent studies, showing that mutations D67N and K70R enhance the pyrophosphorolytic removal (the reverse of the polymerization reaction) of a chain-terminating nucleotide by HIV-1 RT, whereas the mutations T215F and K219Q increase the processivity of DNA synthesis, thus compensating for the higher rate of back reaction of the mutated enzyme.[87,88] The d4T-specific mutation V75T has been shown to act through two different mechanisms: it decreases the affinity of the mutated RT for the triphosphate form of the drug d4TTP and it increases the rate of the pyrophosphorolytic reaction which removes the d4T-terminating residue from the primer. The shift in the balance between polymerization and pyrophosphorolysis towards the latter reaction is then compensated by an increase in processivity of RT, due to an effect of the V75T mutation on the translocation rate of the enzyme.[89]

9.5.4.2 *NNRTIs drug resistance*

The available structural and biochemical data made it clear that, albeit structurally very different, all the known NNRTIs bind at a common site in the HIV-1 RT. The first solved threedimensional structure of a NNRTI/RT complex was the one with Nevirapine.[90] Several other structures for first-generation NNRTIs complexed with RT were solved afterwards. They all bind to a hydrophobic "pocket" which is represented by a large (600–700 Å) solvent-accessible cavity whose inner surface is lined by a number of hydrophobic residues. Comparison of the structures of HIV-1 RT complexed with Nevirapine and other NNRTIs, showed substantially similar modes of interaction. These NNRTIs can be considered to consist of two, hinged, 6-membered rings. These rings adopt similar positions within the binding site, giving rise to a "butterfly"-like spatial arrangement. Based on this analogy, the binding site can be divided into two distinct regions, termed wing I and wing II, each occupied by one of the two ringed elements (or "wings"). The wing I region is lined by residues Y181, Y188, W229, whereas the wing II region is defined by residues V179, V106, L100. Enzyme–inhibitor interaction is primarily stabilized through hydrophobic interactions with the Y181, Y188, L100, V106 residues. In addition, drug-specific electrostatic interactions (charge and hydrogen bonds) contribute to the final strength of binding. The binding pocket communicates with the bulk solvent through a channel lined by the residues K101, V179, E138(p51), which might represent the entry site for the NNRTIs. Thus, it appears that all the NNRTIs adopt a common binding mode within the target site. Comparison of the unliganded RT structure with the RT/NNRTIs complexes shows that, in the inhibitor-free structure, there is a small solvent inaccessible cavity between Y181 and Y188. Upon inhibitor binding, these residues rotate "outwards", filling this cavity and opening up the larger pocket where NNRTIs bind. This movement involves extensive changes in the relative positions of the $\beta4$, $\beta7$ and $\beta8$ strands, which contain the catalytically essential residues D110, D185 and D186. Pre-steady state kinetic analysis of the reaction catalyzed by HIV-1 RT in the absence or in the presence of NNRTIs, showed that the inhibition was due to a great reduction of the polymerization rate k_{pol}.[91] The detailed crystallographic information available for NNRTIs can explain both the mechanism of resistance induced by the different mutations and the high level of cross-resistance characteristic of this class of compounds (**Table 9.4**). The first RT mutations shown to be associated with resistance to NNRTIs were the K103N and Y181C substitutions, which were selected in the presence of pyridinone derivatives, Nevirapine and TIBO R82150.[92] Indeed, the

K103N and Y181C mutations have been observed with virtually all the NNRTIs (**Table 9.4**). Other drug-specific mutations include L100I, V106A and P236L.[93] All the mutated aminoacids are part of the NNRTIs binding pocket and they act by reducing the number of stabilizing interactions between the drug and the mutated residues.[94] One relevant exception is the mutation K103N, which lines the proposed entry channel for the NNRTIs. Enzymological and crystallographic studies have shown that this mutation imposes a thermodynamic barrier for the formation of the RT-inhibitor complex,[94] most likely through stabilization of the unbound form of the enzyme.[83]

9.6 Inhibitors of Hepatitis B DNA Polymerase

Hepatitis B is a chronical viral infection concerning more than 350 million people worldwide. Complications of Hepatitis B virus (HBV) infection are cirrhosis, hepatocellular carcinoma and end-stage liver disease, accounting for approximately one million deaths each year. Even though HBV is a DNA virus, its DNA pol also possesses reverse transcription activity (See Chapter 6). This dual identity of the HBV DNA pol prompted the researchers to exploit inhibitors of the HIV-1 RT and herpes simplex DNA pol as potential anti-HBV agents. This approach was successful, and to date some nucleoside analogs have been approved for the clinical treatment of HBV infections. These drugs are listed in **Table 9.5** and their structures are shown in **Figure 9.7**. As can be seen, anti-HBV agents include two drugs (Lamivudine and Tenofovir) which have been previously approved for the treatment of HIV infections (see **Table 9.2**). Tenofovir and Adefovir belong to the phosphonate analogs class (see Section 9.3.3.4), the same of the anti-herpetic drug Cidofovir. The biochemical details of the mechanism of action of these compounds are limited by the lack of robust *in vitro* systems for studying the enzymology of the HBV DNA pol (Chapter 6). Given the good degree of conservation of the active site residues between HIV RT and HBV DNA pol, structural models have been built,[95]

Table 9.5   **Currently approved drugs for the treatment of Hepatitis B infection**

Drug	Commercial name	Licensed	Known resistance mutations
Lamivudine	(Epivir-HBV, Zeffix, or Heptodin)	1998	M204V/I/S, L80V, V173L, L180M
Adefovir Dipivoxil	(Hepsera)	2002	N236T, A181T/V
Entecavir	(Baraclude)	2005	T184G, S202I, M250V
Telbivudine	(Tyzeka, Sebivo)	2006	M204I
Tenofovir	(Viread)	2008	

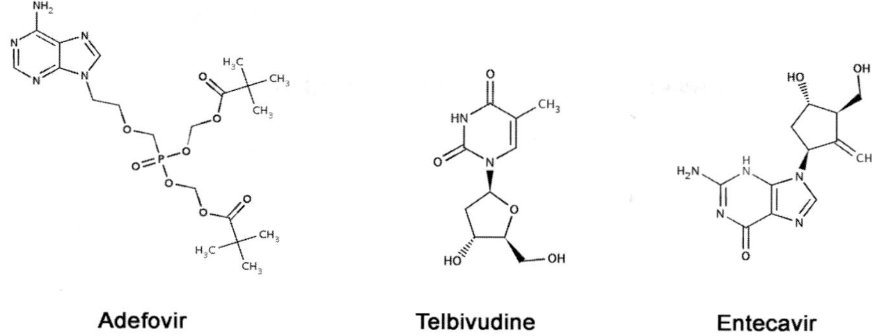

Adefovir   Telbivudine   Entecavir

Figure 9.7   Inhibitors of HBV DNA polymerase.

which helped to clarify the patterns of drug resistance mutations selected during the therapy (**Table 9.5**).

Below are some detailed descriptions of drugs that inhibit HBV DNA polymerase.

Lamivudine. Lamivudine (3TC) has already been discussed in Section 9.5; here we will briefly discuss specific issues related to HBV DNA pol drug resistance towards this drug. Lamivudine has been widely used for the treatment of HBV infections since 1998, but it results in selection of drug resistance mutations (**Table 9.5**) in 24% of treated patients after one year and in 70% of treated individuals after four years of monotherapy.[96] Molecular modeling[95] suggested that 3TC can bind the active site of HBV DNA pol, thanks to an induced fit mechanism, involving the movement of the M204 side chain to allow proper positioning of the oxathiolane ring. This also immediately provides an explanation for the observed high frequency of drug resistance mutations at position M204 during monotherapy. Substitution of Met at position 204 with bulkier sidechains such as Val or Ile reduces the space available for the accommodation of the oxathiolane ring, resulting in steric hindrance. Thus, this mechanism of resistance is analogous to the one described for the M184V/I mutation of HIV-1 RT (see Section 9.5.4.1). Another mutation conferring low-level resistance to 3TC is the L180M mutation. The L180 residue is oriented towards the hydrophobic pocket which accommodates the 3TC oxathiolane ring. Replacing Leu with Met at position 180 results in a shift of this residue away from the 3TC binding site, likely weakening the network of interactions. Combination of the L180M and M204V/I/L mutations results in high-level 3TC resistance.

(R)-9-(2-phosphonylmethoxypropyl)adenine (Tenofovir, PMPA) and its prodrug fumarate salt bis(isopropoxy-carbonyloxymethyl) esther (bis(POC)PMPA, *Viread®*), once phosphorylated to triphosphate they interfere as chain terminator

with the reverse transcriptase reaction of HIV-1 and -2, various retroviruses and HBV.[97]

Phosphonylmethoxyethyladenine (Adefovir, PMEA), whose dipivoxil derivative (bis-POM-PMEA, *Hepsera*®) is used as a prodrug against HBV,[98] after intracellular conversion to the diphosphate form, acts as an alternative substrate for HBV reverse transcriptase and, once incorporated into the DNA, acts as a chain terminator, preventing DNA elongation. Adefovir dipivoxil proved effective against HBV strains that had developed resistance to Lamivudine. Also, resistance to adefovir seems to develop at a much slower rate and lower frequency than for Lamivudine. Molecular modeling studies[95] suggested that Adefovir binds the DNA pol active site in a fashion similar to the natural substrate dATP. It does not interact with the hydrophobic pocket exploited by 3TC, thus explaining its activity against Lamivudine-resistant mutants.

Entecavir. Entecavir is a guanosine nucleoside analog, originally developed against herpes simplex virus infections. However, while it was only moderately effective against herpesviruses, entecavir was later discovered to be a potent inhibitor of HBV DNA pol.[99,100] Like other nucleoside analogs, entecavir needs to be converted intracellularly to its $5'$-triphosphate form, which binds potently to the HBV DNA pol. Modelling studies[95] suggested that it almost perfectly mimics the natural substrate dGTP in its interaction with the enzyme active site, thus explaining its very high potency of inhibition. The partial sensitivity to the M204V mutation was explained by a possible weak steric clash between the exocyclic alkene of Entecavir and the Ile side chain at position 204.

Telbivudine. Telbivudine is the L-nucleoside form of thymidine (see also Section 9.4) used against HBV infections.[101] It selects for the same resistance mutation (M204I) than 3TC, but molecular modeling studies indicated a different mode of interaction.[95] Telbivudine triphosphate binds differently from both 3TC and the natural substrate dTTP. In particular, it does not induce a rearrangement of the M204 residue, as seen with Lamivudine. Thus, mutations at position 204 do not cause a steric clash, but rather weaken the network of stabilizing interactions between the active site and the sugar ring of Telbivudine.

9.7 Chapter Summary

DNA pols are the key enzymes involved in the duplication of the genetic information of all living organisms. This fact immediately qualifies DNA pols as an attractive chemotherapeutic target for the control of the proliferation of cancer cells and microbial pathogens. In particular viruses, that have very compact genetic information, all encode a replicase. Thus, inhibitors of DNA pols and RT's are the cornerstone of all current antiviral chemotherapeutic approaches. The inhibitors of herpes simplex virus DNA pol have been the first antiviral drugs developed and the knowledge gained in their design and synthesis has set the standards for the following development of the first inhibitors of HIV-1 RT. With the ever increasing development of powerful computational and *in silico* techniques, coupled with high resolution X-ray crystallography, novel approaches to the design of antiviral drugs have been made possible, leading to the development of several novel classes of RT inhibitors. The enormous advance in knowledge-based drug design makes it now theoretically possible to develop inhibitors for any DNA pol, even if its crystal structure is unknown, as in the case of the HBV DNA pol. While novel targets and strategies are also being developed, the inhibitors of viral replicases will still constitute the core of antiviral therapies for a long time.

References

1. Sanger F, Nicklen S, Coulson AR. 1977. *Proc Natl Acad Sci U S A* 74: 5463–7.
2. Cozzarelli NR. 1977. *Annu Rev Biochem* 46: 641–68.
3. Kirchmair J, Markt P, Distinto S, Wolber G, Langer T. 2008. *J Comput Aided Mol Des* 22: 213–28.
4. Dror O, Shulman-Peleg A, Nussinov R, Wolfson HJ. 2004. *Curr Med Chem* 11: 71–90.
5. Klepeis JL, Lindorff-Larsen K, Dror RO, Shaw DE. 2009. *Curr Opin Struct Biol* 19: 120–7.
6. De Clercq E. 2002. *Mini Rev Med Chem* 2: 163–75.
7. Furgiuele FP, Pentecost GF, Klein M, Mc GJ. 1962. *Trans Am Ophthalmol Soc* 60: 243–59.
8. Prusoff WH, Chen MS, Fischer PH, Lin TS, Shiau GT, *et al.* 1979. *Pharmacol Ther* 7: 1–34.
9. Prusoff WH, Mancini WR, Lin TS, Lee JJ, Siegel SA, Otto MJ. 1984. *Antiviral Res* 4: 303–15.
10. De Clercq E. 2004. *J Clin Virol* 30: 115–33.
11. Balzarini J, De Clercq E. 1989. *Methods Find Exp Clin Pharmacol* 11: 379–89.
12. De Clercq E. 2004. *Biochem Pharmacol* 68: 2301–15.
13. Miura S, Izuta S. 2004. *Curr Drug Targets* 5: 191–5.
14. Kufe D, Herrick D, Crumpacker C, Schnipper L. 1984. *Cancer Res* 44: 69–73.
15. Falke D, Ronge K, Arendes J, Muller WE. 1979. *Biochim Biophys Acta* 563: 36–45.
16. Crumpacker CS, Schnipper LE, Kowalsky PN, Sherman DM. 1982. *J Infect Dis* 146: 167–72.
17. Muller WE, Zahn RK, Arendes F, Falke D. 1979. *J Gen Virol* 43: 261–71.
18. Ono K, Ohashi A, Ogasawara M, Matsukage A, Takahashi T, *et al.* 1981. *Biochemistry* 20: 5088–93.
19. Trousdale MD, Nesburn AB, Su TL, Lopez C, Watanabe KA, Fox JJ. 1983. *Antimicrob Agents Chemother* 23: 808–13.
20. Elion GB. 1983. *J Antimicrob Chemother* 12 Suppl B: 9–17.
21. Elion GB. 1982. *Am J Med* 73: 7–13.
22. Derse D, Cheng YC, Furman PA, St Clair MH, Elion GB. 1981. *J Biol Chem* 256: 11447–51.
23. Parris DS, Harrington JE. 1982. *Antimicrob Agents Chemother* 22: 71–7.
24. Field HJ, Darby G, Wildy P. 1980. *J Gen Virol* 49: 115–24.
25. Schnipper LE, Crumpacker CS, Marlowe SI, Kowalsky P, Hershey BJ, Levin MJ. 1982. *Am J Med* 73: 387–92.
26. Crumpacker CS, Kowalsky PN, Oliver SA, Schnipper LE, Field AK. 1984. *Proc Natl Acad Sci U S A* 81: 1556–60.

27. Reid R, Mar EC, Huang ES, Topal MD. 1988. *J Biol Chem* 263: 3898–904.
28. Herdewijn P, De Clercq E, Balzarini J, Vanderhaeghe H. 1985. *J Med Chem* 28: 550–5.
29. De Clercq E, Balzarini J, Bernaerts R, Herdewijn P, Verbruggen A. 1985. *Biochem Biophys Res Commun* 126: 397–403.
30. Bernaerts R, De Clercq E, Balzarini J, Herdewijn P, Verbruggen A. 1985. *Antiviral Res* Suppl 1: 51–6.
31. De Clercq E, Bernaerts R, Balzarini J, Herdewijn P, Verbruggen A. 1985. *J Biol Chem* 260: 10621–8.
32. Lea AP, Bryson HM. 1996. *Drugs* 52: 225–30; discussion 31.
33. Oberg B. 1982. *Pharmacol Ther* 19: 387–415.
34. Crumpacker CS. 1992. *Am J Med* 92: 3S–7S.
35. Lietman PS. 1992. *Am J Med* 92: 8S–11S.
36. Verri A, Montecucco A, Gosselin G, Boudou V, Imbach JL, *et al.* 1999. *Biochem J* 337 (Pt 3): 585–90.
37. Spadari S, Maga G, Verri A, Focher F. 1998. *Expert Opin Investig Drugs* 7: 1285–300.
38. Semizarov DG, Arzumanov AA, Dyatkina NB, Meyer A, Vichier-Guerre S, *et al.* 1997. *J Biol Chem* 272: 9556–60.
39. Verri A, Focher F, Priori G, Gosselin G, Imbach JL, *et al.* 1997. *Mol Pharmacol* 51: 132–8.
40. Focher F, Maga G, Bendiscioli A, Capobianco M, Colonna F, *et al.* 1995. *Nucleic Acids Res* 23: 2840–7.
41. Maga G, Focher F, Wright GE, Capobianco M, Garbesi A, *et al.* 1994. *Biochem J* 302: 279–82.
42. Maga G, Verri A, Bonizzi L, Ponti W, Poli G, *et al.* 1993. *Biochem J* 294 (Pt 2): 381–5.
43. Yamaguchi T, Saneyoshi M, Shudo K. 1993. *Nucleic Acids Symp Ser*: 135–6.
44. Spadari S, Maga G, Focher F, Ciarrocchi G, Manservigi R, *et al.* 1992. *J Med Chem* 35: 4214–20.
45. Balzarini J, De Clercq E, Baumgartner H, Bodenteich M, Griengl H. 1990. *Mol Pharmacol* 37: 395–401.
46. Maga G, Spadari S. 2002. *Curr Drug Metab* 3: 73–95.
47. Sharma PL, Nurpeisov V, Hernandez-Santiago B, Beltran T, Schinazi RF. 2004. *Curr Top Med Chem* 4: 895–919.
48. Balzarini J, Herdewijn P, De Clercq E. 1989. *J Biol Chem* 264: 6127–33.
49. Peter K, Gambertoglio JG. 1998. *Pharm Res* 15: 819–25.
50. Stretcher BN. 1995. *Clin Pharmacokinet* 29: 46–65.
51. Wattanagoon Y, Na Bangchang K, Hoggard PG, Khoo SH, Gibbons SE, *et al.* 2000. *Antimicrob Agents Chemother* 44: 1986–9.
52. Hoggard PG, Lloyd J, Khoo SH, Barry MG, Dann L, *et al.* 2001. *Antimicrob Agents Chemother* 45: 976–80.
53. Ho HT, Hitchcock MJ. 1989. *Antimicrob Agents Chemother* 33: 844–9.
54. Gao WY, Agbaria R, Driscoll JS, Mitsuya H. 1994. *J Biol Chem* 269: 12633–8.
55. Rodriguez Orengo JF, Santana J, Febo I, Diaz C, Rodriguez JL, *et al.* 2000. *P R Health Sci J* 19: 19–27.
56. Kewn S, Veal GJ, Hoggard PG, Barry MG, Back DJ. 1997. *Biochem Pharmacol* 54: 589–95.
57. Moore KH, Barrett JE, Shaw S, Pakes GE, Churchus R, *et al.* 1999. *Aids* 13: 2239–50.
58. Kewn S, Hoggard PG, Henry-Mowatt JS, Veal GJ, Sales SD, *et al.* 1999. *AIDS Res Hum Retroviruses* 15: 793–802.
59. Robbins BL, Greenhaw J, Fridland A. 2000. *Nucleosides Nucleotides* 19: 405–13.
60. Faletto MB, Miller WH, Garvey EP, St Clair MH, Daluge SM, Good SS. 1997. *Antimicrob Agents Chemother* 41: 1099–107.
61. Staszewski S. 1999. *Int J Clin Pract Suppl* 103: 35–8.
62. Gao WY, Shirasaka T, Johns DG, Broder S, Mitsuya H. 1993. *J Clin Invest* 91: 2326–33.

63. Back DJ. 1999. *Infection* 27: S42–4.
64. Hübscher U, Spadari S. 1994. *Physiol. Rev.* 74: 259–304.
65. Spadari S, Maga G, Verri A, Focher F. 1997. *Exp. Opin. Invest. Drugs* 7: 1285–300.
66. Borst P, Evers R, Kool M, Wijnholds J. 2000. *J Natl Cancer Inst* 92: 1295–3.02.
67. Schuetz JD, Connelly MC, Sun D, Paibir SG, Flynn PM, *et al.* 1999. *Nat Med* 5: 1048–51.
68. Gupta S, Gollapudi S. 1993. *J Clin Immunol* 13: 289–301.
69. Campiani G, Ramunno A, Maga G, Nacci V, Fattorusso C, *et al.* 2002. *Curr Pharm Des* 8: 615–57.
70. Ren J, Esnouf R, Garman E, Somers D, Ross C, *et al.* 1995. *Nature Struct. Biol.* 2: 293–302.
71. Esnouf R, Ren J, Ross C, Jones Y, Stammers D, Stuart D. 1995. *Nature Struct. Biol.* 2: 303–8.
72. Adkins JC, Noble S. 1998. *Drugs* 56: 1055–64; discussion 65–6.
73. Clotet B. 1999. *Int J Clin Pract Suppl* 103: 21–5.
74. Grob PM, Wu JC, Choen KA, Ingraham RH, Shih CK, *et al.* 1992. *AIDS Res. Hum. Retroviruses* 8: 145–52.
75. Bardsley-Elliot A, Perry CM. 2000. *Paediatr Drugs* 2: 373–407.
76. Scott LJ, Perry CM. 2000. *Drugs* 60: 1411–44.
77. Staszewski S. 1999. *Int J Clin Pract Suppl* 103: 10–5.
78. Haas DW, Fessel WJ, Delapenha RA, Kessler H, Seekins D, *et al.* 2001. *J Infect Dis* 183: 392–400.
79. Vigouroux C, Gharakhanian S, Salhi Y, Nguyen TH, Adda N, *et al.* 1999. *Diabetes Metab* 25: 383–92.
80. Larder BA. 1994. *J Gen Virol* 75: 951–7.
81. Duwe S, Brunn M, Altmann D, Hamouda O, Schmidt B, *et al.* 2001. *J Acquir Immune Defic Syndr* 26: 266–73.
82. Tantillo C, Ding J, Jacobo-Molina A, Nanni RG, Boyer PL, *et al.* 1994. *J Mol Biol* 243: 369–87.
83. Ren J, Stammers DK. 2008. *Virus Res* 134: 157–70.
84. Sarafianos SG, Marchand B, Das K, Himmel DM, Parniak MA, *et al.* 2009. *J Mol Biol* 385: 693–713.
85. Sluis-Cremer N, Arion D, Parniak MA. 2000. *Cell Mol Life Sci* 57: 1408–22.
86. Sarafianos SG, Das K, Ding J, Boyer PL, Hughes SH, Arnold E. 1999. *Chem Biol* 6: R137–46.
87. Meyer PR, Matsuura SE, So AG, Scott WA. 1998. *Proc Natl Acad Sci U S A* 95: 13471–6.
88. Meyer PR, Matsuura SE, Mian AM, So AG, Scott WA. 1999. *Mol Cell* 4: 35–43.
89. Selmi B, Deval J, Boretto J, Canard B. 2003. *Antivir Ther* 8: 143–54.
90. Smerdon SJ, Jager J, Wang J, Kohlstaedt LA, Chirino AJ, *et al.* 1994. *Proc Natl Acad Sci U S A* 91: 3911–5.
91. Spence RA, Kati WM, Anderson KS, Johnson KA. 1995. *Science* 267: 988–93.
92. Balzarini J. 1999. *Biochem Pharmacol* 58: 1–27.
93. Deeks SG. 2001. *J Acquir Immune Defic Syndr* 26: S25–33.
94. Maga G, Amacker M, Ruel N, Hubscher U, Spadari S. 1997. *J Mol Biol* 274: 738–47.
95. Sharon A, Chu CK. 2008. *Antiviral Res* 80: 339–53.
96. Ghany MG, Doo EC. 2009. *Hepatology* 49: S174–84.
97. Reynaud L, Carleo MA, Talamo M, Borgia G. 2009. *Ther Clin Risk Manag* 5: 177–85.
98. Jones J, Shepherd J, Baxter L, Gospodarevskaya E, Hartwell D, *et al.* 2009. *Health Technol Assess* 13: 1–172, iii.
99. Langley DR, Walsh AW, Baldick CJ, Eggers BJ, Rose RE, *et al.* 2007. *J Virol* 81: 3992–4001.
100. Scott LJ, Keating GM. 2009. *Drugs* 69: 1003–33.
101. Hadziyannis SJ, Vassilopoulos D. 2008. *Expert Rev Gastroenterol Hepatol* 2: 13–22.